PETER H. DENHAM MEMORIAL PRIZE

2009/2010
for
John & Iris Barrington Leigh

ALBERTA HIGH SCHOOL MATHEMATICS COMPETITION

The Alberta High School Math Competitions 1957–2006

A Canadian Problem Book

Library of Congress Catalog Card Number 2009933078
ISBN 978-0-88385-830-1
Printed in the United States of America
Current Printing (last digit):
10 9 8 7 6 5 4 3 2 1

The Alberta High School Math Competitions 1957–2006

A Canadian Problem Book

Compiled and Edited by

Andy Liu

University of Alberta

Published by
The Mathematical Association of America
in collaboration with the Canadian Mathematical Society

MAA PROBLEM BOOKS SERIES

Problem Books is a series of the Mathematical Association of America consisting of collections of problems and solutions from annual mathematical competitions; compilations of problems (including unsolved problems) specific to particular branches of mathematics; books on the art and practice of problem solving, etc.

Aha! Solutions, V, Martin Erickson

The Alberta High School Math Competitions 1957–2006, compiled and edited by Andy Liu

The Contest Problem Book VII: American Mathematics Competitions, 1995–2000 Contests, compiled and augmented by Harold B. Reiter

The Contest Problem Book VIII: American Mathematics Competitions (AMC 10), 2000–2007, compiled and edited by J. Douglas Faires & David Wells

The Contest Problem Book IX: American Mathematics Competitions (AMC 12), 2000–2007, compiled and edited by David Wells & J. Douglas Faires

First Steps for Math Olympians: Using the American Mathematics Competitions, by J. Douglas Faires

A Friendly Mathematics Competition: 35 Years of Teamwork in Indiana, edited by Rick Gillman

The Inquisitive Problem Solver, Paul Vaderlind, Richard K. Guy, and Loren C. Larson

International Mathematical Olympiads 1986–1999, Marcin E. Kuczma

Mathematical Olympiads 1998–1999: Problems and Solutions From Around the World, edited by Titu Andreescu and Zuming Feng

Mathematical Olympiads 1999–2000: Problems and Solutions From Around the World, edited by Titu Andreescu and Zuming Feng

Mathematical Olympiads 2000–2001: Problems and Solutions From Around the World, edited by Titu Andreescu, Zuming Feng, and George Lee, Jr.

Problems from Murray Klamkin: The Canadian Collection, V, edited by Andy Liu and Bruce Shawyer

The William Lowell Putnam Mathematical Competition Problems and Solutions: 1938–1964, A. M. Gleason, R. E. Greenwood, L. M. Kelly

The William Lowell Putnam Mathematical Competition Problems and Solutions: 1965–1984, Gerald L. Alexanderson, Leonard F. Klosinski, and Loren C. Larson

The William Lowell Putnam Mathematical Competition 1985–2000: Problems, Solutions, and Commentary, Kiran S. Kedlaya, Bjorn Poonen, Ravi Vakil

USA and International Mathematical Olympiads 2000, edited by Titu Andreescu and Zuming Feng

USA and International Mathematical Olympiads 2001, edited by Titu Andreescu and Zuming Feng

USA and International Mathematical Olympiads 2002, edited by Titu Andreescu and Zuming Feng

USA and International Mathematical Olympiads 2003, edited by Titu Andreescu and Zuming Feng

USA and International Mathematical Olympiads 2004, edited by Titu Andreescu, Zuming Feng, and Po-Shen Loh

MAA Service Center
P. O. Box 91112
Washington, DC 20090-1112
1-800-331-1622 fax: 1-301-206-9789

In memory of
James Geoffrey Butler,
Murray Seymour Klamkin
and
Robert Barrington Leigh

In Gratitude to
The Nickle Family Foundation,
The Peter H. Denham Memorial Fund
and
The Canadian Mathematical Society

Contents

Introduction **1**

Ancient Period: 1957–1966 **9**

Medieval Period: 1967–1983 **23**
- Questions with Multiple Choices . . . 23
- Problems requiring Full Solutions . . . 65

Modern Period: 1983–2006 **77**
- First Round Questions . . . 77
- Second Round Problems . . . 127

Solutions **143**
- Ancient Period: 1957–1966 . . . 143
- Medieval Period: 1967–1983 . . . 145
  - Answers to Questions with Multiple Choices . . . 145
  - Answers to Problems requiring Full Solutions . . . 145
- Modern Period: 1983–2006 . . . 147
  - First Round Answers . . . 147
  - First Round Statistics . . . 148
  - Second Round Answers . . . 160
  - First Round Solutions . . . 165
  - Second Round Solutions . . . 198

Winners: 1983–2006 **255**
- First Round Winners . . . 255
  - Individual Prizes . . . 255
  - Team Prizes . . . 264
- Second Round Winners . . . 278

About the Editor **283**

Introduction

The Competition

Although there are older mathematics competitions in Canada for high school students, Alberta was the first to have a province-wide contest. It started in 1957 as the *Alberta Matriculation Prize and Scholarship Examination.* The number of problems, which all required full solutions, varied from year to year. Some of them resembled examination questions rather than contest problems.

In 1967, the competition was renamed the *Alberta High School Mathematics Prize Examination.* Questions with multiple choices were introduced, to be written in the same session as the problems requiring full solutions.

In 1969, the *Canadian Mathematical Olympiad* (CMO) came into being. It was introduced principally in anticipation of Canada's participation in the *International Mathematical Olympiad* (IMO), which had been initiated by **Romania** in 1959. However, it was not until 1981, when the **United States** hosted the event, that Canada entered a team. The Alberta contest acquired an additional role, as a qualifying round for the CMO. From 1981 to 1983, the University of Alberta hosted the CMO Committee.

In 1983, the *Alberta High School Mathematics Prize Examination Board* was constituted to oversee the administration of our provincial competition. The "multiple-choice" part of the contest was now written in a separate sitting in November, and inherited the name of the **Alberta High School Mathematics Prize Examination**. Team prizes were also awarded to stock school libraries with good mathematics books. The "full-solution" part of the contest became the *Alberta High School Mathematics Scholarship Examination*, written in February by top performers in the November contest. Scholarships were awarded annually to the top three contestants, as well as the top contestant in each of Grades 10 and 11 excluding the top three.

In November 1983, participating schools were asked to name three students to constitute their school teams. Unexperienced in this, many schools named students who had high marks in mathematics classes, but they did not always turn out to be the best performers in the contest. At the request of the schools, this practice was discontinued and the top three performers from each school were automatically named to the school team. As a result, the larger schools in urban centers had an even greater advantage over their smaller rural cousins. Zone prizes, for both teams and individuals, were then established to ensure some geographical distribution. A new team prize was introduced in 1985, to be awarded to the top school which had not won any team prizes before and not won any other team prize that year.

In 1988, the Alberta High School Mathematics Prize Examination Board was renamed the **Alberta High School Mathematics Competition Board**, to emphasize that the competition is *not* an examination. Along with the change of the name of the Board, we also changed the names of the contests. The Alberta High School Mathematics Prize Examination was renamed the *First Round* of the *Alberta High School Mathematics Competition* (A.H.S.M.C). Whereas it formerly consisted of 20 questions with multiple choices, the number had been reduced to 16. The Alberta High School Mathematics Scholarship Examination was became the *Second Round* of the A.H.S.M.C.. It still consisted of 5 problems requiring full solutions.

In 1995, Canada hosted the IMO for the first and only time so far. The members of the AHSBC Board formed the backbone of the Problem Selection Committee for this prestigious competition. We will have more about this later.

In 1997, a new competition called the **Canadian Open Mathematics Challenge** was introduced. It officially replaces the Provincial Contests as the means for qualifying for the CMO. Nevertheless, the AHSMC remains the flagship of mathematics contests within our province, and the top performers in the Second Round are often invited to write the CMO.

Two IMO Summer Training Camps for our National IMO Team were held in Alberta, in 1998 and 2003 respectively. Both began at the University of Calgary, and moved onto the Canadian Rockies, the Kananaskis Field Station of the U of C in 1998 and the Banff International Research Station of the Pacific Institute for the Mathematical Sciences in 2003.

The Sponsors

Right from the very beginning, the Alberta High School Mathematics Competition was generously sponsored by the **Nickle Family Foundation** of **Calgary**, the **Canadian Mathematical Congress** (now **Society**), the **Mathematics Council** of the **Alberta Teachers' Association**, the **Department of Mathematics and Statistics** of the **University of Calgary**, and the **Department of Mathematics** (now **Mathematical and Statistical Sciences** of the **University of Alberta**.

The AHSMC Board would like to acknowledge the immense contributions made by these and subsequent sponsors. Our task would have been daunting without their continual support.

In 1983, the Board was most fortunate in attracting two new sponsors. The **Peter H. Denham Memorial Fund for Mathematics** was set up by **Ross Denham** of the Faculty of Business, the University of Alberta, in memory of his late father who was a pioneer of Edmonton. **W. H. Freeman**, an enterprising publishing company in San Francisco, donated many innovative books to the Board for prizes. The Board also received a one-time donation from **Dover Publications Inc.**, but because of the tie to Freeman, the Board did not pursue further relationship with Dover.

In 1997, the **Nickle Family Foundation** tripled their contribution and made the First Prize in the Second Round a generous $1500. The Board is most grateful to the Foundation. At about the same time, **Greenwoods' Bookshoppe** of Edmonton also became a sponsor, giving the Board a 20% discount on books which are ordered for prizes in the First Round.

After its inception, the **Pacific Institute for the Mathematical Sciences** was welcomed by the Board as a sponsor of our activities. PIMs pays for an annual award dinner for the winners of the Second Round, to be held in Calgary and Edmonton in alternate years, beginning in 1999. PIMs also provides a special award for any contestant who is in Grade 9 or below, and whose performance at least matches that of the winner of either the Grade 10 or Grade 11 fellowship winner.

In 2003, W. H. Freeman & Co. terminated its sponsorship. It was a very amicable parting after twenty years of partnership. A K Peters, another publisher which specializes in innovative mathematics books, replaced Freeman as our new sponsor. The company was based in Natick, Massachusetts and has since moved to Wellesley, Massachusetts.

In 2006, the Nickle Family Foundation also decided to terminate its sponsorship after supporting us through our formative years for half a century. We are fortunate that **ConocoPhillips of Canada**, a Calgary company in the petroleum industry, took over from the Foundation.

The People

Board Members

The Board has been fortunate in being able to draw on the expertise and dedication of many individuals. They have devoted considerable time and effort to this endeavor and other related contest activities.

From the early years, the list of names from the University of Calgary includes **Allan Gibbs, Richard Guy, Tony Holland, Harold Lampkin** and **Jonathan Schaer**. The late **Leo Moser** is acknowledged as the father of the Alberta contest. Others from the University of Alberta are **Ken Andersen, Bill Bruce, Geoffrey Butler, Graham Chambers, Jim Fisher, Herb Freedman, Jack Macki, Jim Muldowney, Jim Pounder, Roy Sinclair, Sudarshan Sehgal** and **Jim Timourian**. These data are compiled to the best of our knowledge, but we are sure that they are incomplete. We apologize to those whose names have been inadvertently omitted.

Geoffrey Butler chaired the the CMO Problem Committee from 1981 to 1983, and served as the Leader of the Canadian teams in the IMO from 1981 to 1984. He was the first Chairman of the A.H.S.M.C. Board. In 1985, he succumbed to cancer at the prime of his life.

Over the years, a nucleus of Board members emerged, consisting of **Bill Sands** and **Bob Woodrow** from the University of Calgary, along with **Alvin Baragar, Murray Klamkin** and **Andy Liu** from the University of Alberta.

Bill is best known for his editorship of *Crux Mathematicorum* for ten long, hard but rewarding years. He served as the Leader Observer in the 2002 IMO in Scotland, and will go as the Team Leader in the 2007 IMO in Vietnam. He is currently the chair of the Subcommittee which oversees the Canadian participation in the IMO.

Bob had served alongside Bill by editing the "Olympiad Corner" in *Crux Mathematicorum*, and is still providing this valuable service. He is the Minister of External Affairs for the Board, seeking and interacting with sponsors among his other valuable contributions to the Board.

Alvin was the Minister of Internal Affairs for the Board. He was our contact person for the Government, the Alberta Teachers' Association, school boards and schools. In 1995, he retired from the University of Alberta after 33 years on the faculty, and stepped down as Chairman of the Board, a post which he had held with distinction for ten years. He still serves on the Board.

Murray needed no introduction in the community of mathematics competitions and problem solving. He was an acknowledged expert in the field of inequalities in general, and the Triangle Inequality in particular. He had left his mark on every major mathematical journal that has a problem section. He was the long-time editor of the problem section in *SIAM Review* (Society of Industrial and Applied Mathematics, not Thailand). His passing in 2004, at the age of 83, marked the end of an era.

Andy served as the Team Leader in the 2000 IMO in South Korea and in the 2003 IMO in Japan. He is currently the Chairman of the Board, a post he had held before.

Other members who joined the Board after its inception were **Claude Laflamme** of the University of Calgary, plus **John Bowman, Dragos Hrimiuc** and **Alexander Litvak** of the University of Alberta. Apart from Dragos who is still on the Board, the others' stay were all too brief.

Ted Lewis served as the Chairman of the Board after the retirement of Alvin Baragar, and held the post for eight years before he in turn retired from the University of Alberta, though he stayed on the Board for another two years. He was succeeded by **Peter Minev**, but after two years, he stepped down as the Chairman and resigned from the Board altogether, as he was appointed Associate Chairman of the Department of Mathematical and Statistical Sciences at the University of Alberta. **Viktoria Mineva**, Peter's wife, went as the Deputy Leader Observer to the 2000 IMO in South Korea.

The University of Alberta, being the host institution of the Board, absorbs quite a bit of incidental expenses as well as providing a congenial environment for the smooth functioning of the Board. More importantly, it offers invaluable secretarial services.

In 1985, **Helga Dmytruk** became the first secretary formally appointed to the Board. Unfortunately, after a period of ten years of devoted service, her office staff position at the Department of Mathematical and Statistical Sciences was eliminated. The A.H.S.M.C. Board was indebted to **Leona Guthrie**, Manager of the General Office of the Department, for looking after its affairs while it searched for a new secretary. Several filled in on short-term basis, but the Board was really fortunate when **Linda Drysdale** was finally appointed to the position. A senior member of the General Office staff, she handled her new portfolio with great confidence and efficiency.

The Students

Over the years, many outstanding students distinguished themselves in the Alberta High School Mathematics Competition and in higher level contests. We can only feature a selected few here. It must also be emphasized that many outstanding students are not drawn to contest activities.

Two Alberta students stood out in the early years. In 1971, **Simon Tu** of Harry Ainlay High School, Edmonton, finished third in the CMO. **John Savard** of St. Joseph High School, Edmonton, had the same accomplishment a year later.

Robert Morewood of Medicine Hat was a top performer in the A.H.S.M.C. in the late 1970s. He had the misfortune of graduating a year before Canada entered the IMO. He made up for this by going as the Deputy Leader Observer to the 2002 IMO in Scotland, as the Leader Observer to the 2003 IMO in Japan, and as the Leader himself to the 2006 IMO in Slovenia.

In 1981, Alberta sent two members, **Arthur Baragar** and **John Bowman**, both of Old Scona Academic High School, Edmonton, to our first ever national team at the IMO. In 1998, Arthur was the Leader Observer at the IMO in Taiwan, and in 2002, he was the Team Leader at the IMO in Scotland. The son of Board member Alvin Baragar, Arthur is currently on the faculty at the University of Nevada at Las Vegas. John is currently a faculty member at the University of Alberta, and had served for one year on the A.H.S.M.C. Board.

Two students of prominence in the mid 1980s were **Sam Maltby** of Bishop Carroll High School, Calgary, and **Graham Denham** of Old Scona Academic High School, Edmonton. In 1987, Graham won the scholarship named after his grandfather.

The early 1990s were marked by the emergence of **Jason Colwell**. He received schooling at home, and moved from Medicine Hat to Edmonton to benefit from the resources available in a major urban center. He was registered with Old Scona Academic High School, though he continued to study on his own. He had earned a number of laurels for his alma mater in the Alberta High School Mathematics Competition. At age thirteen, he entered the University of Alberta and enrolled in the Honours Mathematics program. His academic record was remarkable even if his tender age was not taken into consideration. Later, at age seventeen, he became the youngest ever graduate of the University of Alberta at age seventeen. He is currently on the faculty at the University of California at San Diego.

The years 1993–1997 may be dubbed as the **Chun-Kisman Era**, because these two students dominated the mathematics competition scene within Alberta, and made significant impact beyond.

Byung-Kyu Chun was in Grade 10 in Edwin Parr High School, Athabasca, in 1993/1994. A year later, he moved to Edmonton and attended Harry Ainlay High School. Tempted by the many elective courses available only at a major urban center, he decided to take Grade 10 a second time. In 1996, he lifted the Sun Life Cup which went with the First Prize in the CMO, and he won the Second Prize in both 1995 and 1997. He represented Canada in the IMO three times, winning Bronze Medals in Canada (1995) and India (1996), and a Silver Medal in Argentina (1997).

Derek Kisman attended Queen Elizabeth Junior/Senior High School, Calgary. He represented Canada in the 1996 IMO in India, winning a Silver Medal.

The years 1998–2003 may be dubbed as the **Barrington Leigh-Fink Era**.

Robert Barrington Leigh of Old Scona Academic High School, Edmonton, won the Third Prize in the CMO in 2203. He represented Canada in the IMO twice, winning Bronze Medals in Scotland (2002) and Japan (2003). He also won a Silver Medal in the 2003 International Physics Olympiad.

Alexander Fink attended Queen Elizabeth Junior/Senior High School, Calgary. He won a Silver Medal in the 2002 IMO Scotland.

Both Robert and Alexander finished in the top ten, on more than one occasion, in the prestigious William Lowell Putnam Mathematics Competition among North American university students. Unfortunately, Robert's young and promising life came to a tragic end in 2006.

Two students of prominence in recent years were **David Rhee** of McNally High School, Edmonton, and **Boris Braverman** of Sir Winston Churchill High School, Calgary.

David won the Third Prize in the 2004 CMO, and topped it with a First Prize in 2006. He represented Canada in the IMO three times, winning an Honorable Mention in Greece (2004), a Bronze Medals in Mexico (2002) and a Silver Medal in Slovenia (2003).

Boris won a Gold Medal in the 2006 International Physics Olympiad.

Immediately after his return from Mexico in 2005, David set off on a 33-day European tour, the highlight of which was an intensive week-long Mathematics Summer Seminar in Mir Town, Belarus. He was invited on the strength of his excellent performance in the International Mathematics Tournament of the Towns. Boris also received an invitation, but too late for him to act on it. Two other Canadian students, both from Toronto, made it to the Seminar.

Also present was **Jerry Lo** of Taipei, the winner of the Second Round of the A.H.S.M.C. in 2004, when he was an international student in Vernon Barford Junior High School in Edmonton. He was with David for the whole tour.

The Teachers

Our work would not have borne fruit without the enthusiastic support of the many teachers in our school systems. Their selfless dedication to the well-being of their students is deeply appreciated, with or without explicit recognition.

In 1994, the Alberta High School Mathematics Competition Board introduced a book prize to recognize teachers for their support of their students' extra-curricular mathematical activities. Five winners were selected each year.

1994/1995. **Mr. Ed Heilman**, J. Percy Page High School, Edmonton, **Dr. A. Kapoor**, Tempo School, Edmonton, **Mr. Don Layton**, Forest Lawn High School, Calgary, **Mr. Hank Marcuk**, Archbishop MacDonald High School, Edmonton, and **Mr. Jack Rogers**, Henry Wise Wood High School, Calgary.

1995/1996. **Mr. Dennis Cantrill**, Sir Winston Churchill High School, Calgary, **Mr. Dennis Denham**, Acme School, Acme, **Ms. Marge Hallonquist**, Archbishop Jordan High School, Sherwood Park, **Ms. Elizabeth Mowat**, Strathcona High School, Edmonton, and **Mr. Dennis Remillard**, Olds Junior/Senior High School, Olds.

1996/1997. **Mrs. Corlene Balding**, École Secondaire Beaumont, Beaumont, **Mr. Bruce Kutcher**, Crowsnest Consolidated High School, Coleman, **Mr. Mark Milner**, Dr. E. P. Scarlett High School, Calgary, **Mr. Lorne Pascoe**, Old Scona Academic High School, and **Mr. Herbert Schabert** of Lorne Jenken High School, Barrhead.

1997/1998. **Mr. Patrick Ancelin**, James Fowler High School, Calgary, **Mr. Martyn Chapman**, Hilltop High School, Whitecourt, **Mr. Bob Frizzell**, Strathcona High School, Edmonton, **Mr. Lorne Lindenberg**, Harry Ainlay High School, Edmonton, and **Mrs. Pat McManus**, Bishop Carroll High School, Calgary.

1998/1999. **Mr. Randy Broemling**, Salisbury High School, Sherwood Park; **Ms. Marj Farris**, La Crete Public School, La Crete; **Mr. Perry Kulmatyski**, Richard F. Staples Secondary School, Westlock; **Mr. Ross Marion**, Victoria High School, Edmonton and **Mr. Ted Venne**, Father Patrick Mercredi Catholic High School, Fort McMurray.

1999/2000 **Mr. Stan Bold**, Henry Wise Wood High School, Calgary; **Ms. Linda Jaffray**, Eckville Jr/Sr High School, Eckville; **Mr. Ken May**, Winston Churchill High School, Lethbridge; **Mr. Lawrene Tomko**, St. Francis Xavier High School, Edmonton and **Ms. Hazel Williams**, Western Canada High School, Calgary.

At this point, the award program was temporarily suspended. We had simply run out of eligible candidates, though we were certain we had not run out of deserving candidates. It cannot be overemphasized that the selection is largely based on the data submitted by the teachers themselves, and it is important that they should do so. This is not blowing one's own horn, but a simple statement of service rendered. We will reintroduce the award program at some point in the future.

It should be mentioned that **Western Canada High School** and **Sir Winston Churchill High School**, both of Calgary, have established themselves as power-houses, each entering annually over one hundred students in the First Round and finishing regularly in the top three. Both are also well presented among the top performers in the Second Round.

The 1995 International Mathematical Olympiad

As promised earlier, we now give a more detailed account of the 36th International Mathematical Olympiad in July of 1995. The first one was organized in Romania way back in 1959, and this event was held annually since then, except in 1980. By 1995, the total number of contestants rose to 412, representing 73 countries and regions.

The contest took place on July 19 and 20. Each day, the contestants were given a paper consisting of three problems, and allowed four and a half hours to attempt them. Except in the first few years, the problems were weighted equally at 7 points each. The perfect score of 42 was attained by 14 students in 1995.

According to the rules, roughly one half of the contestants would receive Bronze, Silver or Gold Medals, roughly one quarter would receiver Silver or Gold Medals, and roughly one twelfth would receive Gold Medals. This year, the minimum scores for the three medal categories were 19, 29 and 37 respectively.

At most 6 students may represent a country or region each year. The competition is strictly individual, with no collaboration between team members. However the total of their individual scores is often taken to be the team score, and the teams are ranked accordingly, although no such ranking is officially recognized, and no prizes are awarded to teams. Of the six Canadian team members, two received Silver Medals and three Bronze Medals, including our own Byung-Kyu Chun.

The contest problems are submitted by participating countries and regions. Each may contribute up to 6 problems. The host country puts together a Problem Selection Committee to work on the submissions. The primary task is to clean up the mathematics whenever possible, by making the proposed solutions more elegant, or coming up with alternative solutions. The secondary task is to clean up the language. By this we mean removing ambiguity and not vulgarity (of which there was none).

Of the 13 people on this Committee in 1995, 8 had direct connection to the Alberta High School Mathematics Competition Board. They were Board members **Alvin Baragar, Murray Klamkin, Claude Laflamme, Andy Liu, Bill Sands** and **Bob Woodrow**, and former winners **Graham Denham** and **Sam Maltby**. Graham was in a graduate program in mathematics at the University of Michigan, USA. Sam was in a graduate program in mathematics at the University of Warwick, UK. Murray Klamkin was joined by **Richard Guy**, Professor Emeritus at the University of Calgary, as the Distinguished Members of this committee.

The Problem Selection Committee chose 28 of the 112 submitted problems. These were turned over to the Problem Interpolation Group (known fondly as the PIG) who then put together a "Short-listed Problem Book". A lot of work was involved even at this late stage, including a substantial amount of technical typesetting, as well as translation into French, our other official language. The PIG consisted of Alvin Baragar, Graham Denham, Claude Laflamme, Andy Liu and Sam Maltby.

This book was presented to the Jury, which consisted of the Leaders of the participating teams, upon their arrival at the University of Waterloo. As Chairman of the Problem Selection Committee, Andy Liu was an ex-officio of the International Jury Committee. Murray Klamkin was an observer of this Committee, and his expert advice was called upon on several occasions.

After a three-day deliberation, the Jury chose from the 28 problems the final 6 to be put on the papers. These were then translated into various languages, as all contestants wrote in the languages of their choice. They were housed, along with the Deputy Leaders of the teams, at York University, where the competition was written. It is traditional to keep some physical distance between those writing the contest and those making it up. Afterwards, the Waterloo entourage merged into York.

The scripts were first marked by the Leaders and Deputy Leaders. This was a linguistic necessity in many cases. The teams then defended their marking before a panel of Canadian coordinators. Alvin Baragar and Bill Sands acted as coordinating captains, each with ultimate responsibility for one problem, and Sam Maltby was a coordinator for a third. **Arthur Baragar**, another former winner of the Alberta High School Mathematics Competition, was at the time a visiting scholar at the University of Waterloo. He also acted as a coordinator.

This international event was the swan song of **Alvin Baragar** before his retirement. As Executive Secretary of the Problem Selection Committee, he had pulled his weight incessantly, far beyond the call of duty. His many contributions are largely responsible for the success in all facets of the 36th International Mathematical Olympiad with which he was involved.

Acknowledgements

I would like to thank my co-publishers **Graham Wright** of the Canadian Mathematical Society and **Don Albers** of the Mathematical Association of America, as well as my copy editor **Beverly Ruedi** of the MAA. I am grateful to **David Langstroth** of the CMS for administrative assistance.

Ancient Period: 1957–1966

Problems: 1957

1. Express $\dfrac{\frac{1}{1+x}}{1-\frac{1}{1+x}}+\dfrac{\frac{1}{1+x}}{\frac{x}{1-x}}+\dfrac{\frac{1}{1-x}}{\frac{x}{1+x}}$ as a simple fraction.

2. Solve the equation $\sqrt{16x+1}-2(\sqrt[4]{16x+1})=3$.

3. In the center of the flat rectangular top of a building which is 21 meters long and 16 meters wide, a flagpole is to be erected, 8 meters high. To support the pole, four cables are needed. The cables start from the same point, 2 meters below the top of the pole, and end at the four comers of the top of the building. How long is each of the cables?

4. Solve the equation $x+2=0$.

5. Solve the equation $x^2+x+1=0$.

6. Solve the equation $x^3+x=10$.

7. Given that $ax^2+bx+c=0$ has the roots m and n, prove that

   (a) $m+n=-\frac{b}{a}$;

   (b) $mn=\frac{c}{a}$.

8. From a 12×18 sheet of tin, we wish to make a box by cutting a square from each corner and turning up the sides. Determine the size of the squares which yields the largest box.

9. Solve the system of equations $x+y+z=3$, $x+2y+3z=8$ and $x+3y+4z=11$.

10. Solve the system of equations $x+y+z=1$, $2x+3y+4z=2$ and $3x+5y+7z=4$.

11. A certain sample of radium is decreasing according to the equation $A=3\cdot 2^{-\frac{t}{1800}}$, where t is in years and A is in milligrams.

    (a) How much radium was in the sample at $t=0$?

    (b) How much will there be 900 years later?

12. Two circles, each of radius 1, are such that the center of each lies on the circumference of the other. Find the area common to both circles.

13. A man takes a trip from A to B at an average speed of 40 kph and returns at an average speed of 60 kph. What is his average speed for the entire trip?

14. Four men dined at a hotel. They checked their hats in the cloakroom. Each of the four came away wearing a hat belonging to one of the other three. In how many different ways could this have happened?

15. Given a cylinder of height 6 cm and base radius 2 cm, prove that a spider can go from any point on the surface to any other point on the surface along a path of total length less than 9 cm.

16. Prove that the difference between $\frac{n}{n} + \frac{n-1}{n} + \frac{n-2}{n} + \cdots + \frac{1}{n}$ and the infinite sum $\frac{n}{n+1} + \frac{n}{(n+1)^2} + \frac{n}{(n+1)^3} + \cdots$ is equal to $\frac{n-1}{2}$.

17. If two sides of a quadrilateral are parallel to each other, prove that the straight line joining their midpoints passes through the point of intersection of the diagonals.

18. In a certain school all students study Mathematics, Physics and French. Forty percent prefer both Mathematics and Physics to French. Fifty percent prefer Mathematics to French and sixty percent prefer Physics to French. If all students have definite preferences between subjects, what percentage prefer French to both Mathematics and Physics?

Problems: 1958

1. Evaluate $\frac{1}{7} + \frac{2}{3} + \frac{5}{8}$.

2. Express $\dfrac{c}{a + \frac{b}{c}} + \dfrac{a+c}{a - \frac{b}{c}}$ as a simple fraction.

3. Evaluate $9!(\frac{1}{8!1!} + \frac{1}{7!2!} + \frac{1}{6!3!} + \frac{1}{5!4!})$.

4. Solve the equation $(x-2)(x-3) = 1$.

5. Solve the equation $\frac{1}{x-2} + \frac{1}{x-3} = 1$.

6. Solve the system of equations $3x + 2y + z = 1$, $4x - y = 2$ and $x - y + 2z = 3$.

7. You are traveling along a road at 4 kph parallel to a double track railroad. Two trains meet you, each being of the same length. The first takes 40 seconds to pass you, the second 30 seconds. Prove that it takes 4 minutes for the second train to completely pass the first.

8. Find the greatest common divisor of 3910, 8551 and 11475.

9. A quartet is chosen by lot from all the high school students in a certain school. There are 40 students in Grade X, 30 in Grade XI and 20 in Grade XII. In how many ways can the quartet consist of

   (a) students from Grade X alone;

   (b) at least three students from Grade XII?

10. Let $f(n+1) = 5f(n) - 6f(n-1)$ for a function f defined for all positive integers n. Prove that $f(n) = 3^n - 2^n$ is a solution of this equation.

11. For $f(n) = 3^n - 2^n$, find the sum $f(1) + f(2) + \cdots + f(9)$.

12. Use Newton's Binomial Theorem to compute the cube root of 26 to three decimal places.

13. State the Remainder Theorem for polynomials.

14. Solve the equation $4x^3 - 8x^2 - 3x + 9 = 0$.

15. Let P, R and S be any three points on a circle such that the extension of RS meets the tangent at P at the point T. Prove that $\frac{PS}{PR} = \frac{ST}{PT} = \frac{PT}{RT}$.

16. (a) Prove that the sum of the first n positive integers is $\frac{n(n+1)}{2}$.

    (b) Prove that the sum of the cubes of the first n positive integers is $\frac{n^2(n+1)^2}{4}$.

17. Identify the curves given by the equations $x^2 + 4y^2 - 6x - 16y + 21 = 0$ and $y^2 - x^2 - \frac{8xy}{3} = 0$.

18. Solve the system of equations $x^2 + 4y^2 - 6x - 16y + 21 = 0$ and $y^2 - x^2 - \frac{8xy}{3} = 0$.

19. At $r\%$ interest compounded semi-annually, how much would p dollars deposited on each of the dates: January 1, 1940, 1941, 1942, 1943 and 1944, be worth altogether on January 1, 1945?

20. Prove the Sine Law for triangles.

21. Prove that $\sin(A + B) = \sin A \cos B + \cos A \sin B$.

22. From the top of a tower, the angle of depression of a point A, in the same horizontal plane as the base of the tower, is x. On the top of the tower is a flagstaff whose length is equal to one-quarter of the tower. Prove that the tangent of the angle which the flagstaff subtends at A is equal to $\frac{\sin x \cos x}{4+\sin 2x}$.

Problems: 1959

1. Solve the equation $2x - 6 + \sqrt{3x - 2} = 0$.

2. Solve the equation $x^3 + 3x^2 - 4x - 12 = 0$.

3. Solve the system of equations $\frac{6}{x} + \frac{15}{y} = 4$, $\frac{18}{y} - \frac{16}{z} = 1$ and $\frac{14}{x} + \frac{12}{z} = 5$.

4. Given that the roots of $ax^2 + bx + c = 0$ are m and n, find the condition on a, b and c so that

    (a) $m = \frac{1}{n}$;

    (b) $m = -n$.

5. Prove that $\frac{n!}{(n-2)!2!} + \frac{(n+1)!}{(n-1)!2!} = n^2$.

6. $ABCD$ is a kite-shaped figure with $AB = AD$ and $CB = CD$. Find the point P the sum of whose distances from the four vertices is as small as possible.

7. Six papers are set in an examination, two of them in mathematics. In how many different orders can the papers be given, provided only that the two mathematics papers are not consecutive?

8. Prove that $\sin(A+B)\sin(A-B) = \sin 2A - \sin 2B$.

9. A vertical flagpole stands on a hillside which makes an angle A with the horizontal. At a distance k down the slope from the pole, it subtends an angle B. Prove that the height of the pole is given by $\frac{k \sin B}{\cos(A+B)}$.

10. If 1 is added to the product of four consecutive positive integers, is the sum always the square of an integer?

11. If 41 is added to the product of two consecutive positive integers, is the sum always a prime number?

12. What is the coefficient of r^2 in the expansion $\frac{1}{\sqrt{r^2 - 2r\cos\theta + 1}}$ in powers of r?

13. The first three terms of an arithmetic progression are m, $4m-1$ and $5m+3$. What is the sum of the first $4m$ terms?

14. Explain why and how it is possible to attach definite meanings to zero, negative and fractional powers of a positive number.

15. How many different combinations of 2-cent, 3-cent, 4-cent and 5-cent stamps yield a total postage of 25 cents?

16. A line through C intersects the line AB at O, with $OA = a$ and $OB = b$. A point P moves along OC, and P_1 and P_2 are two positions of P such that the $\angle AP_1B$ and $\angle AP_2B$ are equal.

    (a) Prove that $OP_1 \cdot OP_2 = ab$.

    (b) At what distance of P from O is $\angle APB$ a maximum?

17. What is wrong with the following argument?

$$
\begin{aligned}
(x+1)^2 &= x^2 + 2x + 1, \\
(x+1)^2 - (2x+1) &= x^2, \\
(x+1)^2 - (2x+1) - x(2x+1) &= x^2 - x(2x+1), \\
(x+1)^2 - (x+1)(2x+1) + \frac{(2x+1)^2}{4} &= x^2 - x(2x+1) + \frac{(2x+1)^2}{4}, \\
\left((x+1) - \frac{2x+1}{2}\right)^2 &= \left(x - \frac{2x+1}{2}\right)^2, \\
x + 1 - \frac{2x+1}{2} &= x - \frac{2x+1}{2}, \\
1 &= 0.
\end{aligned}
$$

18. PQ is a chord passing through the focus $(a, 0)$ of a parabola with vertex at the origin. The slope of PQ is m.

    (a) Prove that the coordinates of its midpoint are $a(1 + \frac{2}{m^2})$ and $\frac{2a}{m}$.

    (b) Prove that the locus of the midpoints of all chords passing through the focus is again a parabola.

    (c) Find the focus and directrix of the parabola in (b).

19. Given m vertical and n horizontal lines, prove that the number of rectangles which can be formed having segments of these lines as sides is $\binom{m}{2}\binom{n}{2}$.

Problems: 1960

1. Find the distance between the points $(-1,-5)$ and $(-13,0)$.

2. Evaluate $(2+\sqrt{3})^4 + (2-\sqrt{3})^4$.

3. Find the area of an equilateral triangle of side 1.

4. One root of $2hx^2 + (3h-6)x - 9 = 0$ is the negative of the other.

   (a) Find the value of h.

   (b) Solve the equation.

5. Find a cubic equation with integral coefficients which has as roots the numbers $\frac{2}{3}, -2$ and -1.

6. Solve the equation $x^3 - 4x^2 + x + 6 = 0$.

7. (a) Prove that, for any positive number n, the numbers $x = (1+\frac{1}{n})^{n+1}$ and $y = (1+\frac{1}{n})^n$ satisfy the equation $x^y = y^x$.

   (b) Do the formulae in (a) give all the positive solutions of $x^y = y^x$?

8. Given that $f(x) = \dfrac{1}{x + \frac{1}{x+\frac{1}{x}}}$ and $g(x) = x - \dfrac{1}{x}$, determine

   (a) $f(g(x))$;

   (b) $g(f(x))$.

9. Prove that $\log_b xy = \log_b x + log_b y$.

10. Prove that $\log_b x^n = n \log_b x$.

11. From A, a pilot flies $12\sqrt{2}$ km in the direction N30°W to position B, and then $12\sqrt{2}$ km in the direction S60°E to position C. How far and in what direction must he now fly to again reach A?

12. Prove that $\frac{1-\cos x+\sin x}{1+\cos x+\sin x} = \tan\frac{x}{2}$.

13. There are ten points in the plane, no three in the same straight line.

    (a) How many lines are determined by the points?

    (b) How many of the lines pass through a chosen point?

    (c) How many triangles are determined by the points?

    (d) How many of the triangles have a chosen point as a vertex?

    (e) How many of the triangles have two chosen points as vertices?

14. In how many ways can the word PYRAMID be spelt out, using adjacent letters of the arrangement below?

D I M A R Y P Y R A M I D
D I M A R Y R A M I D
D I M A R A M I D
D I M A M I D
D I M I D
D I D
D

15. Prove that if one side of a triangle is greater than another, the angle opposite the greater side exceeds the angle opposite the shorter side.

16. Prove that if $AC = 2BC$ in triangle ABC, then $\angle B$ is more than twice $\angle A$.

17. A ball is dropped from a height of 6 meters. Each time it strikes the ground after falling from a height of h meters, it rebounds to a height of $\frac{2h}{3}$ meters.

    (a) How far has the ball travelled when it hits the floor for the fifth time?

    (b) What is the total distance travelled by the ball before it comes to rest?

18. Three numbers are in geometric progression. If A, G and H are their arithmetic, geometric and harmonic means, respectively, prove that $G^2 = AH$.

19. Numerical calculation seems to show that the relation $8 - \sqrt{62} = \frac{\sqrt[3]{2}}{10}$ is at least approximately true. Determine whether the relation is exact.

Problems: 1961

1. Express $1 + \dfrac{2}{x + \frac{3}{x+\frac{4}{x}}} - \dfrac{3}{x - \frac{3}{x+\frac{4}{x}}}$ as a simple fraction.

2. Express $\sqrt{\dfrac{a-b}{a+b}} + \sqrt{\dfrac{a+b}{a-b}} - \sqrt{c^2a^2 - c^2b^2}$ as a simple fraction.

3. Solve the equation $x - 7\sqrt{x-4} - 12 = 0$.

4. Solve the equation $\sqrt{3x+9} - \sqrt{x+5} = \sqrt{2x+8}$.

5. Solve the equation $2^x = 10$.

6. Solve the inequality $\frac{3x-2}{x} > 1$.

7. In triangle ABC, the longest side BC is of length 20 and the altitude from A to BC is of length 12. A rectangle $DEFG$ is inscribed in ABC, with D on AB, E on AC and both F and G on BC. Determine the maximum area of $DEFG$.

8. Prove that the largest triangle which can be inscribed in a circ]e is equilateral.

9. Find the locus of a point P such that $AP^2 - BP^2 = d^2$, where A and B are two fixed points and d is a real number.

10. Prove that the point of intersection of the lines $2y + x + 1 = 0$ and $y - 3x + 4 = 0$ lies on the line $A(2y+x+1) + B(y-3x+4) = 0$, where A and B are real numbers.

11. Identify the curve given by the equation $(x^2 + y^2)(x^2 + y^2 - 1) = 0$.

12. Let x take on a set of values which form a geometric progression. Prove that the corresponding values of $y = \log_b x$ form an arithmetic progression, where b is a positive real number not equal to 1.

13. Let $f(x) = 4 - x^2$. Consider $\frac{f(x)-f(1)}{x-1}$.

    (a) Interpret this expression geometrically as x takes on values successively nearer to 1.

    (b) Determine its limiting value.

14. Let $S_n = 1 + \frac{1}{2} + \frac{1}{4} + \frac{1}{8} + \cdots + \frac{1}{2^n}$. What is the least value of n such that $S_n > \frac{127}{64}$?

15. Convert the recurring decimal 0.147147147... into a fraction.

16. The equation $ax^2 + bx + c = 0$ has two real roots. Prove that, if a is very "small" while b and c are of "moderate" size, then one of the roots is close to $-\frac{c}{b}$ while the other is very large numerically.

17. Prove that $\binom{n}{0} + \binom{n}{1} + \binom{n}{2} + \cdots + \binom{n}{n} = 2^n$ for all non-negative integers n.

18. Prove that $\binom{n}{0} - \binom{n}{1} + \binom{n}{2} - \cdots + (-1)^n \binom{n}{n} = 0$ for all positive integers n.

19. Nine students are to be assigned to three rooms, three students to a room.

    (a) In how many ways can this be done?

    (b) What if two particular students refuse to be assigned to the same room?

20. Fifteen passengers rode on a railway line which leads to 25 towns. If no two persons get off at the same town, what is the total number of ways in which they can get off?

21. Solve the equation $\sin x = \sin 2x$.

22. Solve the equation $\tan x = \cot 2x$.

23. In triangle ABC, $BC = a > b = AC$ and the difference between $\angle A$ and $\angle B$ is x. Given a, b and x, describe an Euclidean construction of triangle ABC.

Problems: 1962

1. Determine d such that when $\dfrac{\sqrt{1-4x^2} - \frac{x}{2}\frac{1}{\sqrt{1-4x^2}}(-8x)}{1-4x^2} = \dfrac{1}{(1-4x^2)^d}$, where $-\frac{1}{2} < x < \frac{1}{2}$.

2. (a) Determine a and b such that $a(3x+5) + b(2x+3) = 12x + 19$ for every x.

   (b) Determine A and B such that for every x except $-\frac{3}{2}$ and $-\frac{5}{3}$, $\frac{12x+19}{(3x+5)(2x+3)} = \frac{A}{2x+3} + \frac{B}{3x+5}$.

3. Solve the equation $x^2 - x - 20 = 0$.

4. Solve the equation $\frac{1}{x-2} + \frac{1}{x+2} = \frac{4}{x^2-4}$.

5. Solve the equation $\sqrt{4-3x} - x = 12$.

6. For any two real numbers x and y, each greater than 1, prove that $\log_y x = \frac{1}{\log_x y}$.

7. Solve the equation $2^x - 4 \cdot 2^{-x} + 3 = 0$.

8. Using only an instrument with two straight edges meeting at a right angle, construct the center of a given circle.

9. Solve the equation $\cos 2x = \cos x$.

10. Solve the equation $\sin 5x - \sin x = \cos 3x$.

11. Solve the equation $\sec x - 2\cos x - \tan x = 0$.

12. (a) Let $S_n = \frac{1}{1\cdot 2} + \frac{1}{2\cdot 3} + \frac{1}{3\cdot 4} + \cdots + \frac{1}{n(n+1)}$. Express S_n as a simple fraction in terms of n.

    (b) Determine the smallest n such that S_n in (a) is greater than $\frac{100}{101}$.

13. Let m and n be the roots of the equation $ax^2 + bx + c = 0$. Find a quadratic equation with coefficients expressed in terms of a, b and c which has $m+2$ and $n+2$ as roots.

14. Prove that $1 + 2r + 3r^2 + \cdots + nr^{n-1} = \frac{1-r^n}{(1-r)^2} - \frac{nr^n}{(1-r)}$, where r is a real number not equal to 1.

15. An after-dinner speaker anticipates delivering 35 speeches during the next five years. So as not to become bored with his jokes, he decides to tell exactly three jokes in every speech, and in no two speeches to tell exactly the same three jokes.

    (a) What is the minimum number of jokes that will accomplish this?

    (b) What is the minimum number if he decides never to tell the same joke twice?

16. In triangle ABC, side BC, $\angle B$ and $\angle C$ are given. Prove that the length of the altitude from A to BC is $\frac{BC}{\cot B + \cot C}$.

17. Prove that $\tan A + \tan B + \tan C = \tan A \tan B \tan C$, where A, B and C are the angles of any triangle.

18. Prove that, for any two positive numbers whose sum is 8, their product is a maximum when they are equal.

19. Prove that, among all rectangles with a given area, the square has the least perimeter.

20. The incircle of triangle ABC has center I and touches BC at P. One of the excircles of triangle ABC has center U and touches BC at Q.

    (a) Prove that $BP = CQ$.

    (b) Prove that B, C, I and U lie on a circle.

21. (a) For the quartic equation $Ax^4 + Bx^3 + Cx^2 + Bx + A = 0$, prove that if r is a root, so also is $\frac{1}{r}$.

    (b) If $t = x + \frac{1}{x}$, express $x^2 + \frac{1}{x^2}$ in terms of t.

    (c) Solve the equation $x^4 - x^3 - 10x^2 - x + 1 = 0$ and verify that its roots occur in pairs as indicated in (a).

22. Let ABC be any triangle.

 (a) Prove that $\sin A + \sin B + \sin C = 2\cos\frac{C}{2}(\cos\frac{A-B}{2} + \sin\frac{C}{2})$.

 (b) For a fixed value of $\angle C$, what is the relation between $\angle A$ and $\angle B$ if $\sin A + \sin B + \sin C$ is to have its maximum value?

 (c) Prove that the maximum value of $\sin A + \sin B + \sin C$ is $\frac{3\sqrt{3}}{2}$.

23. Let S be a finite set of points in the plane. There will be a smallest distance d between some pair of them which may, of course, occur between several pairs. Prove that, for any point P of S, there cannot be more than six other points of S whose distance from P is d.

Problems: 1963

1. Express $\dfrac{1}{x-\frac{1}{x}} - \dfrac{1}{x+\frac{1}{x}} - \dfrac{2x}{x^2-\frac{1}{x^2}}$ as a simple fraction.

2. Express $a - 2ax + 4ax^2 - \frac{8ax^3}{1+2x}$ as a simple fraction.

3. Express $\dfrac{\frac{2x}{1-x^2}}{2+\frac{2x^2}{1-x^2}}$ as a simple fraction.

4. Simplify $(2+5x)^2 + (5-2x)^2 - 13x^2$.

5. Express $\frac{a}{(2-\sqrt{3})^2} + \frac{b}{(3+2\sqrt{2})^2}$ as a simple fraction.

6. Solve the system of equations $x + y = 80$ and $x^2 + y^2 = 3250$.

7. If the sum of two numbers is 80, find the largest possible value of their product.

8. If the sum of two positive numbers is equal to N, what is the smallest value of the sum of their reciprocals?

9. Two cars, A and B, cover a distance of 200 km, each at constant speed, but with B traveling at a constant speed $\frac{25}{6}$ kph faster and hence requiring 12 minutes less time. Determine the speeds of the cars.

10. Solve the equation $\sqrt{8x} - \sqrt{6x+1} = 1$.

11. Two straight lines $y = 3x + 7$ and $y = 5x - 4$ meet in one point. Prove that all other lines passing through the same point have equations of the form $y = k(3x+7) + (1-k)(5x-4)$.

12. A hall is 20 meters long. The end walls are 10 meters by 10 meters. In the middle of one of them, one meter from the floor, is a fly. In the middle of the other, one meter from the ceiling, is a spider. The spider, being hungry, wishes to take the shortest route possible to crawl from where it is to where the fly is. What is the length of the shortest route?

13. (a) Prove that the sum of the first n positive integers is $\frac{n(n+1)}{2}$.

 (b) Prove that the sum of the cubes of the first n positive integers is $\frac{n^2(n+1)^2}{4}$.

(c) Simplify $n^3(n+1)^3 - n^3(n-1)^3$.

(d) Prove that the sum of the fifth powers of the first n positive integers is $\frac{n^3(n+1)^3}{6} - \frac{n^2(n+1)^2}{12}$.

14. Prove that the perpendicular distance of any point (a, b) such that $3a + 4b > 10$ from $3x + 4y = 10$ is $\frac{3a+4b-10}{5}$.

15. Let S be the sum of the first n terms of the series $a + 4ax + 9ax^2 + \cdots + n^2ax^{n-1}$.

(a) Prove that $S - xS = a + 3ax + 5ax^2 + \cdots + (2n-1)ax^{n-1} - n^2ax^n$.

(b) Determine $(1-x)^2S$.

16. From a point P on the circumference of a circle, a distance PT of 10 meters is laid out along the tangent. The shortest distance from T to the circle is 5 meters. A straight line is drawn through T cutting the circle at X and Y. The length of TX is $\frac{15}{2}$ meters.

(a) Determine the radius of the circle.

(b) Determine the length of XY.

17. The equation $x^4 - 19x^2 + 20x - 4 = 0$ may be rewritten as $(x^2 + sx + p)(x^2 - sx + q) = 0$ for constants s, p and q.

(a) Prove that $(\frac{20}{s})^2 = (s^2 - 19)^2 + 16$.

(b) Verify that $s = 4$ satisfies the equation in (a).

(c) Solve the original quartic equation.

18. The sides of a triangle a, b and c are related by $c^2(a+b) = a^3 + b^3$.

(a) Prove that one angle is exactly 60°.

(b) Express the area of the triangle in terms of a and b.

19. An isosceles triangle has an interior angle of 36° between two sides, each one meter long. One of the angles at the base is bisected by a line from that vertex to the opposite side. This line is x meters long.

(a) Prove that the base is also x meters long.

(b) Prove that one of the segments of the divided side is also x meters long.

(c) Prove that $x + x^2 = 1$.

20. (a) Prove that $\sin 2 - x = 2\sin x \cos x$.

(b) Prove that $4\sin 18° \sin 54° = 1$.

(c) Prove that $\sin(A+B) - \sin(A-B) = 2\cos A \sin B$.

(d) Prove that $\sin 54° - \sin 18° = \frac{1}{2}$.

(e) Prove that $\sin 18° = \frac{\sqrt{5}-1}{4}$.

21. (a) Prove that $\frac{2}{\tan 2x} = \frac{1}{\tan x} - \tan x$.

(b) Express $\frac{1}{\tan x} - \tan x - 2\tan 2x - 4\tan 4x$ as a simple fraction.

22. Prove that $\frac{2}{\sin 2x} = \frac{1}{\tan x} + \tan x$.

23. (a) Prove that $\frac{\sin(n+1)x - \sin nx}{\cos(n+1)x + \cos nx} = \tan\frac{x}{2}$.

(b) Prove that $1 + 2\cos x + 2\cos 2x + 2\cos 3x + \cos 4x = \frac{\sin 4x}{\tan\frac{x}{2}}$.

Problems: 1964

1. If the distance s meters that a bomb falls vertically in t seconds is given by the formula $s = \frac{16t^2}{1+\frac{3t}{50}}$, how many seconds are required for a bomb released at an altitude of 20000 meters to reach ground level?

2. (a) Find three consecutive positive even integers such that the square of the largest is equal to the sum of the squares of the other two.

   (b) Prove that this is impossible for consecutive positive odd integers.

3. A man has 15878 equilateral triangular pieces of mosaic, all of side length one cm. He constructs the largest possible mosaic in the shape of an equilateral triangle.

   (a) What is the side length of the mosaic?

   (b) How many pieces will he have left over?

4. A, B, C and D are four points in a plane. The midpoints of AB, BC, CD and DA are P, Q, R and S, respectively.

   (a) Prove that $PQRS$ is a parallelogram.

   (b) How is this result modified if the four points A, B, C and D are not all in one plane?

5. Solve the equation $9 \cdot 10^{2x} - 6 \cdot 10^x + 1 = 0$.

6. Explain how logarithms may be used to compute the fifth root of a real number.

7. Let p, q and r be three positive numbers such that $p + q + r = 12$.

   (a) If p is held fixed while q and r are allowed to vary, prove that product pqr is greatest when $q = r$.

   (b) What is the greatest possible value of pqr?

8. Find the value of the constant k so that the equation $kx^2 + 6x - 4 = 0$ has two equal roots.

9. Find a quadratic equation whose roots are the reciprocals of the roots of $x^2 + x + 4 = 0$.

10. Prove that $\sin 2A + \sin 2B + \sin 2C = 4 \sin A \sin B \sin C$, where A, B and C are the angles of any triangle.

11. Divide 100 loaves among five men so that the shares received shall be in arithmetic progression, and so that one-seventh of the sum of the largest three shares shall be equal to the sum of the smallest two shares. Individual loaves may be subdivided if necessary. What are the shares of the five men?

12. Assume that $\sqrt{2 + \sqrt{2 + \sqrt{2 + \sqrt{2 + \cdots}}}}$, where the number of 2's and radical signs are infinite, is a meaningful expression and has a definite real value. Prove that this value is 2.

13. (a) Find the square root of the complex number $4 - 6\sqrt{5}i$.

    (b) Represent graphically the given number and also the square roots.

14. A circle is folded along a chord AB and a line is drawn from B to cut the two circular arcs at P and Q, respectively. Prove that triangle PAQ is isosceles.

15. Prove that $1 - \frac{1}{2} + \frac{1}{4} - \cdots + (-\frac{1}{2})^{n-1} = \frac{2}{3}(1 - (-\frac{1}{2})^n)$.

16. Two cyclists are 20 km apart on a straight road and, at the same moment, begin cycling towards each other at a speed of 10 kph. At the instant they begin moving, a fly which can travel at 20 kph leaves the nose of one of them and flies towards the other. As soon as it arrives at the second nose, it turns around and flies back to the first, continuing to go backwards and forwards until the cyclists meet. How far has the fly flown?

17. Use Newton's Binomial Theorem to compute the cube root of 63.9 to four decimal places.

18. In expression $(a^2 - 2b^2)^{\frac{3}{2}}$ may be expanded so that the first term is a^3. Find the fifth term.

19. Four cards are drawn at random from an ordinary deck of 52. What is the probability that exactly three of these will be clubs?

Problems: 1965

1. Determine the value of $(\sqrt{2}^{\sqrt{2}})^{\sqrt{2}}$.

2. Prove that, for any positive number n, the numbers $x = (1+\frac{1}{n})^{n+1}$ and $y = (1+\frac{1}{n})^n$ satisfy the equation $x^y = y^x$.

3. Prove that $1 + \frac{2x}{x^2-x+1} = (1 - \frac{2}{x^3+1})(1 + \frac{2}{x-1})$ if $x \neq 1$ and $x^3 + 1 \neq 0$.

4. Find constants a and b such that the equation $\sqrt{(x+1)^2 + y^2} - \sqrt{(x-1)^2 + y^2} = 1$ may be rewritten in the form $(\frac{x}{a})^2 - (\frac{y}{b})^2 = 1$.

5. Solve the equation $2x - 5 = \sqrt{2x + 1}$.

6. Find the sum of the squares of the roots of the equation $x^3 - px = q$ in terms of p and q.

7. Solve the equation $x^3 - 13x = 12$.

8. Two parallel walls at some distance apart are perpendicular to the ground-level and two ladders are placed one against each wall so that the other ends touch the bases of the opposite walls. The ladders touch each other at some point between the wall, h meters above the ground. The top of the ladder of length m meters is at a height of a meters above the ground. The height of the top of the ladder of length n meters is b meters above the ground.

    (a) Prove that $h = \frac{ab}{a+b}$.

    (b) Find an equation involving a, m, n and h but not b.

9. (a) Prove that there do not exist positive integers m and n such that $10^m = 2^n$.
   (b) Prove that $\log_{10} 2$ is not a rational number.

10. Prove that $\binom{2n}{n}$ is an even number where n is any positive integer.

11. Prove that $\displaystyle\binom{2n}{n-1} + \binom{2n}{n+1} = \frac{2n}{n+1}\binom{2n}{n}$.

12. A motorized column is advancing over flat country at the rate of 15 kph. It is 1 km long. A dispatch rider is sent from the rear to the front on a motorcycle traveling at a constant speed. He returns immediately at the same speed and his total time is 3 minutes. How fast is he going?

13. Through a point R outside a circle with center 0 and radius r, a line is drawn cutting the circle in two distinct points P and Q. Prove that $RP - RQ = OR^2 - r^2$.

14. Prove that $\sin(A + B)\sin(A - B) = \sin 2A - \sin 2B$.

15. (a) Prove that $\frac{1}{n^2} - \frac{1}{(n+1)^2} = \frac{2n+1}{n^2(n+1)^2}$.
    (b) Prove that $\frac{3}{1^2 2^2} + \frac{5}{2^2 3^2} + \cdots + \frac{2n-1}{n^2(n-1)^2} = 1 - \frac{1}{n^2}$.
    (c) Express $\frac{1}{n(n^2-1)}$ as a sum of fractions with simpler denominators.
    (d) Express $\frac{1}{2(2^2-1)} + \frac{1}{3(3^2-1)} + \cdots + \frac{1}{n(n^2-1)}$ as a simple fraction in terms of n.

16. The radius of the base of a right circular cone is 2 meters and the slant height from the edge of the base to the vertex is 6 meters. Find the total surface area of the cone.

17. The radius of the base of a right circular cone is 2 meters and the slant height from the edge of the base to the vertex is 6 meters. From a point A on the edge of the base one may proceed to a point B halfway up the cone towards the vertex. Consider the point C directly opposite B, also halfway up the cone. Find the shortest distance from A to C on the surface of the cone.

Problems: 1966

1. Factor $x^5 + x^4 + x^3 + x^2 + x + 1$ as far as possible into polynomials with integral coefficients.

2. Prove that the roots of the equation $bx^3 + a^2x^2 + a^2x + b = 0$ are in geometric progression.

3. If the product of two positive numbers is 36, prove that their sum is at least 12.

4. ABC is a triangle with $AB = AC$. D is a point on the extension of AB, and E is a point on CA or its extension, such that $\angle BEC = \angle BDC$. Prove that $BE = CD$.

5. Solve the inequality $\frac{1}{x-1} + \frac{1}{x+1} > \frac{1}{2}$.

6. Solve the system of equations $x + y + z = 7, 3x + 2y - z = 3$ and $x^2 + y^2 + z^2 = 21$.

7. There are ten guests at a party. Assume that all acquaintances are mutual and that no one is considered an acquaintance of himself or herself. Prove that two of the guests are acquainted with the same number of guests at the party.

8. Indicate the region in the xy-plane for which $x + y$ takes values between -2 and 2 inclusive.

9. The line $y = 3x + b$ meets the parabola $2y = x^2 + 2x$ in two distinct points P and Q.

   (a) What restriction does this place on b?

   (b) Prove that the x coordinate of the midpoint of PQ is independent of b.

10. Prove that the area of a triangle inscribed in a parallelogram is at most one-half the area of the parallelogram.

11. Let S_n denote the sum of the first n terms of the series $1 + \frac{2}{2} + \frac{3}{4} + \cdots + \frac{n}{2^{n-1}} + \cdots$.

   (a) Calculate S_5.

   (b) Prove that $4 - S_n = (n + 2)(\frac{1}{2})^{n-1}$.

   (c) Find the "sum to infinity" of this series.

12. P is a point on the side CD of a parallelogram $ABCD$. AP and BC, extended if necessary, meet at Q. AD and BP, extended if necessary, meet at R. Prove that $\frac{1}{BQ} + \frac{1}{AR} = \frac{1}{AD}$.

13. Let $F(x, y)$ be a function such that for any x and y, $F(x, y) = F(y, x)$ and $F(x, y) = F(x, x - y)$. Prove that $F(x, y) = F(-x, -y)$.

14. A triangle has sides 20 cm, 20 cm and 5 cm. Determine the lengths of its interior angle bisectors.

15. (a) Prove that two consecutive integers have no common divisors other than ± 1.

   (b) Suppose $n + 1$ positive integers are taken, all different and none greater than $2n$. Prove that at least two of them have no common divisors other than ± 1.

16. (a) Express $\frac{x}{1-x} - \frac{x}{1+x}$ as a simple fraction.

   (b) Prove that $\frac{x}{1-x} = \frac{x}{1+x} + \frac{2x^2}{1+x^2} + \frac{4x^4}{1+x^4} + \cdots$ for $-1 < x < 1$.

17. Prove that, however large the positive number N may be, one can always find a number whose logarithm to base 10 is greater than N.

18. Prove that $\sin A + \sin B + \sin C = 4 \cos \frac{A}{2} \cos \frac{B}{2} \cos \frac{C}{2}$, where A, B and C are the angles of any triangle.

Medieval Period: 1967–1983

Questions with Multiple Choices

Problems: 1967

1. When the base of a triangle is increased 10%, and the altitude to this base is decreased 10%, the change in area is

   (a) 1% increase (b) $\frac{1}{2}$% increase (c) 0%

   (d) $\frac{1}{2}$% decrease (e) 1% decrease

2. If $\frac{4^x}{2^{x+y}} = 8$ and $\frac{9^{x+y}}{3^{5y}} = 243$, then xy is

   (a) $\frac{12}{5}$ (b) -4 (c) 4 (d) 12 (e) 6

3. The value of $\frac{(4-\sqrt{5})(2+\sqrt{5})}{7+\sqrt{5}}$ is

   (a) $\frac{8}{7} - \sqrt{5}$ (b) $\frac{4-\sqrt{5}}{11}$ (c) $\frac{8+4\sqrt{5}}{11}$ (d) $\frac{5}{\sqrt{5}-1}$ (e) $\frac{1+\sqrt{5}}{4}$

4. The graph of the equation $x^2 - 4y^2 = 0$ is

   (a) a point (b) a pair of straight lines

   (c) a parabola (d) an ellipse (e) none of these

5. If $x - y < x$ and $x + y < y$, then

   (a) $y < x$ (b) $0 < x < y$ (c) $x < y < 0$ (d) $x < 0, y < 0$ (e) $x < 0 < y$

6. The number of values of x satisfying the equation $\frac{2x^2-10x}{x^2-5x} = x - 3$ is

   (a) zero (b) one (c) two (d) three (e) more than three

7. The radius of the circle whose equation is $x^2 + y^2 - 16x - 10y + 64 = 0$ is

   (a) 4 (b) 5 (c) 6 (d) 8 (e) 10

8. One root of the equation $\frac{1}{(1+x)^3} + \frac{1}{(1-x)^3} + \frac{1}{2} = 0$ is $i = \sqrt{-1}$. Then the number of real roots is

(a) zero (b) one (c) two (d) three (e) four

9. The sides of a triangle are 8, 13 and 15 centimeters. In square centimeters, its area is

(a) 52 (b) $20\sqrt{2}$ (c) 60 (d) $30\sqrt{3}$ (e) none of these

10. If for all x we have $1 = ax^2 + (bx + c)(x + 1)$, then

(a) $c + a + 2b = 0$ (b) $a + b + 2c = 0$ (c) $b + c + 2a = 0$

(d) $ab = c^2$ (e) $bc = a^2$

11. When $x^3 + k^2x^2 - 2kx - 6 = 0$ is divided by $x + 2$, the remainder is 10. Then k must be

(a) 2 (b) -2 (c) 2 or -3 (d) 2 or -1 (e) none of these

12. A student wrote that the product of $a + i$ and $b - i$ was $a + b + i$ where $i = \sqrt{-1}$. If this answer was correct, then the minimum value of ab is

(a) 2 (b) 1 (c) 0 (d) -1 (e) -2

13. The converse of the statement "If $a = 0$, then $ab = 0$" is

(a) If $a \neq 0$, then $ab \neq 0$. (b) If $a \neq 0$, then $ab = 0$.

(c) If $a = 0$, then $ab \neq 0$. (d) If $ab = 0$, then $a = 0$.

(e) If $ab = 0$, then $a = 0$ or $b = 0$.

14. Let ABC be a triangle with $\angle A < \angle C < 90° < \angle B$. Consider the external angle-bisectors at A and B, each measured from the vertex to the opposite side (extended). If each of these line segments is equal to AB, then $\angle A$ is

(a) 6° (b) 9° (c) 12° (d) 15° (e) none of these

15. The sum and the product of two numbers are each equal to $s + \frac{1}{s} + 2$ where $s > 1$. Then the difference between the squares of the reciprocals of the numbers is

(a) 1 (b) 2 (c) $(\frac{s-1}{s-2})^2$ (d) $\frac{s-1}{s+1}$ (e) at least 1

16. The distance that a body falls from rest varies as the square of the time of falling. If it falls from rest at a distance of 256 meters in 4 seconds, then during the tenth second it falls a distance, in meters, of

(a) 288 (b) 304 (c) 320 (d) 336 (e) 384

17. The roots of the equation $x^3 + 3px^2 + q^2x + r^3 = 0$ are in arithmetic progression. Then we must have

(a) $pq^2 = 2p^3 + r^3$ (b) $p = 0$ (c) $q^4 = 3pr^3$

(d) $3p + r^3 = 2q^2$ (e) $3pr = q^2$

18. In calm weather, an aircraft can fly from one city to another 200 kilometers north of the first and back in exactly 2 hours. In a steady north wind, the round trip takes 5 minutes longer. The speed of the wind, in kilometers per hour, is

(a) 8 (b) 16 (c) 32 (d) 35 (e) 40

19. The length of the common chord of two intersecting circles is 16 meters. If the radii are 10 meters and 17 meters, then the distance, in meters, between the centers is

(a) 27 (b) 21 (c) $\sqrt{389}$ (d) 15 (e) none of these

20. The number of positive integers less than 500 that are divisible by neither 3 nor by 5 is

(a) 269 (b) 267 (c) 265 (d) 234 (e) 201

21. The system of equations $x + (k-2)y = 1$ and $(k+2)x - 3y = 1$ can be solved for x and y in terms of k, provided that

(a) $k \neq 1$ (b) $k \neq 0$ (c) $k \neq -1$

(d) $k \neq 1, k \neq -1$ (e) none of these

22. The smoke trail of a steamship sailing due east at 30 knots is in a direction 60° west of north. It overtakes a freighter sailing east at 10 knots, whose smoke trail is in a direction 30° west of north. The wind must be blowing from the direction

(a) 30° east of north (b) 135° west of north (c) due north

(d) due south (e) 150° west of north

23. When the last digit of a certain six-digit number N is transferred to the first position, the other digits moving one place to the right, the new number is exactly one-third of N. The sum of the six digits is

(a) 28 (b) 27 (c) 26 (d) 25 (e) 24

24. In the figure below, $CA = CF$ and $\angle B = \angle C$. Then we must have

(a) $AE = EF$ (b) $AE = AD$ (c) $AD = \frac{1}{2}CF$ (d) $AE = EB$ (e) $AE = DF$

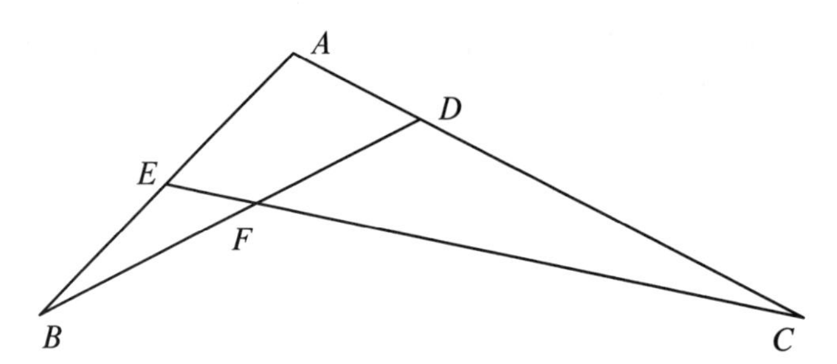

25. The guests at a party play as follows. Each player in turn names a real number, not 0 or 1. No number is to be repeated, and either the sum or the product of every pair of successive numbers must be 1, The greatest number of turns there can be is

(a) 2 (b) 4 (c) 6 (d) 8 (e) no limit

Problems: 1968

1. The equation $x^2 + x + 2 = 0$ has

(a) two positive roots (b) two negative roots

(c) one positive and one negative roots

(d) no real roots (e) none of these

2. The equation $(a + b)x = 3$ in x will have no solution

(a) if $a = b$ (b) if $a = -b$ (c) if $a + b = 3$

(d) if $a + b = -3$ (e) under any circumstance

3. The most general parallelogram which has equal diagonals is a

(a) rhombus (b) square (c) rectangle (d) trapezium (e) none of these

4. What is the value of $5^{\log_5 6}$?

(a) 1 (b) 5 (c) 6 (d) $\log_5 6$ (e) none of these

5. Which of the following constructions is impossible, using only an umnarked ruler and compass?

(a) to trisect a given angle (b) to trisect a given line

(c) to bisect a given line (d) to bisect a given angle

(e) none of these

6. Given that $\log_{10} 2 = x$ and $\log_{10} 3 = y$, then $\log_{10} 15 = ?$

(a) $1 + x + y$ (b) $1 - x - y$ (c) $1 + x - y$ (d) $1 - x + y$ (e) none of these

7. Let S be the set of points (x, y) in the plane satisfying both $x^2+y^2 \leq 1$ and $x^2+y^2 \geq r^2$. A value of r such that S is the empty set is

(a) 1 (b) -1 (c) $\frac{1}{2}$ (d) $-\frac{1}{2}$ (e) none of these

8. If S, T and V are any sets, then $(S \cap T) \cup (S \cap V)$ is the same set as

(a) S (b) $T \cup V$ (c) $T \cap V$

(d) $S \cap (T \cup V)$ (e) none of these

9. What is the greatest number of possible points of intersection of three lines in the plane, of different slopes?

(a) 2 (b) 4 (c) 6 (d) 8 (e) none of these

10. What is the greatest number of possible points of intersection of three circles in the plane, of different radii?

(a) 2 (b) 4 (c) 6 (d) 8 (e) none of these

11. O is the center of a circle, AB is a diameter and C a point on this circle. If $\angle COB = 40°$, then $\angle CAB =$

(a) 30° (b) 40° (c) 60° (d) 80° (e) none of these

12. A metal disc has one face marked "1" and the other face marked "2". A second metal disc has one face marked "2" and the other marked "3". Assume that, when tossed, the two faces of a disc are equally likely to turn up. If both discs are tossed, what is the probability that 4 is the sum of the numbers turning up?

(a) $\frac{1}{4}$ (b) $\frac{1}{2}$ (c) $\frac{3}{4}$ (d) $\frac{1}{3}$ (e) none of these

13. Which of the following statements about $\frac{1+\sqrt{2}}{1-\sqrt{2}}$ is true?

(a) it is irrational (b) it is rational (c) it is imaginary

(d) it is positive (e) none of these

14. What is the longest rod that can be put in a rectangular box of dimensions $6 \times 3 \times 2$?

(a) 6 (b) $3\sqrt{5}$ (c) $2\sqrt{10}$ (d) 7 (e) none of these

15. If $i = \sqrt{-1}$, then i^6 is

(a) 1 (b) -1 (c) i (d) $-i$ (e) none of these

16. If $i = \sqrt{-1}$, then $\frac{1+i}{1-i}$ is

(a) 1 (b) -1 (c) i (d) $-i$ (e) none of these

17. The solution set of the inequality $x^2 - x - 2 < 0$ is the interval

(a) $-2 \leq x \leq 1$ (b) $-2 < x < 1$ (c) $-2 \leq x < 1$

(d) $-2 < x \leq 1$ (e) none of these

18. If $a > 0, b > 0$ and $a > b$, which of the following is false?

(a) $\frac{1}{a} > \frac{1}{b}$ (b) $a^2 > b^2$ (c) $a^3 > b^3$ (d) $\sqrt{a} > \sqrt{b}$ (e) none of these

19. The solution set in the plane of the equation $y^2 = xy$ is

(a) a parabola (b) a rectangular hyperbola

(c) a pair of straight lines (d) a circle

(e) none of these

20. The distance between the two points represented by the complex numbers $1 - 2i$ and $2i - 2$ is

(a) 5 (b) $-1 + 4i$ (c) $\sqrt{17}$ (d) $3 - 4i$ (e) none of these

21. At the end of a party, everyone shakes hands with everyone else. Altogether there are 28 handshakes. How many people are there at the party?

(a) 8 (b) 14 (c) 20 (d) 56 (e) none of these

22. How many odd multiples of "3" are there between 100 and 200?

(a) 15 (b) 17 (c) 33 (d) 49 (e) none of these

23. Let $aabb$ be a four-digit number in base r, with $b = 0$. This number is divisible by

(a) r only (b) $r + 1$ only

(c) $r - 1$ only (d) more than one of these

(e) none of these

24. Let $\{a_1, a_2, a_3, \ldots\}$ be an infinite sequence of real numbers with $a_k \geq 1$ for $k = 1, 2, 3, \ldots$. Define $U_n = a_1 a_2 \cdots a_n$. Consider the following statements:

$$\textbf{I: } U_n = U_{n+1}; \textbf{ II: } U_n < U_{n+1}; \textbf{ III: } U_n > U_{n+1}.$$

The true ones are

(a) I only (b) I and II only (c) I and III only

(d) III only (e) none of these

25. A rectangular floor 24×40 is covered by squares of sides 1. A chalk line is drawn from one corner to the diagonally opposite corner. How many tiles have a chalk line segment on them?

(a) 40 (b) 56 (c) 63 (d) 64 (e) none of these

Problems: 1969

1. O is the center of a circle. $CDOE$ is a rectangle, with C lying on this circle. If $DE = 5$ and $CE = 3$, the diameter of the circle is

(a) $4\sqrt{2}$ (b) 8 (c) 10

(d) $10\sqrt{5}$ (e) cannot be determined

2. A man spends $\frac{1}{3}$ of his money and loses $\frac{1}{3}$ of the remainder. He then has $12. How much money had he at first?

(a) $56 (b) $27 (c) $108 (d) $112 (e) none of these

3. For all real numbers a and b,

(a) $a^2 + b^2 \geq 2ab$ (b) $a^2 + b^2 > 2ab$ (c) $a^2 + b^2 < 2ab$

(d) $a^2 + b^2 \leq 2ab$ (e) none of these hold

4. The total number of subsets that can be formed from a set containing six elements is

(a) 4 (b) 8 (c) 16 (d) 32 (e) none of these

5. A gambler visited three gambling houses. At the first he doubled his money, and then spent $30. At the second he tripled his remaining money and then spent $54. At the third he quadrupled his remaining money and then spent $72, and he then had $48 left. How much money did he start with?

(a) $29 (b) $30 (c) $31 (d) $32 (e) $33

6. Let r be the result of doubling both the base and the exponent of a^b, $b \neq 0$. If r equals the product of a^b and x^b, then x equals

(a) a (b) $2a$ (c) $4a$ (d) 2 (e) 4

7. The symbol $|a|$ means a if a is a positive number or zero, and $-a$ if a is a negative number. For all real values of x the expression $\sqrt{x^4 + x^2} =$

(a) x^3 (b) $x^2 + x$ (c) $|x^2 + x|$ (d) $x\sqrt{x^2+1}$ (e) $|x|\sqrt{1+x^2}$

8. In the base ten number system, the number 526_{10} means $5 \cdot 10^2 + 2 \cdot 10 + 6$. If in the base r number system the equation $1000_r - 440_r = 340_r$ holds, then r is

(a) 2 (b) 5 (c) 7 (d) 8 (e) 12

9. While three watchmen were guarding an orchard, a thief slipped in and stole some apples. On his way out he met the three watchmen one after another, and to each in turn he gave a half of the apples he then had and two besides. Thus he managed to escape with one apple. How many had he stolen originally?

(a) 16 (b) 22 (c) 32 (d) 76 (e) none of these

10. Men are digging a hole. If one man can dig the hole in one hour, and a second man can dig the hole in one-and-a-half hours, how many minutes must they work together to dig a hole?

(a) 16 (b) 36 (c) 46 (d) 56 (e) none of these

11. If the radius of a circle is increased by one unit, the ratio of the new circumference to the new diameter is

(a) $\pi + 2$ (b) $\frac{2\pi+1}{2}$ (c) π (d) $\frac{2\pi-1}{2}$ (e) $\pi - 2$

12. A square and an equilateral triangle have equal perimeters. The area of the triangle is $9\sqrt{3}$ square centimeterss. Expressed in centimeters, the diagonal of the square is

(a) $\frac{9}{2}$ (b) $2\sqrt{5}$ (c) $4\sqrt{2}$ (d) $\frac{9\sqrt{2}}{2}$ (e) none of these

13. A hungry hunter came upon two shepherds, one of whom had three loaves of bread, and the other five, all of the same size. The loaves were divided equally among the three and the hunter paid 8 cents for his share. How should the shepherds divide the money?

(a) 1 and 7 (b) 2 and 6 (c) 3 and 5 (d) 4 and 4 (e) none of these

14. The average of a set of 50 numbers is 38. If two numbers of the set, namely 45 and 55 are discarded, the average of the remaining set of numbers is

(a) 38.5 (b) 37.5 (c) 37 (d) 36.5 (e) 36

15. A circle is inscribed in an equilateral triangle, and a square is inscribed in the circle. The ratio of the area of the triangle to the area of the square is

(a) $\sqrt{3}$:1 (b) $\sqrt{3}$:$\sqrt{2}$ (c) $3\sqrt{3}$:2 (d) 3:$\sqrt{2}$ (e) 3:$2\sqrt{2}$

16. Every day at noon, a ship leaves New York for Lisbon and at the same instant, a ship leaves Lisbon for New York. Each trip lasts exactly 8 days. How many ships from Lisbon will each ship from New York meet?

(a) 11 (b) 13 (c) 15 (d) 17 (e) none of these

17. Points P and Q are both on the line segment AB and on the same side of its midpoint. P divides AB in the ratio 2:3, and Q divides AB in the ratio 3:4. If $PQ = 2$, then the length of AB is

(a) 60 (b) 70 (c) 75 (d) 80 (e) 85

18. Given two equiangular polygons P_1 and P_2 with different numbers of sides, each angle of P_1 is $x°$ and each angle of P_2 is $kx°$, where k is an integer greater than 1. The number of possibilities for the pair (x, k) is

(a) infinite (b) finite but more than two

(c) two (d) one (e) zero

19. Given that the following have the same perimeters, which has the largest area?

(a) square (b) equilateral triangle (c) circle

(d) regular pentagon (e) two or more of these are the same

20. The angles at A, B, C, D and E of a pentagon $ABCDE$ are in the ratio 5:3:8:5:6. The largest of these angles has the value

(a) 90° (b) 110° (c) 130° (d) 150° (e) none of these

21. If $\binom{n}{8} = \binom{n}{24}$, then $n =$

(a) 8 (b) 16 (c) 24 (d) 32 (e) none of these

22. The number of solutions of $2^{2x} - 2^{2y} = 55$ in which x and y are integers is

(a) zero (b) one (c) two (d) three (e) more than three

23. In racing over a given distance d at uniform speed, A can beat B by 20 meters, B can beat C by 10 meters, and A can beat C by 28 meters. In meters, $d =$

(a) 58 (b) not determined by the above information

(c) 100 (d) 116 (e) 120

24. If $x_{k+1} = x_k + \frac{1}{2}$ for $k = 1, 2, \ldots, n-1$ and $x_1 = 1$, find $x_1 + x_2 + \cdots + x_n$.

(a) $\frac{n+1}{2}$ (b) $\frac{n+3}{2}$ (c) $\frac{n^2-1}{2}$ (d) $\frac{n^2+n}{4}$ (e) $\frac{n^2+3n}{4}$

25. Three disks labelled one to three are put in a bag. Three other disks labelled one to three are put in a second bag. A disk is drawn from each bag and the two disks thus drawn are stacked in a pile on a table. This is repeated two more times. What is the probability that at least one of the stacks will contain disks with the same number?

(a) $\frac{1}{6}$ (b) $\frac{1}{3}$ (c) $\frac{1}{2}$ (d) $\frac{2}{3}$ (e) none of these

Problems: 1970

1. The number $10a + b$, where a and b are digits, is divisible by nine only if

(a) $a + b = 7$ (b) $a + b = 8$ (c) $a + b = 9$

(d) $a + b = 10$ (e) none of these

2. If $a \neq b$ and $ax + b^2 = a^2 - bx$, then $x =$

(a) $a + b$ (b) $a - b$ (c) $b - a$ (d) $a^2 + b^2$ (e) none of these

3. For the function $f(x) = x^{50} - 2a^{47}x^3 + a^{50}$, the following is a factor.

(a) $x - a$ (b) $x - a^2$ (c) $x + a$ (d) $x + a^2$ (e) none of these

4. Which of the following inequalities are true for all positive x?

(a) $x + \frac{1}{x} < 2$ (b) $x + \frac{1}{x} \leq 2$ (c) $x + \frac{1}{x} > 2$

(d) $x + \frac{1}{x} \geq 2$ (e) none of these

5. A steamer was able to go 20 kilometers per hour downstream and 15 kilometers per hour upstream. On a return trip, the steamer took 5 hours longer coming up than going down. In kilometers, the total distance travelled by the steamer is

(a) 500 (b) 600 (c) 700 (d) 800 (e) none of these

6. Suppose n is a positive integer. Then $\frac{n^2+(n+2)^2}{2}$ is

(a) sometimes an integer (b) always a perfect square

(c) sometimes a perfect square (d) never a perfect square

(e) none of these

7. The solution set of the inequality $x^2(x-1)^2 \leq 0$ consists of

(a) an interval (b) two intervals

(c) an interval and a point (d) an interval and two points

(e) none of these

8. The expression $\sqrt[8]{x^8} + \sqrt[7]{x^7}$ is always equal to

(a) x (b) $2x$ (c) $2x^2$ (d) 0 (e) none of these

9. The binary operation $*$ between two integers m and n is defined by $m * n = m^2 + n^2$. Which of the following do not hold?

(a) commutative law (b) associative law (c) $m * n$ is an integer

(d) $m * n \geq 0$ (e) all of the previous hold

10. Given the four quadrants labelled in the usual manner, the solution set of the simultaneous inequalities $x^2 + y < 0$ and $x^2 + y^2 > 4$ lies entirely in quadrants

(a) I and II (b) II and III (c) III and IV

(d) IV and I (e) none of these

11. Two cyclists race on a circular track. The first can ride around the track in 6 seconds, and the second in 4 seconds. If they start off at the same point, the second cyclist can overtake the first in

(a) 12 seconds (b) 14 seconds (c) 16 seconds

(d) 18 seconds (e) none of these

12. An equilateral triangle is inscribed upside-down in a larger equilateral triangle. The ratio of the area of the smaller to the area of the larger is

(a) $\frac{1}{12}$ (b) $\frac{1}{6}$ (c) $\frac{1}{4}$ (d) $\frac{1}{3}$ (e) none of these

13. If $f(n) = n^2$, where n is an integer, and $n \neq k$, then $\frac{f(f(n))-f(f(k))}{f(n)-f(k)} =$

(a) $n^2 + k^2$ (b) $n^2 - k^2$ (c) $\frac{n^2+k^2}{n^2-k^2}$ (d) $\frac{n^2-k^2}{n^2+k^2}$ (e) none of these

14. Which of the following inequalities hold for all real x and y?

(a) $\sqrt{x^2+y^2} < x + y$ (b) $\sqrt{x^2+y^2} \leq x + y$ (c) $\sqrt{x^2+y^2} < |x| + |y|$

(d) $\sqrt{x^2+y^2} \leq |x| + |y|$ (e) none of these

15. The equation of the line through the origin and perpendicular to the line $y = 3x + 1$ is

(a) $y = 3x$ (b) $y = -3x$ (c) $y = \frac{x}{3}$ (d) $y = -\frac{x}{3}$ (e) none of these

16. Given that $\log_a b = c$, then $\log_{a^2} b =$

(a) c^2 (b) $\frac{c}{2}$ (c) $2c$ (d) $\sqrt{c}$ (e) none of these

17. A polynomial whose graph passes through the points (3,0), (4,2) and (0,6) is

(a) $y = x - 6$ (b) $y = x^4 + x^3 + 3x - 5$ (c) $y = x^2 - 5x + 6$

(d) $y = x^2 - 9$ (e) none of these

18. Let $f(x) = 3x^2 + Kx + 1$, where K is a real constant. If r and s are the roots of $f(x)$, which of the following is impossible?

(a) $r = s$ (b) $rs = 1$ (c) $r + s = 1$

(d) $r > 0$ and $s > 0$ (e) none of these

19. If A is the area of an equilateral triangle of side length S, then the area of an equilateral triangle of side length $2S$ is

(a) $2A$ (b) $4A$ (c) A^2 (d) $2A^2$ (e) none of these

20. The following figures all have the same area. Which has the smallest perimeter or circumference?

(a) circle (b) square (c) equilateral triangle

(d) regular pentagon (e) regular hexagon

21. The value of K so that the polynomial $x^4 + Kx^3 - 7x^2 - 27x - 18$ will be divisible by $x - 3$ is

(a) 1 (b) -1 (c) 2 (d) -2 (e) none of these

22. The domain of the function $f(x) = \frac{\sqrt{x+1}}{x}$ is

(a) a single point
(b) an infinite interval
(c) an infinite interval and a single point
(d) an infinite interval with a single point deleted
(e) none of these

23. Let $\{a_1, a_2, a_3, \ldots\}$ be a sequence of numbers such that the sum of the first n terms is $\frac{n(n+1)}{2}$. Then $a_k =$

(a) k (b) 1 (c) $k+1$ (d) $k-1$ (e) none of these

24. If a, b and c are positive integers such that $a+b=2c$, then $2^a 2^b =$

(a) 2^c (b) 2^{c^2} (c) 4^c (d) 4^{c^2} (e) none of these

25. A man drives from Edmonton to Calgary at a speed of 30 miles per hour (mph). At what speed must he drive from Calgary to Edmonton so that the average speed for the whole trip is 40 mph?

(a) 45 mph (b) 50 mph (c) 55 mph (d) 60 mph (e) none of these

Problems: 1971

1. Given that x is inversely proportional to y, y is inversely proportional to z and z is inversely proportional to v, the relation between x and v, for some constant k, is

(a) $x = kv$ (b) $x = \frac{k}{v}$ (c) $x = kv^2$ (d) $x = \frac{k}{v^2}$ (e) none of these

2. Given the function $f(x) = 2^{x-1}$, $f(-1) =$

(a) $\frac{1}{4}$ (b) 1 (c) -1 (d) $-\frac{1}{4}$ (e) none of these

3. A triangle with sides of length 12, 13 and 5

(a) is a right triangle (b) is an acute triangle (c) is an obtuse triangle

(d) does not exist (e) none of these

4. If $p = \frac{x+1}{x}$ where $\frac{1}{100} < x < 100$, for what value of x does p have its smallest value?

(a) 101 (b) $\frac{1}{100}$ (c) $\frac{101}{100}$ (d) 100 (e) none of these

5. For what values of M and N is the equation $M^{\log N} = N^{\log M}$ true? Both logarithms are base 10.

(a) no values of M and N (b) all negative values of M and N

(c) all values of M and N (d) all positive values of M and N

(e) none of these

6. A triangle with sides of length 5, 16 and 8 is

(a) a right triangle (b) an obtuse triangle (c) non-existent

(d) an acute triangle (e) none of these

7. At what time between four and five o'clock is the minute hand exactly two minutes ahead of the hour hand?

(a) 4:21 (b) 4:22 (c) 4:23 (d) 4:24 (e) none of these

8. Let n be an integer. Define $f(n)$ to be the number of positive integers not exceeding n. Then $f(n) + f(-n) =$

(a) 0 (b) $-n$ (c) n (d) $2n$ (e) none of these

9. Let $x = 1 - 2t$ and $y = 1 + 2^{-t}$. Which of the following is true for all t?

(a) $y = \frac{1}{x-1}$ (b) $y = \frac{x-2}{x-1}$ (c) $y = \frac{2-x}{x-1}$ (d) $y = \frac{x}{x-1}$ (e) none of these

10. A perfect number is a positive integer such that it is equal to the sum of all positive integers smaller than it which divide it. Which of the following is perfect?

(a) 4 (b) 6 (c) 8 (d) 10 (e) none of these

11. Let S be the set of points defined by the inequalities $0 \leq y \leq 1$, $0 \leq x \leq 1$ and $y \leq x + \frac{1}{2}$. The area of the region determined by S is

(a) $\frac{1}{8}$ (b) $\frac{3}{8}$ (c) $\frac{5}{8}$ (d) $\frac{7}{8}$ (e) none of these

12. A gear of radius 1 revolves around a fixed gear of radius 2. During one complete revolution, the smaller gear will rotate

(a) 360° (b) 540° (c) 720° (d) 900° (e) none of these

13. For any integer n, the expression $n(n + 1)(n + 2)$ cannot assume the value

(a) 0 (b) 731 (c) 1320 (d) 7980 (e) none of these

14. The equation of the perpendicular bisector of the line segment with end points (1,5) and $(-3, 2)$ is

(a) $8x - 6y + 29 = 0$ (b) $8x + 6y + 29 = 0$ (c) $8x + 6y - 29 = 0$

(d) $8x - 6y - 29 = 0$ (e) none of these

15. Given the quadratic equation $x^2 + 2bx + 3c = 0$, the absolute value of the difference of the roots is

(a) $\frac{\sqrt{b^2-4ac}}{2a}$ (b) $\frac{\sqrt{b^2-4ac}}{a}$ (c) $\frac{2\sqrt{b^2-4ac}}{a}$ (d) $\frac{\sqrt{b^2-3ac}}{2a}$ (e) none of these

16. A triangle ABC with $\angle A = 30°$ is inscribed in a circle. The radius of the circle is

(a) BC (b) $\frac{AB+AC}{2}$ (c) $\frac{BC}{2}$ (d) AB (e) none of these

17. If $y^2 - 1 + \log_{10} x = 0$, the values of x for which y takes on real values are

(a) $x \leq 10$ (b) $0 < x \leq 10$ (c) $0 < x$ (d) all x (e) no x

18. The number 123456789012345678901234567890 is not divisible by

(a) 2 (b) 3 (c) 5 (d) 9 (e) none of these

19. Let ABC be a triangle such that $AB = 5$ and $AC = 7$. Let AH be the altitude from A to BC. If $AH = 1$, $BC =$

(a) $\sqrt{24}$ (b) $2\sqrt{24}$ (c) $(1+\sqrt{2})\sqrt{24}$

(d) $\sqrt{2}\sqrt{24}$ (e) none of these

20. Let x and y be positive numbers. Let $a = \frac{x+y}{2}$, $b = \sqrt{xy}$ and $c = \sqrt{\frac{x^2+y^2}{2}}$. Which of the following is always true?

(a) $a \leq b \leq c$ (b) $c \leq b \leq a$ (c) $b \leq c \leq a$

(d) $b \leq a \leq c$ (e) none of these

Problems: 1972

1. A quadrilateral $ABCD$ is inscribed in a circle of radius 5. If the center of the circle lies on AB, and the length of AD is 7, then diagonal BD has length

(a) 2 (b) 3 (c) $2\sqrt{6}$ (d) $\sqrt{51}$ (e) none of these

2. A three-digit number abc is chosen. The difference between bca and abc is calculated and found to lie between 400 and 500. The number equals

(a) 400 (b) 404 (c) 429 (d) 495 (e) none of these

3. A pentagon has angles 110°, 143°, 87° and 52°. Its remaining angle equals

(a) 148° (b) 110° (c) 178° (d) 90° (e) none of these

4. In 9 years, John will be $\frac{3}{5}$ as old as his father was 10 years ago. In 10 years, John will be $\frac{1}{2}$ as old as his father was 1 year ago. John's age is

(a) 8 (b) 18 (c) 12 (d) 16 (e) none of these

5. The expression $i^{2073} =$

(a) i (b) $-i$ (c) 1 (d) -1 (e) none of these

6. Let a and b be the roots of $x^2 - 7x + 3 = 0$. Then $a^3 + b^3 =$

(a) 91 (b) -18 (c) 324 (d) 360 (e) none of these

7. The long hand of a clock points exactly at a minute and the short hand points exactly two minutes ahead of the long hand. The time is

(a) 3:17 (b) 4:26 (c) 7:36 (d) 11:58 (e) none of these

8. The expression $\cos^4\theta - \sin^4\theta =$

(a) $2\cos^2\theta - 1$ (b) $\cos\theta$ (c) $2\sin^2\theta - 1$ (d) $\sin\theta$ (e) none of these

9. What is the next number after 1, 1, 2, 6, 15?

(a) 9 (b) 31 (c) 16 (d) 28 (e) 42

10. Which of the following numbers divides $9^5 + 33^5 + 39^{10} - 4^{10}$?

(a) 4 (b) 7 (c) 3 (d) 2 (e) none of these

11. In triangles ABC and DEF, the sides AB, BC, DE and EF have the same length, and $\angle ABC = 2\angle FDE$. The ratio of the areas of ABC to that of DEF is

(a) 2:3 (b) 1:2 (c) 2:1 (d) 1:1 (e) none of these

12. If $f(x) = \log_2(\frac{1}{x^2})$, then $f(-\frac{1}{4}) =$

(a) 0 (b) 4 (c) -4 (d) -3 (e) none of these

13. The sum of the first twenty odd integers is

(a) 420 (b) 800 (c) 400 (d) 190 (e) none of these

14. The expression $|\frac{2+5i}{3-4i}| =$

(a) 7 (b) -7 (c) $\frac{5}{\sqrt{29}}$ (d) $\frac{\sqrt{29}}{5}$ (e) none of these

15. Let p be an integer. Then $2x^2 + px + 3 = 0$ has real roots for

(a) all p (b) no p (c) $p \leq 24$ (d) $p \geq 24$ (e) none of these

16. If $g(t) = -g(-t)$ for some function g, which of the following are true?

$$\textbf{I: } g(t) > 0 \text{ for } t > 0; \textbf{ II: } g(t) + g(-t) = 0; \textbf{ III: } g(0) = 0;$$
$$\textbf{IV: } g(t) = t; \textbf{ V: } (g(t))^2 + 2g(-t) + 1 \geq 0.$$

(a) all (b) I, II and III only (c) II, III and V only

(d) I, IV and V only (e) none

17. The expression $i^{4421} + i^{3663} =$

(a) 2 (b) -2 (c) $2i$ (d) $-2i$ (e) none of these

18. A triangle with sides of length 5, 7 and 9 is

(a) nonexistent (b) obtuse (c) acute (d) right (e) none of these

19. A quadrilateral with sides 16, 38, 7 and 12 is

(a) impossible (b) convex (c) concave (d) a trapezoid (e) none of these

20. Which of the following is true if x and y are any real numbers?

(a) $x^2 + 9y^2 \leq 6xy$ (b) $x^2 + 9y^2 \geq x$ (c) $x^2 + 9y2 \geq 1$

(d) $x^2 + 9y^2 \geq 9xy$ (e) none of these

Problems: 1973

1. A triangle ABC is inscribed in a circle of radius 3. If $BC = 2$ and $AC = 6$, then $AB =$

(a) 5 (b) 6 (c) $4\sqrt{2}$ (d) $\sqrt{2} + 2\sqrt{3}$ (e) none of these

2. The line $-x + 3y = 9$ meets the parabola $y^2 = 4x$ in

(a) no points (b) one point (c) two points (d) four points (e) none of these

3. Of the following numbers, select the largest.

(a) $2 + 3\sqrt{3}$ (b) $3 + 2\sqrt{2}$ (c) 4 (d) $\sqrt{39}$ (e) $2\sqrt{10}$

4. Let u and v be the roots of $x^2 - 5x + 3 = 0$. Then $u^2 + v^2 =$

(a) 9 (b) 15 (c) 19 (d) 25 (e) none of these

5. If a quadrilateral is circumscribed about a circle, then

   (a) the sum of two diagonally opposite angles is 180°
   (b) it must contain a right angle
   (c) it must have two equal sides
   (d) the sum of two opposite sides is half the perimeter of the quadrilateral
   (e) none of the above

6. If $*$ is commutative and associative, and if $a * b = c$ and $c * a = a$, then $b * a * c * a * b =$

   (a) a (b) b (c) c (d) $b * b$ (e) none of these

7. What is the last digit of 728^{4921}?

   (a) 2 (b) 4 (c) 6 (d) 8 (e) none of these

8. What positive integer is $\sqrt{3}^{\sqrt{3}^{\sqrt{3}}} \cdot \sqrt{3}$?

   (a) 3 (b) 9 (c) 27 (d) 1 (e) none of these

9. The Edmonton-Calgary Airbus can fly the 189 miles in 40 minutes on a calm day. With a headwind, the time was 45 minutes. What was the speed of the wind in miles per hour?

   (a) 40.5 (b) 38.2 (c) 31.5 (d) 60.5 (e) none of these

10. AB is a diameter of a circle with center C, and D is another point on the circle such that $\angle BCD = 72°$. What is $\angle BAD$?

    (a) 18° (b) 24° (c) 30° (d) 36° (e) none of these

11. A quadrilateral with sides 5, 3, 5 and 7 in that order must necessarily be a

    (a) parallelogram (b) non-isosceles trapezoid

    (c) isosceles trapezoid (d) rhombus (e) none of these

12. A pentagon $MNPQR$ has $MN = 2, NP = 7, PQ = 4, QR = 5$ and $RM = 1$. The sum of the lengths of the diagonals, $MP + MQ + NQ + NR + PR$, cannot possibly equal

    (a) 53 (b) 33 (c) 18 (d) 16 (e) none of these

13. The number 999,999,999,999 is divisible by

    (a) 23 (b) 77 (c) 101 (d) 162 (e) none of these

14. Consider two face diagonals of a cube which meet at a vertex. They form an angle of

(a) 90° (b) 60° (c) 45° (d) 72° (e) none of these

15. If $f(x) = 2^x x^2$, then $f(1) =$

(a) 0 (b) 1 (c) 2 (d) 4 (e) none of these

16. The athletics banquet at Dudgeon High School costs 450 dollars, and the committee decided that the cost would be shared equally by all those attending. When 75 of those eligible to attend did not, the cost to each attending was 50 cents higher than it would otherwise have been. How many were eligible to attend?

(a) 200 (b) 225 (c) 300 (d) 325 (e) none of these

17. Find the sum $1 + 3 + 5 + \cdots + 199$ of the first 100 odd numbers.

(a) 10,200 (b) 10,201 (c) 10,001 (d) 10,000 (e) none of these

18. A given quadrilateral can be inscribed in a circle if

(a) the sum of its angle is 360°
(b) the sum of any two opposite angles is 180°
(c) it has at least two right angles
(d) the sum of opposite sides is half the perimeter
(e) none of these

19. Find the next number in the sequence $-3, 1, 5, 9, 31, 53, 75, 97, 101$ and 501.

(a) 301 (b) 700 (c) 505 (d) -3 (e) none of these

20. The altitude of a regular tetrahedron of edge length 1 equals

(a) $\frac{\sqrt{3}}{2}$ (b) $\frac{\sqrt{6}}{3}$ (c) $\frac{\sqrt{3}}{3}$ (d) $\frac{2\sqrt{3}}{3}$ (e) none of these

Problems: 1974

1. In a regular pentagon $ABCDE$, $\angle ACB =$

(a) 18° (b) 27° (c) 36° (d) 45° (e) 54°

2. Find the smallest positive integer n for which $1 + 2 + 3 + \cdots + n > 5,000$.

(a) 10 (b) 99 (c) 100 (d) 101 (e) 1,000

3. Let $a = \frac{1}{2}(\sqrt{5} + \sqrt{7}), b = \sqrt{6}$ and $c = \sqrt[4]{35}$. Then

(a) $a < b < c$ (b) $b < a < c$ (c) $c < a < b$ (d) $c < b < a$ (e) none of these

4. John and Susan are both younger than 5 years old. Three times John's age equals twice the age Susan will be 5 years from now. Susan's age is

(a) 0 (b) 1 (c) 2 (d) 3 (e) 4

5. Consider the curve $C : y = \sin x, -\infty < x < \infty$. The line $y = \frac{1}{2}$ intersects C

(a) once (b) twice (c) never

(d) infinitely often (e) five times

6. Two chords of a circle of radius 2 are perpendicular to each other. The largest possible value of their total length is

(a) $2\sqrt{2}$ (b) 8 (c) $4\sqrt{2}$ (d) $6\sqrt{2}$ (e) none of these

7. The length of the segment joining any top vertex of a cube of side 1 to the center of the bottom face is

(a) $\sqrt{3}$ (b) $\sqrt[3]{2}$ (c) $\sqrt{2}$ (d) $2\sqrt{2}$ (e) $\frac{1}{2}\sqrt{6}$

8. The graph of the equation $x^2 + \frac{y^2}{9} = 1$ is

(a) the empty set (b) (0,0) (c) two straight lines

(d) an ellipse (e) a hyperbola

9. If $0 < x + y < 3$ and $1 < x - y < 2$, then

(a) $1 < x < 5$ (b) $|x| < 1$ (c) $x < 1$ (d) $\frac{1}{2} < x < \frac{5}{2}$ (e) none of these

10. The expression $4^{\log_4 3} =$

(a) 3 (b) 4 (c) 64 (d) $\sqrt[3]{4}$ (e) none of these

11. A regular polygon of 300 sides, F_1, is inscribed in a circle, as is another regular polygon, F_2, with 600 sides. The perimeters P_1 and P_2 of F_1 and F_2, respectively, satisfy

(a) $P_1 = P_2$ (b) $P_1 < P_2$ (c) $P_1 > P_2$

(d) $P_1 + P_2 = P_1 - P_2$ (e) none of these

12. A right triangle with area 12 and hypotenuse 5 has a perimeter of

(a) 49 (b) 7 (c) 12 (d) 37 (e) none of these

13. If $\tan\theta = \frac{1}{2}$, then $\sin 2\theta =$

(a) $\frac{1}{\sqrt{2}}$ (b) $\frac{1}{\sqrt{5}}$ (c) $\frac{4}{5}$ (d) $\frac{2}{\sqrt{5}}$ (e) none of these

14. The function $f(x) = x^4 - 4x^2 + 4$ achieves a minimum at the value $x =$

(a) 0 (b) $\sqrt{2}$ (c) 2

(d) $2\sqrt{2}$ (e) does not attain a minimum

15. If b and c are odd integers, which of the following could be the roots of $x^2+bx+c = 0$?

(a) 5,7 (b) 4,7 (c) $3+2i, 3+4i$

(d) $5+\sqrt{7}, 5-\sqrt{7}$ (e) none of these

16. Two non-overlapping circles of radius 5 are inscribed in a circle of radius 10. If A_1 is the area of part of the large circle outside the two circles, and A_2 is the total area of the two circles, then

(a) $A_1 > A_2$ (b) $A_1 = A_2$ (c) $A_1 < A_2$

(d) $A_1 + A_2 = A_1 - A_2$ (e) $A_1 = \pi A_2$

17. Given a cloth divided into 4 horizontal stripes, suppose we wish to create different flags by colouring the stripes. If we can use the colour red, blue, white, green and yellow, how many different flags can we make, if adjacent stripes must be different colours?

(a) 80 (b) 210 (c) 320 (d) 400 (e) none of these

18. A point A is chosen outside a circle with center C. A tangent from A meets the circle at B, while AC meets the circle at P and Q, with A closer to P than to Q. Given $AB = 10$ and $AP = 2$, the radius of the circle must equal

(a) 12 (b) $12\sqrt{2}$ (c) $2\sqrt{10}$

(d) 24 (e) not enough information

19. In base-4 arithmetic, the product of 332 and 32 is

(a) 100210 (b) 31210 (c) 30211 (d) 21301 (e) none of these

20. A parallelogram is circumscribed about a circle. It is necessarily a

(a) rectangle (b) square

(c) parallelogram with a 60° angle

(d) rhombus (e) none of these

Problems: 1975

1. The sum of three consecutive positive integers is always

(a) odd (b) even (c) a perfect square

(d) divisible by 3 (e) none of these

2. Which of the following holds true?

(a) $\log_3 2 < \log_2 3$ (b) $\log_3 2 = \log_2 3$ (c) $\log_3 2 > \log_2 3$

(d) $\log_3 2 = 1$ (e) $\log_2 3 = 1$

3. In triangle ABC, $\angle A = 60°$ 4and $\angle B = 30°$. If $a = BC, b = CA$ and $c = AB$, which of the following is true?

(a) $a = b$ (b) $b = 2a$ (c) $c = 2a$ (d) $c = 2b$ (e) none of these

4. "The operation $*$ is commutative" means

(a) $x * 1 = 1$ (b) $x * (y * z) = (x * y) * z$

(c) $x * y = y * x$ (d) $x * x = x$ (e) none of these

5. In base-3 arithmetic, if $x = 0.1102$, then $x^2 =$

(a) 0.11021102 (b) 0.10102 (c) 0.01222111 (d) 0.010211 (e) none of these

6. A square is inscribed in a circle which is inscribed in an equilateral triangle. If each side of the triangle has length 6, what is the length of each side of the square?

(a) $\frac{1}{2}\sqrt{6}$ (b) $\sqrt{3}$ (c) $\sqrt{6}$ (d) $2\sqrt{3}$ (e) none of these

7. Which is larger, the volume of a sphere of radius 1 or the volume of a right circular cone of height 1 and base radius 2?

(a) they are equal (b) these volumes do not exist

(c) the sphere (d) the cone (e) none of these

8. The expression $\frac{a^4+a^2b^2+b^4}{a^2+ab+b^2} =$

(a) $a^2 + ab + b^2$ (b) $a^2 + ab - b^2$ (c) $a^2 - ab - b^2$

(d) $a^2 - ab + b^2$ (e) none of these

9. Five years from now Bill will be twice as old as he was two years after he was half as old as he will be in one year from now. His age must be

(a) 16 (b) 13 (c) 8 (d) 41 (e) cannot be determined

10. The repeating decimal $1.131313\cdots$ is the same as

(a) $\frac{112}{99}$ (b) $\frac{113}{99}$ (c) $\frac{100}{99}$ (d) $\frac{1131313}{1000000}$ (e) none of these

11. A jar contains 15 balls, of which 10 are red and 5 are black. If 3 balls are chosen at random, the probability that all three will be red is

(a) 0 (b) $\frac{2}{3}$ (c) $\frac{4}{9}$ (d) $\frac{8}{27}$ (e) none of these

12. The square $ABCD$ has side length 1. E, F and G are points on AB, BC and CD, respectively, such that AG is perpendicular to EF. If $EF = \sqrt{\frac{7}{6}}$ and $BF = \sqrt{\frac{1}{7}}$, then $AG =$

(a) $\sqrt{\frac{7}{6}}$ (b) $\sqrt{7}$ (c) $\sqrt{\frac{6}{7}}$ (d) $\sqrt{\frac{8}{7}}$ (e) none of these

13. Of the following numbers, which is the largest?

(a) 2^{4^3} (b) 2^{3^4} (c) 4^{2^3} (d) 3^{2^4} (e) 3^{4^2}

14. A circle of radius 3 has center C. Let A be at a distance 5 from C and AB be a tangent to the circle. Let AC meet the circle at D, and let E be a point on AB such that ED is perpendicular to AC. Then $AD =$

(a) 1 (b) 2 (c) $\sqrt{2}$ (d) $\sqrt{3}$ (e) none of these

15. The system of equations $2x - 3y = 4, 2y - 4x = 8$ has

(a) ten solutions (b) two solutions (c) one solution

(d) no solutions (e) none of these

16. Suppose a_1 is an integer not divisible by 3 and $a_1^2 + a_2^2 + \cdots + a_n^2$ is divisible by 3, where $a_2, \ldots, a_n$ are integers. Then n is

(a) arbitrary (b) at least 3 (c) at most 2 (d) always odd (e) none of these

17. In a quadrilateral $ABCD, AB = 9, BC = 2, CD = 5$ and $DA = 3$. X and Y are points on AB and CD, respectively. Then XY cannot equal

(a) 2 (b) 4 (c) 7 (d) 10 (e) none of these

18. Assume the earth is a perfect sphere and a wire is stretched tightly around the equator. The wire is lengthened one meter and then expanded uniformly so as to form a somewhat larger circle. By approximately how many meters will the new radius be larger than the old one?

(a) 0.016 (b) 0.032 (c) 0.16 (d) 0.32 (e) 1

19. School X has 100 students and school Y has 50 students. These schools are to be replaced by a single school Z. If the students live in the immediate vicinities of their respective schools, X or Y, where should Z be placed so as to minimize the total distance travelled by all the students.

(a) half way in between (b) one third of the way from X to Y

(c) at X (d) at Y (e) none of these

20. An urn contains 40 red balls, 27 green balls, 26 blue balls and 7 white balls. What is the smallest number of balls that must be drawn, at random, to guarantee that at least 15 balls have the same colour?

(a) 86 (b) 50 (c) 43 (d) 39 (e) none of these

Problems: 1976

1. What is the value of $5^{\log_5 6}$?

(a) 1 (b) 5 (c) 6 (d) $\log_6 5$ (e) none of these

2. Let S be the set of points (x, y) in the plane satisfying both $x^2+y^2 \leq 1$ and $x^2+y^2 \geq r^2$. A value of r such that S is the empty set is

(a) 1 (b) -1 (c) $\frac{1}{2}$ (d) $-\frac{1}{2}$ (e) none of these

3. If S, T and V are sets, then $(S \cap T) \cup (S \cap V)$ is the same set as

(a) S (b) $T \cup V$ (c) $T \cap V$

(d) $S \cap (T \cup V)$ (e) none of these

4. A metal disc has one face marked "1" and the other face marked "2". A second metal disc has one face marked "2" and the other marked "3". Assume that, when tossed, the two faces of a disc are equally likely to turn up. If both discs are tossed, what is the probability that 4 is the sum of the numbers turning up?

(a) $\frac{1}{4}$ (b) $\frac{1}{2}$ (c) $\frac{3}{4}$ (d) $\frac{1}{3}$ (e) none of these

5. Which of the following statements about $\frac{1+\sqrt{2}}{1-\sqrt{2}}$ is true?

(a) it is irrational (b) it is rational (c) it is imaginary

(d) it is positive (e) none of these

6. If $i = \sqrt{-1}$, then $i^6 =$

(a) 1 (b) -1 (c) i (d) $-i$ (e) none of these

7. The solution set of the inequality $x^2 - x - 2 < 0$ is the interval

(a) $-2 \leq x \leq 1$ (b) $-2 < x < 1$ (c) $-2 \leq x < 1$

(d) $-2 < x \leq 1$ (e) none of these

8. At the end of a party, everyone shakes hands with everyone else. Altogether there are 28 handshakes. How many people are there at the party?

(a) 8 (b) 14 (c) 20 (d) 56 (e) none of these

9. Let $[x]$ denote the largest integer not exceeding x. Which of the following statements are always true?

I: $[x + y] = [x] + [y]$; **II:** $[2x] = 2[x]$; **III:** $[-x] = -[x]$.

(a) I only (b) II only (c) III only (d) all (e) none

10. If the sum of the first n positive integers is $\frac{n(n+1)}{2}$, the sum of the first n positive odd integers is

(a) $\frac{n(n+1)}{4}$ (b) $\frac{n(2n+1)}{2}$ (c) n^2 (d) $n^2 - 4$ (e) none of these

11. Suppose $d = x^2 - y^2$ where x and y are two odd integers. Which of the following statements are always true?

I: d is odd; **II:** d is divisible by 4; **III:** d is a perfect square.

(a) I only (b) II only (c) III only

(d) II and III only (e) none of these

12. If each term of the sequence $\{a_1, a_2, \ldots, a_n\}$ is either 1 or -1, then $a_1 + a_2 + \cdots + a_n$ is always

(a) 0
(b) 1
(c) odd if n is odd, even if n is even

(d) odd if n is even, even if n is odd

(e) none of these

13. If x is a real number satisfying the equation $x^{x^{x^{\cdot^{\cdot^{\cdot}}}}} = 2$, then $x =$

(a) ∞ (b) 2 (c) $\sqrt[4]{2}$ (d) $\sqrt{2}$ (e) none of these

14. The number of pipes of inside diameter 1 unit that will carry the same amount of water as one pipe of inside diameter 6 units of the same length is

(a) 6π (b) 6 (c) 12 (d) 36 (e) none of these

15. The expression $2^{-(2k+1)} - 2^{-(2k-1)} + 2^{-2k} =$

(a) 2^{-2k} (b) $2^{-(2k-1)}$ (c) $-2^{-(2k+1)}$ (d) 0 (e) none of these

16. Let P be the product of any 3 consecutive odd integers. The largest integer dividing all such P is

(a) 15 (b) 6 (c) 5 (d) 3 (e) none of these

17. If $|x - \log y| = x + \log y$ where x and $\log y$ are real, then

(a) $x = 0$ (b) $y = 1$ (c) $x = 0$ and $y = 1$

(d) $x(y - 1) = 0$ (e) none of these

18. Each of a group of 50 girls is blonde or brunette and is blue or brown eyed. If 14 are blue-eyed blondes, 31 are brunettes and 18 are brown eyed, the number of brown-eyed brunettes is

(a) 7 (b) 9 (c) 11 (d) 13 (e) none of these

19. After finding the average of 35 scores, a student carelessly included the average with the 35 scores and found the average of these 36 numbers. The ratio of the second average to the true average was

(a) 1:1 (b) 35:36 (c) 36:35 (d) 2:1 (e) none of these

20. If the line $y = mx + 1$ intersects the ellipse $x^2 + 2y^2 = 1$ exactly once, then $m^2 =$

(a) $\frac{1}{2}$ (b) $\frac{2}{3}$ (c) $\frac{3}{4}$ (d) $\frac{4}{5}$ (e) none of these

Problems: 1977

1. If $a > b > 0$, then

(a) $\frac{a+1}{a} > \frac{b+1}{b}$ (b) $\frac{a+1}{a} \geq \frac{b+1}{b}$ (c) $\frac{a+1}{a} \leq \frac{b+1}{b}$

(d) $\frac{a+1}{a} < \frac{b+1}{b}$ (e) none of these

2. Let AB be a diameter of a circle of radius 1 and let C be a point on the circumference such that $AC = BC$. Then $AC =$

(a) 2 (b) $\frac{1}{2}$ (c) $\sqrt{2}$ (d) $\frac{1}{\sqrt{2}}$ (e) none of these

3. Out of 100 people, 60 report that they receive the daily news by watching television, whereas 70 read the newspaper. Of those that read the newspaper, 70% also watch television. The number not receiving any news by television or newspaper is

(a) 15 (b) 19 (c) 23 (d) 27 (e) none of these

4. The expression $64^9 32^{-0.08} =$

(a) 64 (b) 32 (c) 24 (d) 8 (e) none of these

5. Let $f(x)$ be a non-constant polynomial with real coefficients. If $f(x) = f(x-1)$ for all x, then $f(x)$

(a) cannot exist (b) has either no or infinitely many roots

(c) has exactly 1 root (d) has exactly 2 roots (e) none of these

6. Suppose k is a real number such that $0 < k < 1$. Of the two roots of $kx^2-3x+k = 0$,

(a) both are zero (b) one is positive and one is negative

(c) both are positive (d) both are negative (e) none of these

7. Let ℓ be a line in the plane passing through the points (1,1) and (3,5). Then ℓ passes through the point $(2, y)$ where $y =$

(a) 4 (b) 2 (c) 3 (d) 5 (e) none of these

8. ABC is an equilateral triangle with sides of length 1. D and E are points on AC and AB, respectively, such that $BCDE$ is a trapezoid. If the area of triangle ADE is equal to the area of this trapezoid, then $DE =$

(a) $\frac{1}{2}$ (b) $\frac{1}{3}$ (c) $\frac{1}{\sqrt{2}}$ (d) $\frac{\sqrt{2}-1}{\sqrt{2}}$ (e) $\frac{\sqrt{3}-1}{\sqrt{3}}$

9. The inequality $(x+1)(x-1) \geq x^2$ is valid

(a) for all real x (b) for no real x (c) for all $x > 1$

(d) for all $x < 0$ (e) none of these

10. Suppose a bowl contains 3 red balls and 3 yellow balls. The probability that two balls drawn out without replacement will both be red is

(a) $\frac{1}{2}$ (b) $\frac{1}{4}$ (c) $\frac{1}{3}$ (d) $\frac{1}{6}$ (e) none of these

11. The compound fraction $\frac{1}{1+\frac{1}{1+\frac{1}{1+1}}} =$

(a) 1 (b) $\frac{1}{2}$ (c) $\frac{2}{3}$ (d) $\frac{3}{5}$ (e) $\frac{5}{8}$

12. The expression $\frac{xy-x^2}{xy-y^2} - \frac{xy}{x^2-y^2} =$

(a) $\frac{x^3}{y^3-yx^2}$ (b) $\frac{x^2}{y^2-x^2}$ (c) $\frac{x^2+y^2}{x^2-y^2}$

(d) $\frac{x^4+xy^3}{(x^2-y^2)(xy-y^2)}$ (e) none of these

13. If the radius of a sphere is increased 100%, the volume is increased by

(a) 100% (b) 200% (c) 300% (d) 400% (e) none of these

14. The expression $x^4 + 16 =$

(a) $(x^2+4)(x^2+4)$ (b) $(x^2+4)(x^2-4)$

(c) $(x^2-4x+4)(x^2+4x+4)$ (d) $(x^2-2x\sqrt{2}2+4)(x^2+2x\sqrt{2}+4)$

(e) none of these

15. The price of a book has been reduced by 20%. To restore it to its former value, the last price must be increased by

(a) 25% (b) 10% (c) 15% (d) 20% (e) none of these

16. $OABC$ is a rectangle such that B is on the circle with center O and radius 10. If $OA = 5$, then $AC =$

(a) $5\sqrt{2}$ (b) $5\sqrt{3}$ (c) 8 (d) 12 (e) none of these

17. The lengths of the medians of a right triangle which are drawn from the vertices of the acute angles are $\sqrt{73}$ and $2\sqrt{13}$. The length of the third median is

(a) $\sqrt{73+52}$ (b) $\sqrt{73}+2\sqrt{13}$ (c) 5

(d) 10 (e) none of these

18. A car travels 240 kilometers from one town to another at an average speed of 30 kilometers per hour (kpm). On the return trip the average speed is 60 kph. The average speed for the round tiip is

(a) 35 mph (b) 40 mph (c) 45 mph (d) 50 mph (e) 55 mph

19. The expression $\log_3 6 + \log_3 \frac{3}{2} =$

(a) $\frac{5}{2}$ (b) 3 (c) 2 (d) 1 (e) 0

20. The slope of the line passing through the points (3,4) and (1,9) is

(a) $-\frac{5}{2}$ (b) $\frac{5}{2}$ (c) 5 (d) -2 (e) 6

Problems: 1978

1. Which of the following inequalities are true for all positive numbers x?

(a) $x + \frac{1}{x} > 2$ (b) $x + \frac{1}{x} < 2$ (c) $x + \frac{1}{x} \geq 2$ (d) $x + \frac{1}{x} \leq 2$ (e) none of these

2. A steamer was able to go 20 kilometers per hour upstream and 25 kilometers per hour downstream. On a return trip the steamer took 2 hours longer coming upstream than it took coming downstream. In kilometers, the total distance travelled by the steamer was

(a) 100 (b) 200 (c) 400 (d) 800 (e) 150

3. If n is a positive integer, then $n^2 + 3n + 1$ is

(a) sometimes a square (b) sometimes an even integer

(c) always a square (d) never a square (e) none of these

4. The solution set of the inequality $x^2(x^2 - 1) < 0$ is

(a) an interval (b) an interval and a point

(c) a point (d) two intervals (e) all real numbers

5. D is the point inside an equilateral triangle ABC such that DBC is an isosceles triangle and $\angle CDB = 90°$. Then $\angle ADC =$

(a) 45° (b) 90° (c) 120° (d) 135° (e) none of these

6. The binary operation $*$ between two positive integers m and n is such that $m * n = mn + 1$. Which of the following does not hold?

(a) commutative law (b) associative law (c) $m * n$ is a positive integer

(d) $m * n \geq 2$ (e) $m * n$ is odd whenever m is even

7. Label the four quadrants of the plane in the usual manner. The solution set of the simultaneous inequalities $x^2 - y < 0$ and $x^2 + y^2 < 1$ lies entirely in quadrants

(a) I and II (b) II and III (c) III and IV (d) IV and I (e) none of these

8. If $f(n) = n^2$, where n is an integer, then $\frac{f(f(n+l)) - f(f(n-1))}{f(n+1) - f(n-1)} =$

(a) n^2 (b) $2n^2 + 2$ (c) $n^2 + 1$ (d) $n^4 + 1$ (e) none of these

9. Which of the following inequalities hold for all pairs of real number (x, y)?

(a) $\sqrt{x^2 + y^2} \leq x + y$ (b) $\sqrt{x^2 + y^2} \leq x^2 + y^2$ (c) $\sqrt{x^2 + y^2} \leq xy$

(d) $\sqrt{x^2 + y^2} \leq |x| + |y|$ (e) none of these

10. Two similarly proportioned boxes have their surface areas in the ratio 4:1. Their volumes are in the ratio

(a) 9:1 (b) 8:1 (c) 3:1 (d) 2:1 (e) none of these

11. The roots of the quadratic polynomial $2x^2 + kx + 1$ are r and s. Which of the following are impossible

(a) $r = s$ (b) $r - s = 1$ (c) $r + s = 1$ (d) $r + s = 0$ (e) none of these

12. A hat contains three slips of paper, of which one bears the name John, one bears the name Diana and the other bears both names. If John and Diana each draw a slip, the probability that they both draw slips with their own names on is

(a) $\frac{1}{9}$ (b) $\frac{1}{6}$ (c) $\frac{1}{4}$ (d) $\frac{1}{3}$ (e) none of these

13. The value of k such that $x^6 - kx^4 + kx^2 - kx + 4k + 6$ is divisible by $x - 2$ is

(a) 1 (b) 5 (c) 7 (d) 11 (e) no such k

14. ABC is an equilateral triangle inscribed in a circle of diameter 1. If AD is a diameter of this circle, then $BD =$

(a) $\frac{1}{2}$ (b) 1 (c) 2 (d) $\frac{\sqrt{3}}{2}$ (e) $\frac{1}{\sqrt{2}}$

15. The equation of the line through the point (1,1) that is perpendicular to the line $y = -2x - 3$ is

(a) $y = \frac{2x+1}{3}$ (b) $y = \frac{x+2}{3}$ (c) $y = 2x - 1$ (d) $y = \frac{x+1}{2}$ (e) none of these

16. If $\log_a b = c$, then $\log_a(b^c) =$

(a) bc (b) b^c (c) c^c (d) c^2 (e) $2c$

17. A circle and a square can never have in common exactly

(a) one point (b) two points (c) three points

(d) four points (e) all of these are possible

18. Let $\{a_1, a_2, a_3, \ldots,\}$ be a sequence of real numbers such that the sum of the first n terms is $n^2 + n$. Then $a_n =$

(a) n (b) $2n - 1$ (c) $2n + 1$ (d) 1 (e) none of these

19. The domain of the function $f(x) = \sqrt{1 - \sqrt{1 - x^2}}$ is

(a) a single point (b) an infinite interval with a point deleted

(c) a finite interval (d) an infinite interval (e) none of these

20. A polynomial whose graph passes through the points $(-1, 7)$, $(1, 0)$ and (2,0) is

(a) $x + 8$ (b) $x^2 - 3x + 2$ (c) $x^2 + 9$

(d) $x^3 - x^2 + x - 1$ (e) none of these

Problems: 1979

1. A triangle has sides of lengths 1, 2 and $\sqrt{3}$. Its area is

(a) $\frac{1}{2}$ (b) $\frac{\sqrt{3}}{2}$ (c) 2 (d) $2\sqrt{3}$ (e) none of these

2. The polynomial $4x^4 - kx^2 + 1$ has double roots if $k =$

(a) 1 (b) 2 (c) 3 (d) 4 (e) none of these

3. For what values of a and b is the equation $(a^{\log_{10} b})^{ab} = (b^{\log_{10} a})^{ba}$ true?

(a) all values (b) no values (c) all positive values

(d) all negative values (e) none of these

4. The equation of the line that is perpendicular to the line $x + 2y = 3$ and passes through the point (4,5) is

(a) $x - 2y = 3$ (b) $2x - y = 3$ (c) $2x + y = 3$

(d) $-2x + y = 3$ (e) none of these

5. Let S be the set of points defined by the inequalities $x+y \geq 1$, $x-y \leq 1$ and $y \leq \frac{1}{2}$. The area of the region determined by S is

(a) $\frac{3}{8}$ (b) $\frac{1}{2}$ (c) $\frac{5}{8}$ (d) 1 (e) none of these

6. X is a square of diagonal 1, Y is an equilateral triangle of side 1 and Z is an isosceles right triangle whose equal sides have length 1. Comparing the areas of these figures.

(a) X is larger than both Y and Z (b) Y is larger than both Z and X

(c) Z is larger than both X and Y (d) X, Y and Z all have the same area

(e) none of these

7. A positive integer is squarefree if it not divisible by the square of an integer larger than 1. The number of positive squarefree integers less than 20 is

(a) 0 (b) 3 (c) 9 (d) 13 (e) none of these

8. The solutions of the equation $(\sin\theta + \cos\theta)^2 = 1$ are

(a) all multiples of $\frac{\pi}{2}$ (b) all multiples of π

(c) all odd multiples of $\frac{\pi}{2}$ (d) all even multiples of π

(e) none of these

9. The sum of the first 27 odd positive integers is

(a) 153 (b) 196 (c) 144 (d) 216 (e) none of these

10. The expression $(((\sqrt{2})^{\sqrt{2}})^2)^{\sqrt{2}} =$

(a) 2 (b) $\sqrt{2}$ (c) 4 (d) 8 (e) none of these

11. If r and s are the roots of $x^2 + 7x - 5 = 0$, then $r^2 + s^2 =$

(a) 59 (b) 47 (c) -15 (d) 35 (e) none of these

12. AB and CD are diameters of a circle with center O. The tangents to the circle at A and D meet at P, while the tangents to the circle at B and C meet at Q. If $\angle AOC = 30°$, then $PQ =$

(a) 2 (b) 3 (c) 4 (d) $\frac{1}{2}$ (e) none of these

13. For which values of x is a triangle with sides $x, x+1$ and $x+2$ an acute triangle ?

(a) $x = 1$ (b) $x > 2$ (c) $x < 4$ (d) $x > 3$ (e) none of these

14. Which of the following inequalities is always true for any pair of real numbers (x, y)?

(a) $x + y \leq xy$ (b) $(x + y)^2 \geq xy$ (c) $(x + y) \geq xy$

(d) $(x + y)^2 \geq x + y$ (e) none of these

15. A twelve-hour digital watch displays the hours, minutes and seconds. During one complete day it registers at least one figure 3 for a total time of

(a) 1 hour and 5 seconds (b) 1 hour, 15 minutes and 15 seconds

(c) 2 hours and 24 minutes (d) 3 hours

(e) none of these

16. $ABCD$ is a square of side 1. P4P and Q are points on BD and CD, respectively, such that APQ is an equilateral triangle. Then $DQ =$

(a) $\frac{1}{2}$ (b) $\sqrt{2} - 1$ (c) $\frac{1}{3}$ (d)$2 - \sqrt{3}$ (e) none of these

17. The product of John's and Mary's ages if five more than four times the sum of their ages. If Mary is 4 years younger than John, John'res age is

(a) 13 (b) 11 (c) 9 (d) 7 (e) none of these

18. In a poll of 1000 coffee drinkers, 40% preferred their coffee with neither cream nor sugar and 60% of the remainder preferred their coffee with cream only. After deducting both of these groups, 40% of those left preferred their coffee with sugar only. The rest preferred coffee with cream and sugar and their number was

(a) 144 (b) 216 (c) 96 (d) 172 (e) none of these

19. A sphere and a triangle cannot have in common exactly

(a) 1 point (b) 2 points (c) 3 points (d) 4 points (e) none of these

20. The picture cards are removed from a pack of 52 playing cards. The number of ways of drawing 2 cards from the remaining 40 so that the sum of the numerical values is 10 is

(a) 100 (b) 10 (c) 50 (d) 70 (e) none of these

Problems: 1980

1. There are 5 roads between the towns A and B and 4 roads between B and C. The number of different ways of driving the round trip $A \to B \to C \to B \to A$ without using the same road more than once is

   (a) 32 (b) 240 (c) 400 (d) 16 (e) none of these

2. In the figure below, if the area of the triangle is two-fifths of the area of the parallelogram, then $x =$

   (a) 30 (b) 12 (c) 15 (d) 24 (e) none of these

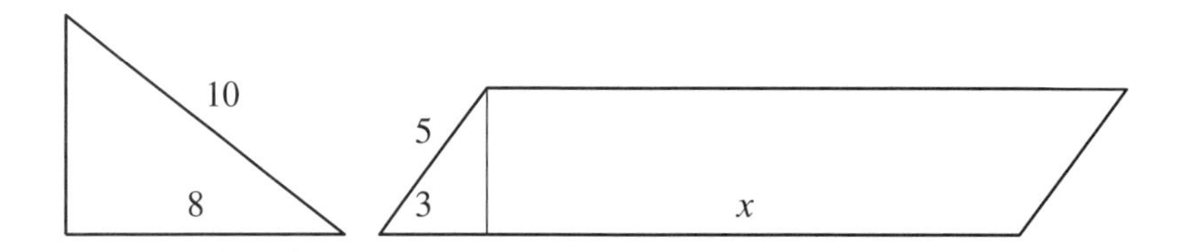

3. A fair die is thrown repeatedly until a 6 is obtained. The probability that this will happen on the third throw is

   (a) $\frac{1}{6}$ (b) $\frac{1}{216}$ (c) $\frac{5}{36}$ (d) $\frac{25}{216}$ (e) none of these

4. Find all real values of k for which $kx^2 + kx + 1$ has no real roots.

   (a) $-2 < k < 6$ (b) $-2 < k < 4$ (c) $0 < k < 2$

   (d) $0 < k < 4$ (e) none of these

5. If the lines $9x + ky = 7$ and $kx + y = 2$ parallel, then $k =$

   (a) 3 (b) ± 3 (c) $\frac{1}{3}$ (d) $\pm\frac{1}{3}$ (e) none of these

6. Let $x = \log_{a^2} b$. Then $\log_b a^x =$

   (a) $\sqrt{b}$ (b) b^2 (c) $\frac{1}{2}$ (d) 2 (e) none of these

7. If $\sin 3x = 0$, then $\sin x =$

   (a) 0 (b) 0, $\pm\frac{1}{2}$ or $\frac{\sqrt{3}}{2}$

   (c) 0 or $\pm\frac{\sqrt{3}}{2}$ (d) 0 or $\pm\frac{1}{2}$ (e) none of these

8. The right triangle ABC has hypotenuse $AB = x^2 + y^2$ and side $BC = 2xy$. Then $AC =$

   (a) $x^2 - y^2$ (b) $(x - y)^2$ (c) $(x^2 - y^2)^2$ (d) $(x + y)^2$ (e) none of these

9. In a certain examination, the average mark the 30 boys in a class was 60. The girls did rather better, their average being 65. If the overall average fro the class was 62, the number of girls who took the examination was

(a) 20 (b) 24 (c) 18 (d) 36 (e) none of these

10. Which of the following is true for all real numbers x and y?

(a) $(x+y)^2 \geq (x-y)^2$ (b) $|x+y| \geq x+y$ (c) $\sqrt{x^2+y^2} \geq x+y$

(d) $xy \geq x+y$ (e) none of these

11. The expression $\log_{10}(144^{144}) =$

(a) $576\log_{10} 2 + 288\log_{10} 3$ (b) $144\log_{10} 2 + 144\log_{10} 3$

(c) $(\log_{10} 144)^{144}$ (d) $2\log_{10} 144$

(e) none of these

12. The line ℓ has the equation $y = -2x-4$ and the line m has the equation $y = 2x+4$. The equation of the line through (0,0) and the point of intersection of ℓ and m is

(a) $y = x$ (b) $y = -2x$ (c) $2x + 3y = 1$ (d) $-y - 2x = 1$ (e) none of these

13. The maximum possible value of $3x - 3x^2$ where x is real is

(a) $\frac{5}{4}$ (b) 0 (c) -1 (d) $\frac{3}{4}$ (e) none of these

14. $ABCD$ is a unit square. M is the midpoint of AD and N is a point inside the square such that MN is parallel to AB. If the segments MN, BN and CN divide the square into three parts of equal area, then $MN =$

(a) $\frac{1}{3}$ (b) $\frac{1}{\sqrt{2}}$ (c) $\frac{1}{4}$ (d) $\frac{1}{\sqrt{3}}$ (e) $\frac{1}{2}$

15. $ABCD$ is a unit square. M is the midpoint of AD and N is a point inside the square such that MN is parallel to AB. If $MN = BN = CN$, then their common value is

(a) $\frac{1}{2}$ (b) $\frac{5}{8}$ (c) $2-\sqrt{2}$ (d) $\frac{1}{\sqrt{2}}$ (e) $\frac{3}{5}$

16. The expression $\sqrt{\frac{\sqrt{5}-2}{\sqrt{5}+2}} =$

(a) $\frac{1}{\sqrt{5}-\sqrt{2}}$ (b) $\frac{1}{\sqrt{2}+5}$ (c) $\sqrt{5}-2$ (d) $\sqrt{5}+2$ (e) $\frac{1}{5-\sqrt{2}}$

17. If $f(x) = x^2 - 5$ and $f(4+a) = f(4a)$, then $a =$

(a) $\frac{4}{3}$ or $-\frac{4}{5}$ (b) 4 or $-\frac{4}{3}$ (c) $\frac{4}{3}$ or $\frac{4}{5}$ (d) -4 or $\frac{4}{3}$ (e) -4 or $-\frac{4}{5}$

18. Two circles and a straight line are drawn in the plane to form exactly N bounded regions. n cannot be

(a) 3 (b) 4 (c) 5

(d) 6 (e) all of these are possible

19. A bowl contains 2 marbles of each of 4 colours. If you randomly remove 3 marbles from the bowl without replacing them, what is the probability that you have removed two of the same colour?

(a) $\frac{1}{4}$ (b) $\frac{1}{7}$ (c) $\frac{3}{7}$ (d) $\frac{1}{2}$ (e) none of these

20. If n is a positive integer such that $1+2+\cdots+n = (n+1)+(n+2)+\cdots+118+119$, then $n =$

(a) 60 (b) 69 (c) 76 (d) 84 (e) 89

Problems: 1981

1. In how many real points do the circle $x^2 + y^2 = 1$ and the ellipse $\frac{x^2}{4} + (y-2)^2 = 1$ intersect?

(a) one (b) two (c) three (d) four (e) zero

2. If $x°$ Fahrenheit is equal to $-x°$ Celsius, then $x =$

(a) 10 (b) -40 (c) $-\frac{16}{9}$ (d) $-\frac{80}{7}$ (e) none of these

3. If $x^2 + (x+1)(k+1) = 0$ has real solutions in x, then

(a) $k = -1$ or $k = -3$ (b) $-1 \leq k \leq 3$ (c) $k \geq 3$ or $k \leq -1$

(d) k is arbitrary (e) none of these

4. The expression $\sec x \csc x =$

(a) $\sec x + \csc x$ (b) $\tan x + \cot x$ (c) $\sin x + \cos x$

(d) $\frac{1}{\sec x + \csc x}$ (e) none of these

5. The sum of the roots of $3x^5 - 30x^4 + 105x^3 - 105x^2 + 72x$ is

(a) 3 (b) -30 (c) 10 (d) -35 (e) none of these

6. For what values of k can the correct solution of $\log(3x+2) + \log(4x-1) = 2\log k$ be obtained by "canceffing the log" and solving $(3x+2) + (4x-1) = 2k$?

(a) no such k (b) $k = 3$ (c) $k = 11$ (d) k arbitrary (e) none of these

7. Seven Canadian coins, none of which has a value greater than 25 cents, add up to 81 cents. The number of nickels is

(a) none (b) one (c) two (d) five (e) none of these

8. Let $ABCD$ be a square of side 1 and P be any point. The minimum total length of line-segments PA, PB, PC and PD is

(a) 2 (b) $2\sqrt{2}$ (c) $2+\sqrt{2}$ (d) 4 (e) none of these

9. The expression $(1+\sqrt{2}+\sqrt{3})(1+\sqrt{2}-\sqrt{3})(1-\sqrt{2}+\sqrt{3})(1-\sqrt{2}-\sqrt{3}) =$

(a) -8 (b) 4 (c) $2\sqrt{3}+2\sqrt{2}$ (d) 0 (e) none of these

10. In how many ways can one choose 5 of the first 10 positive integers so that no two of them are consecutive?

(a) 2 (b) 4 (c) 6 (e) 8 (e) none of these

11. The only integer solution of $(3^x)^{3^x}(2^x)^{2^x} = 3^{18}2^9$ is $x =$

(a) 1 (b) 2 (c) 3 (d) 0 (e) no such solution

12. In each meeting, every pair of people shook hands once. At two successive meetings, the second one having a higher attendance, 100 handshakes in all took place. The second attendance was higher than the first by

(a) 1 (b) situation impossible

(c) 10 (d) 11 (e) none of these

13. $ABCD$ is a square of side 1. E and F are points on BC and CD, respectively. If AEF is an equilateral triangle, then $AE =$

(a) $\sqrt{2}$ (b) $\sqrt{3}$ (c) $\sqrt{6}-\sqrt{3}$ (d) $\sqrt{6}-\sqrt{2}$ (e) none of these

14. What is the sum of all the digits appearing in the first 99 positive integers?

(a) 900 (b) 1800 (c) 4950 (d) 9900 (e) none of these

15. Joe tosses a fair coin three times in succession. Moe tosses three fair coins all at once. The probability that Joe gets more heads than Moe is

(a) $\frac{1}{2}$ (b) $\frac{3}{4}$ (c) $\frac{5}{8}$ (d) $\frac{7}{16}$ (e) none of these

16. In the figure below, $x =$

(a) 1 (b) 2 (c) 3 (d) 4 (e) none of these

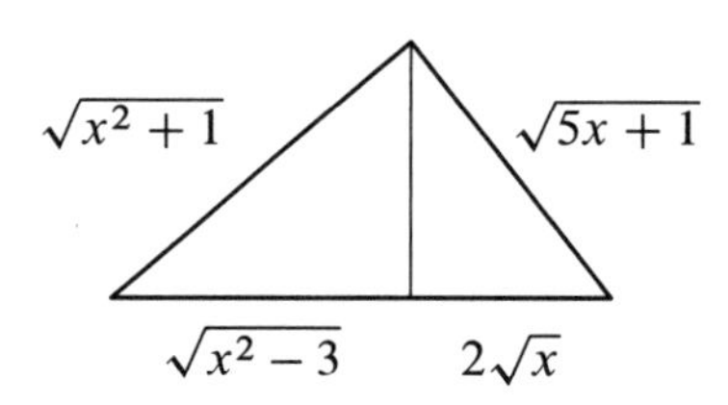

17. Let $f(x, y)$ be a function which is not identically zero and such that $f(x, y) = kf(y, x)$ for all x and y. The possible values of k are

(a) 0 (b) 1 (c) -1 (d) ± 1 (e) none of these

18. The number of four-digit numbers that remain unchanged when their digits are reversed is

(a) 81 (b) 90 (c) 100 (d) 121 (e) none of these

19. The area of the triangle whose vertices have coordinates (0,0), (2,4) and (4,2) is

(a) 12 (b) $6\sqrt{2}$ (c) $2\sqrt{10}$ (d) $2\sqrt{2}+4\sqrt{5}$ (e) none of these

20. The sum $\frac{1}{2x+1} + \frac{1}{(2x+1)(4x+1)} + \frac{1}{(4x+1)(6x+1)} + \cdots + \frac{1}{(2(k-1)x+1)(2kx+1)} =$

(a) $\frac{1}{2k+1}$ (b) $\frac{k}{2k+1}$ (c) $\frac{1}{(2x+1)(4x+1)\cdots(2k+1)}$

(d) $\frac{k}{(2x+1)(4x+1)\cdots(2kx+1)}$ (e) none of these

Problems: 1982

1. A plane takes off from Edmonton, flies north for 500 kilometers, then east for 500 kilometers, next south for 500 kilometers and finally west for 500 kilometers. It will land

(a) north of Edmonton (b) east of Edmonton (b) South of Edmonton

(d) west of Edmonton (e) right at Edmonton

2. Merle has an assortment of pennies, nickels, dimes and quarters but is unable to change a dollar bill. The largest possible amount Merle has in coins is

(a) \$0.94 (b) \$0.99 (c) \$1.19 (d) \$1.24 (e) none of these

3. The angle between the hour hand and the minute hand at twenty minutes past one is

(a) 60° (b) 70° (c) 80° (d) 90° (e) none of these

4. The number of solutions (x, y, z) of the equation $x + y + z = 15$, where x, y and z are integers such that $0 < x < y < z$, is

(a) 12 (b) 13 (c) 14 (d) 15 (e) none of these

5. If r and s are the roots of $7x^2 - 4x + 12$, then $\frac{1}{r} + \frac{1}{s} =$

(a) -3 (b) $-\frac{1}{3}$ (c) $\frac{1}{3}$ (d) 3 (e) none of these

6. In the figure below, $AC = BC$, $BE = EF$ and AG is parallel to FB. Then $\frac{BD}{AD}$ is

(a) less than 2 (b) equal to 2 (c) between 2 and 3

(d) equal to 3 (e) greater than 3

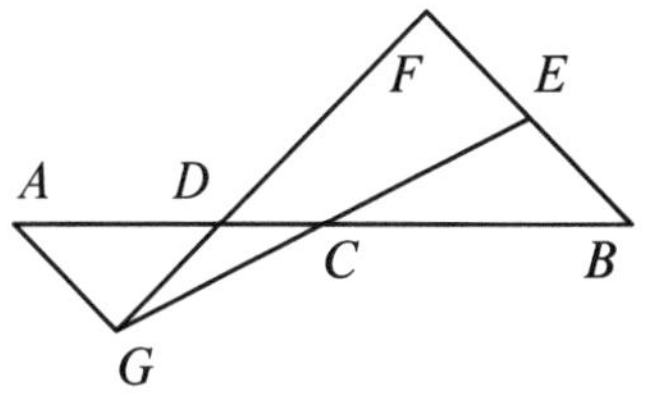

7. Stephen collects bugs. Some of them are spiders (8 legs each) and the rest are beetles (6 legs each). His collection consists of 8 bugs with a total of 54 legs. The number of spiders in his collection is

(a) 1 (b) 3 (c) 5 (d) 7 (e) none of these

8. The minimum value of $x^2 + 2x - 3$ is

(a) -4 (b) -3 (c) -1 (d) 1 (e) none of these

9. Kelly goes to school at a rate of 2 kilometers per hour (kph). In order that the complete journey (home to school to home) is travelled at an average of 4 kph, the rate at which the trip home must be made is

(a) 5 kph (b) dependent on home-school distance

(c) 6 kph (d) 8 kph (e) none of these

10. Let A, B, C, D and E be five points in space. If $AB = 30$, $BC = 80$, $CD = 236$, $DE = 86$ and $EA = 40$, then $CE =$

(a) $\sqrt{236} - \sqrt{86}$ (b) $\sqrt{236^2 - 86^2}$ (c) 150 (d) 236 (e) none of these

11. Let $x = \max[\min(a, b), \min(c, d)]$ and $y = \min[\max(a, c), \max(b, a)]$, where a, b, c and d are distinct real numbers. Then x is always

(a) strictly greater than y (b) greater than or equal to y

(c) less than or equal to y (d) strictly less than y (e) none of these

12. The sum of the positive divisors of 36 is

(a) 9 (b) 43 (c) 91 (d) 666 (e) none of these

13. Peter and Trefor bought 300 grams and 500 grams of jelly beans respectively. They ate them together with Mr. Smith, each eating the same amount. Afterwards, Mr. Smith paid the boys 80 cents. Peter's fair share amounted to

(a) 10 cents (b) 20 cents (c) 30 cents (d) 40 cents (e) none of these

14. If $a = 6\sqrt{3} - 3\sqrt{13}$ and $b = 6\sqrt{10} - 15\sqrt{2}$, then

(a) $a - b \geq 1$ (b) $0 < a - b < 1$ (c) $a - b = 0$

(d) $-1 < a - b < 0$ (e) $a - b \leq -1$

15. Five baskets contain 5, 12, 14, 22 and 29 eggs respectively. In each basket, some of the eggs are chicken eggs while the remaining ones are duck eggs. After one of the baskets is sold, the total number of chicken eggs remaining is equal to twice the total number of duck eggs remaining. The number of eggs in the basket sold is

(a) 5 (b) 12 (c) 14 (d) 22 (e) 29

16. An astronaut 2 meters tall walks once around the equator of a gigantic spherical planet. The top of his head describes a circle. The circumference of this circle is longer than the equator of the planet by

(a) less than 50 meters (b) between 50 and 500 meters

(c) between 500 and 5000 meters (d) between 5000 and 50000 meters

(e) greater than 50000 meters

17. The number of terms in the arithmetic series $8 + 16 + 24 + \cdots$ such that their sum first exceeds 1982 is

(a) 20 (b) 22 (c) 24 (d) 248 (e) none of these

18. Rose and Mary play a series of three games. In each game, Rose's probability of winning is $\frac{2}{3}$. If Rose's probability of winning at least two of the three games is p, then

(a) $p < \frac{1}{2}$ (b) $\frac{1}{2} \leq p < \frac{2}{3}$ (c) $p = \frac{2}{3}$ (d) $\frac{2}{3} < p \leq \frac{3}{4}$ (e) $p > \frac{3}{4}$

19. If $\log_z x = 2$ and $\log_{xy} z = \frac{1}{6}$, then $\log_y z =$

(a) $\frac{1}{4}$ (b) $\frac{1}{3}$ (c) 3 (d) 4 (e) none of these

20. The expression $\frac{\cos x}{1 - \sin x} =$

(a) $\sin x + \tan x$ (b) $\sin x - \tan x$ (c) $\sec x + \tan x$

(d) $\sec x - \tan x$ (e) none of these

Problems: 1983

1. Assume that the vertical distance between floors in a building is constant. The ratio of the vertical distance between the first and the third floors to the vertical distance between the first and the sixth floors is

(a) 2:5 (b) 1:2 (c) 2:1 (d) 5:2 (e) none of these

2. The compound fraction $\frac{2}{3+\frac{2}{3+\frac{2}{3}}} =$

(a) $\frac{4}{9}$ (b) $\frac{22}{39}$ (c) $\frac{2}{3}$ (d) 1 (e) none of these

3. A boy stands at the center of a circle of radius 8 meters. A girl stands at a point 4 meters from the boy. The boy runs to some point on the circle and then to the girl. In meters, the shortest distance the boy must run is

(a) 20 (b) $8 + 4\sqrt{5}$ (c) $8 + 4\sqrt{3}$ (d) $2\sqrt{68}$ (e) none of these

4. The smallest positive integer which leaves a remainder of 2 whether it is divided by 12 or by 15 is

(a) 14 (b) 58 (c) 62 (d) 182 (e) none of these

5. A bird in the hand is worth two in the bush. Five birds in hand and three birds in the bush together are worth $30 more than three birds in hand and five birds in the bush together. A bird in hand is worth

(a) $5.00 (b) $7.50 (c) $10.00 (d) $15.00 (e) none of these

6. In triangle ABC, $\angle A$ is at least $10°$ more than $\angle B$ and $\angle B$ is at least $25°$ more than $\angle C$. The maximum value of $\angle C$ is

(a) $35°$ (b) $40°$ (c) $45°$ (d) $50°$ (e) none of these

7. If $f(x-1) = 3x^2 + 2x - 5$, then $f(x) =$

(a) $3x^2 + 2x - 4$ (b) $3x^2 + 2x - 6$ (c) $3x^2 + 8x$

(d) $3x^2 - 4x - 4$ (e) none of these

8. Let $[x]$ denote the greatest integer not exceeding x. Then the set of all possible values of $[x] + [-x]$ is

(a) $\{0\}$ (b) $\{0, 1\}$ (c) $\{0, -1\}$ (d) $\{0, 1, -1\}$ (e) none of these

9. Let a, b and c be real constants. It is possible for all the points (1,3), (2,1) and (3,5) to he on a graph of the form

(a) $y = a^2x + b$ (b) $y = -a^2x + b$ (c) $y = a^2x^2 + bx + c$

(d) $y = -a^2x^2 + bx + c$ (e) none of these

10. The sum of five numbers is 100. The sum of the first and the second number is 44, the sum of the second and third is 47, the sum of the third and the fourth is 37 and the sum of the fourth and the fifth is 35. The third number is

(a) 16 (b) 18 (c) 19 (d) 21 (e) none of these

11. There were 16 participants in a mathematics contest. Of every two participants, at least one was right-handed. If the eventual winner was left-handed, then the number of right-handed participants was

(a) 8 (b) impossible to detemine

(c) 15 (d) 7 (e) none of these

12. If r and s are the roots of $x^2 + x + 1 = 0$, then $\frac{r}{s} + \frac{s}{r} =$

(a) -3 (b) -1 (c) 1 (d) 3 (e) none of these

13. The smallest positive integer which can be expressed in the form $11x + 8y$, where x and y are integers, is

(a) 1 (b) 3 (c) 8 (d) 19 (e) none of these

14. If $m = \frac{1}{2+\sqrt{3}}$ and $n = \frac{1}{2-\sqrt{3}}$, then $\frac{1}{m+1} + \frac{1}{n+1} =$

(a) 0 (b) 1 (c) $\sqrt{3}$ (d) 2 (e) none of these

15. The expression $\log_{10} 10 \div \log_{10} \sqrt{10} =$

(a) $\log_{10} \sqrt{10}$ (b) $\log_{10}(10 - \sqrt{10})$ (c) 2

(d) $\sqrt{10}$ (e) none of these

16. The minimum value of $\sin^5 x + \cos^5 x$ for all $x > 0$ is

(a) 0 (b) $\frac{1}{2\sqrt{2}}$ (c) 1 (d) 2 (e) none of these

17. A square root of the complex number $8 + 6i$ is

(a) $\sqrt{8} + \sqrt{6}i$ (b) $\sqrt{8} - \sqrt{6}i$ (c) $3 + i$ (d) $3 - i$ (e) none of these

18. The four vertices of a regular tetrahedron consist of two opposite vertices of the top face of a unit cube together with two opposite vertices of the bottom face. The volume of this tetrahedron is

(a) $\frac{2\sqrt{2}}{3}$ (b) $\frac{1}{2}$ (c) $\frac{1}{3}$ (d) $\frac{1}{4}$ (e) none of these

19. Kelly rolls a fair die and scores the number of spots showing on its upper face. Kerry rolls two fair dice and scores the total number of spots showing on their upper faces. The probability that Kelly's score is higher than Kerry's score is

(a) $\frac{10}{216}$ (b) $\frac{20}{216}$ (c) $\frac{25}{144}$ (d) $\frac{1}{3}$ (e) none of these

20. Three unit circles are drawn with centers at (1,1), (3,1) and (5,3). A line passing through (1,1) is such that the total area of the parts of the three circles to one side of the line is equal to the total area of the parts of the three circles to the other side of it. The slope of this line is

(a) $\frac{1}{3}$ (b) $\frac{1}{\pi}$ (c) $\frac{3}{\pi^2}$ (d) $\frac{1}{4}$ (e) none of these

Problems requiring Full Solutions

Problems: 1967

1. Four points are given in the plane.
   (a) Describe an Euclidean construction of a square such that each of the points lies on a different side of the square.
   (b) How many such squares are there?

2. Let $S = \{x_1, x_2, x_3, \dots, x_n\}$ be a set of n positive numbers. Let

$$f(Q) = \frac{1}{a_1(a_1 + a_2)(a_1 + a_2 + a_3)\cdots(a_1 + a_2 + a_3 + \cdots + a_n)}$$

for any permutation $Q = \langle a_1, a_2, a_3, \dots, a_n \rangle$ of S. Prove that

$$\sum f(Q) = \frac{1}{x_1 x_2 x_3 \cdots x_n},$$

where the summation ranges over the $n!$ permutations of S.

3. P is a variable point on the side CD of a rectangle $ABCD$. AP meets the extension of BC at Q, and BP meets the extension of AD at R. Prove that $CQ \cdot DR$ is constant.

4. Prove that the number of distinct ratios $\frac{d_1}{d_2}$, where d_1 and d_2 are positive divisors of a positive integer m, equals the number of positive divisors of m^2.

5. (a) Prove that if the roots of the equation $Ax^2 + Bxy + Cy^2 = 0$ are real, then its graph consists of two straight lines through the origin. The two lines may coincide.

(b) Prove that the real points of intersection, if any, of the graph of $3x + 4y = 5$ and the graph of $25(x^2 + y^2) = a^2(3x + 4y)^2$ lie on a certain circle with center at the origin.

(c) Find the perpendicular distance from the origin to the line $3x + 4y = 5$.

6. Prove that at least two faces of any polyhedron have the same number of edges.

7. Consider the sequence of integers $\{t_n\}$ defined by $t_1 = t_2 = 1$ and $t_{n+1} = 2t_n + t_{n-1}$ for $n \geq 2$. Let s_n denote the sum of the first n terms of this sequence. Prove that the sequence $\{s_n\}$ satisfies the analogous relation $s_{n+1} = 2s_n + s_{n-1}$ for $n \geq 2$.

8. C is an interior point of a line segment AB and P is any point. Prove that

$$\frac{PA^2}{AB \cdot AC} + \frac{PB^2}{BC \cdot BA} = 1 + \frac{PC^2}{CA \cdot CB}.$$

9. At a certain party, no boy dances with every girl, but each girl dances with at least one boy. Prove that there must be two girls g and G, and two boys b and B, such that g dances with b and G with B, but g does not dance with B nor G with b.

10. For $n = 1, 2, 3, \ldots$, let $x_n = 1 + \cos 2t - (\sin 2t)(\tan nt)$. Prove that $x_{n+1} = 4(1 - \frac{1}{x_n}) \cos^2 t$.

Problems: 1968

1. The three sides of a triangle, when extended, divide the plane into seven areas. Prove that no circle can be drawn with an arc in each of the seven areas.

2. M is the midpoint of side BC of triangle ABC. Prove that the length of the median AM is less than the average of the lengths of the sides AB and AC.

3. A 3-digit number in base 10 is n times the sum of its digits.

   (a) Find all such numbers for $n = 17$

   (b) Prove that no such numbers exist for $n = 9$.

4. Prove that $\frac{1}{a} + \frac{1}{d} > \frac{1}{b} + \frac{1}{c} > \frac{4}{a+d}$, where a, b, c and d are distinct positive numbers in an arithmetic progression.

5. Prove by induction or otherwise that $5^n + 2 \cdot 3^{n-1} + 1$ is divisible by 8 for any positive integer n.

6. The roots of the equation $x^2 - 14x - 1 = 0$ are α and β, with $\alpha > \beta$. Without solving the equation, prove that $-\frac{1}{14} < \beta < 0$ and $14 < \alpha < 14 + \frac{1}{14}$.

7. (a) What is the smallest positive number in base 10, with all its digits zeros and ones, which is a multiple of 225?

   (b) $ABCABC$ is a 6-digit number in base eight. Prove that it must have as a factor the number 27 in base 10.

8. Solve the equation $\log_x(7x - 6) = 3$.

9. (a) In a lunar probe attempt, a three-stage rocket is used. If any stage fails, the rocket is automatically destroyed. The probability that the first stage fails is $\frac{1}{4}$, that the second fails is $\frac{1}{5}$, and that the third fails is $\frac{1}{6}$. What is the probability that the probe is successfully launched?
   (b) A box contains 3 white, 2 red, and 1 black balls. A second box contains 2 white and 2 red balls. One ball is selected at random from each box. What is the probability that they are of the same color?

10. $M(a,b)$ and $m(a,b)$ represent the larger and smaller, respectively, of the two real numbers a and b, and $M(a,a) = m(a,a) = a$. Which of the following statements are true and which are false?

    (a) $m(a,b) = m(b,a)$;
    (b) $M(a, M(b,c)) = M(M(a,b),c)$;
    (c) $M(a, m(b,c)) = m(M(a,b),c)$;
    (d) $M(a, m(b,c)) = m(a, M(b,c))$;
    (e) $M(a, m(b,c)) = m(M(a,b), M(a,c))$.

Problems: 1969

1. If k, m and n are integers such that $(x-k)(x-10)+1 = (x-m)(x-n)$, prove that $m = n$.

2. Prove that if x is a real number such that $0 < x < \frac{1}{2}$, then there exists an integer $k > 1$ such that $\frac{k^2}{k^2+1} < 1 + x$.

3. For any three consecutive positive integers, prove that the cube of the largest cannot be the sum of the cubes of the other two.

4. Given a straight-edge 1 meter in length and a compass capable of drawing circles of radii up to 1 meter, describe a construction using these tools only which will connect two points in the plane by a straight line, if the distance between them is bigger than 1 meter but less than 2 meters.

5. Let α and β be the roots of the quadratic equation $x^2 - 14x - 1 = 0$.
   (a) If $\alpha < \beta$, determine α.
   (b) Find a quadratic equation with integer coefficients whose roots are α^2 and β^2.

6. An integer in base 10 consists of digits which are ones and nines only. Prove that it is impossible for such an integer to be a perfect square if it is bigger than ten.

7. Let $x > 1$. Prove that $2(\sqrt{x}-1) > 3(\sqrt[3]{x}-1)$.

8. Prove that the sum of all six three-digit numbers which can be formed by permuting three distinct digits a, b and c is $222(a+b+c)$.

9. STU is a triangle with $ST = 6, TU = 10$ and $US = 8$. $PQRS$ is a rectangle with P on ST, Q on TU and R on US. Determine the dimensions of $PQRS$ which yield the maximum area.

10. Let $x^2 = ab + bc + ca$. Prove that $(x^2+a^2)(x^2+b^2)(x^2+c^2)$ is a perfect square.

Problems: 1970

1. Find the remainder upon dividing the polynomial $x + x^3 + x^9 + x^{27} + x^{81} + x^{243}$ by

   (a) $x - 1$;

   (b) $x^2 - 1$.

2. Determine the perimeter of a right triangle with hypotenuse H and area A.

3. A circle fits into the corner of a square. If the shortest distance from the circle to the corner is 2, find the radius of the circle.

4. Let a, b and c be real numbers. Prove that $(a + b + c)^2 < 3(a^2 + b^2 + c^2)$.

5. A hundred identically looking coins are divided into ten equal stacks. In nine of the stacks, all coins weigh 10 grams each. In the tenth stack, all coins weight 11 grams each. Describe a method to determine in as few weighings as possible on a scale, not a balance, which stack has the "heavy" coins.

6. Prove that there is no four-digit integer which is doubled when its last digit is transferred to the front.

7. Prove that $4^{5555} - 4^{2222}$ is divisible by 7.

8. Let $2^\alpha, 2^\beta$ and 2^γ be the highest powers of two which divide $b - c, c - a$ and $a - b$ respectively, where a, b and c are integers. Prove that at least two of α, β and γ are equal.

9. A and B are two points on the same side of a line ℓ. Determine the point C on ℓ such that $AC + CB$ is minimum.

10. Five points are inside a unit square. Prove that the distance between some two of them is at most $\frac{1}{\sqrt{2}}$.

Problems: 1971

1. The polynomial $f(x)$ has at least two positive roots. Moreover, the sum of any two roots is also a root. Prove that $f(x)$ is identically zero.

2. $ABCD$ is a rectangle. P is the foot of perpendicular from B to the diagonal AC. If $AP = 1$ and $PC = 2$, determine $\frac{AD}{DC}$.

3. Let a and b be integers. Prove that 17 divides $2a+3b$ if and only if 17 divides $9a+5b$.

4. Let a, b and c be numbers such that $x^3 + x + 3 = (x - a)(x - b)(x - c)$ for all x. Determine $a^3 + b^3 + c^3$.

5. Let $f(x)$ be a function such that $f(1) = 1$ and for all x and y, $f(x + y) = f(x) + f(y)$. Prove that $f(x) = x$

   (a) for all integers x;

   (b) for all rational numbers x.

6. Let $a > b > 1$. Determine which of

$$\frac{1 + a + a^2 + \cdots + a^{n-1}}{1 + a + a^2 + \cdots + a^n} \quad \text{and} \quad \frac{1 + b + b^2 + \cdots + b^{n-1}}{1 + b + b^2 + \cdots + b^n}$$

is greater.

7. P is a point on a circle with radius 1. A circle with center P and radius r is such that the area common to both circles is $\frac{\pi}{2}$. Prove that $r < \sqrt{2}$.

8. Each term of a sequence $\{x_n\}$ is one of $1, 2, \ldots, m$. For any n, $x_n \neq x_{n+1}$. Moreover, if $x_i = a = x_k$ and $x_j = b$ for $i < j < k$, then $x_n \neq b$ for any $n > k$.

   (a) Find a sequence of length $2m - 1$.

   (b) Prove that in every such sequence, there must be a term which is not equal to any other term.

9. Prove or disprove that $1^n + 5^n + 8^n + 12^n = 2^n + 3^n + 10^n + 11^n$ for all $n = 0, 1, 2, \ldots, 100$.

10. The perimeter of a quadrilateral, convex or otherwise, is 1. Prove that it can be covered by a rectangle of area less than $\frac{1}{4}$.

Problems: 1972

1. Find the second smallest sequence of five consecutive positive integers such that the first is divisible by 8, the second by 9, the third by 10, the forth by 11 and the fifth by 12.

2. A finite set S of real numbers is such that whenever a is in S, so is $\frac{a}{a+1}$. Determine all such S.

3. Given two non-parallel lines on the plane, find the locus of the point P such that the sum of the distances from P to the lines is 2.

4. Find a polynomial $g(x)$ such that $1 + (x + (x^3 + 1)g(x))^3$ is divisible by $(x^3 + 1)^2$.

5. A chord CD of a circle meets a diameter AB of the circle at G. Determine the radius of the circle if $CG = 2, GD = 5$ and $BG = 1$.

6. A point is inside an equilateral triangle with side length 1. Determine the minimum sum of the distances from this point to the three sides of the triangle.

7. The numbers 1, 2, 3, 4, 5, 6, 7, 8 and 9 are arranged in a 3×3 table such that each row, column and diagonal has the same sum. Determine the only possible number which can occupy the central position.

8. Let $a_1 < a_2 < a_3$ be positive integers. Prove that $a_1^2 + a_2^2 + a_3^2 > a_1a_2 + a_2a_3 + a_3a_1 + 3$.

9. Find all possible bases $1 < r \leq 10$ and $1 < s \leq 10\,0$ such that $23_r = 32_s$.

10. Let $f(x)$ be such that $f(2) = 1$, $f(x) \geq 0$ for $x \geq 0$ and $f(xy) \leq f(x) + f(y)$ for all positive x and y. Prove that there is a positive number N such that for all $x > N$, $\frac{f(x)}{x} < 10^{-1000}$.

Problems: 1973

1. An integer in base 10 has 91 digits, all of which are ones. Prove that it is not a prime number.

2. Given two points A and B such that $AB = 1$, describe a construction, using only a compass, a point C such that $AC = 3$.

3. Prove that $(a+b+c)^2 < 3(a^2+b^2+c^2)$ for all real numbers a, b and c.

4. (a) Express the number $\frac{3}{22}$ in the form $\frac{A}{10} + \frac{B}{10 \cdot 99}$, where A and B are positive integers.

   (b) Prove that any rational number between 0 and 1 can be expressed in the form $\frac{A}{10^n} + \frac{B}{10^n(10^m-1)}$, where A, B, n and m are positive integers.

5. Suppose p lines are drawn in the plane, dividing it into regions. Determine the maximum number of regions if p is

   (a) 1;

   (b) 2;

   (c) 3;

   (d) 4;

   (e) 100.

6. Given are 51 distinct numbers between 1 and 100 inclusive. Prove that at least two of them must be relatively prime.

7. Given that $\sqrt{1+2\sqrt{1+2\sqrt{1+\cdots}}}$ represents a real number, determine its value.

8. A circle is inscribed in a unit square. Determine the radius of a circle that is inscribed between the first circle and a corner of the square.

9. Xavier, Yolanda and Zeke each picks a positive integer between 1 and 10 inclusive. Xavier's number is multiplied by 12, and Yolander's number is added to the product. The sum is then multiplied by 10, and the three original choices are then added. The result is 879. Determine the original numbers.

10. Four of the roots of $x^{10} + ax^8 + bx^6 + cx^4 + dx^2 + 144 = 0$ are $\frac{1}{2}$, 1, 2 and 3. Determine the remaining roots.

Problems: 1974

1. Divide the numbers 1, 2, 3, 4 and 5 into two groups in any way. Prove that at least one group must contain two numbers and their difference.

2. Determine if the product of any n consecutive positive integers is divisible by $n!$.

3. ABC is an equilateral triangle inscribed in a circle. P is an arbitrary point on the arc AC. Prove that $PA + PC = PB$.

4. Prove that $\displaystyle\sum_{k=2}^{n} \frac{1}{k^2-1} = \frac{(3n+2)(n-1)}{4n(n+1)}$ for $n \geq 2$.

5. Determine which of $\sqrt{5}^{\sqrt{3}}$ and $\sqrt{3}^{\sqrt{5}}$ is greater.

6. A right triangle has legs of length a and b. Prove that the bisector of the right angle has length $\frac{ab\sqrt{2}}{a+b}$.

7. At a party, some people shake hands and some do not. Prove that there must exist at least two people who shake hands with the same number of people.

8. For any three consecutive positive integers, prove that the cube of the largest cannot be the sum of the cubes of the other two.

9. (a) How many n-digit numbers are there such that each digit is 1, 2 or 3?

   (b) How many of these numbers use all three of the digits 1, 2 and 3?

10. Let $x_1, x_2, \ldots, x_n$ be positive angles whose sum does not exceed a right angle. Prove that

$$\sin(x_1 + x_2 + \cdots + x_n) \leq \sin x_1 + \sin x_2 + \cdots + \sin x_n.$$

Problems: 1975

1. The four corners of a unit square are cut off to leave behind a regular octagon. Determine the side length of this octagon.

2. Prove that if a prime number is divided by 30, the remainder is prime.

3. Find all integers k such that $x^2 + k(x+1) + 7 = 0$ has only integer solutions.

4. M is the midpoint of side BC of triangle ABC. Prove that $AB + AC > 2AM$.

5. Find an integer $0 < n < 25$ such that $(n-1)^3 + n^3 + (n+1)^3$ is divisible by 102.

6. Determine if $99^n + 100^n > 101^n$ holds for all positive integer n.

7. The side lengths of a right triangle form an arithmetic progression. Prove that they are in the ratio 3 : 4 : 5.

8. Prove that 1 plus the product of four consecutive integers is a perfect square.

9. A group of students write a set of k tests. Suppose that a_i of the students failed at least i tests, $1 \leq i \leq k$. Determine the total number of failed exams.

10. A, B, C and D are four points on a line with $AB = BC = CD = 1$. P is a point such that $PA = 1$ and PA is perpendicular to AD. Prove that $\angle PBA + \angle PCB + \angle PDC = 90°$.

Problems: 1976

1. Prove that it is impossible to find positive integers a and b such that $\frac{21}{23} = \frac{1}{a} + \frac{1}{b}$.

2. A fly sitting at a corner of a room measuring $6 \times 6 \times 2$ wishes to walk to the opposite corner.

   (a) What is the minimum distance the fly must walk?

   (b) How many different paths achieve this distance?

3. Find all real numbers k such that $x^3 + x^2 - 4kx - 4k = 0$ has two of its three roots equal.

4. In a round-robin tournament, each of n players $P_1, P_2, \ldots, P_n$ plays one game with each of the players, with no ties allowed. If w_i and ℓ_i denote respectively the number of games won and lost by player P_i, prove that $w_1^2 + \cdots + w_n^2 = \ell_1^2 + \cdots + \ell_n^2$.

5. An equilateral triangle ABC is divided into triangles ABE, CAD, EDA and a trapezoid $BCDE$. If $DE = 1$, $BE = CD = 2$ and $AD = AE = \sqrt{7}$, determine BC.

6. Given real numbers a, b and c such that $a^2 + b^2 + c^2 = 1$, determine the smallest value of r such that $-\frac{r}{2} \leq ab + bc + ca \leq r$.

7. Determine all integers a, b and c such that $a \log_{10} 2 + b \log_{10} 5 + c \log_{10} 7 = 1$.

8. Determine A and B such that $\cos 3\theta = A\cos^3\theta + B\cos\theta$ holds for every real number θ.

9. Nine lines are drawn parallel to the base of a triangle so that the other two sides are divided into 10 equal segments and the area into 10 portions. If the largest has area 38, determine the area of the triangle.

10. Let a and b be odd integers and let n be a positive integer.

    (a) If $a - b$ is divisible by 2^n, prove that $a^3 - b^3$ is divisible by 2^n.
    (b) If $a^3 - b^3$ is divisible by 2^n, prove that $a - b$ is divisible by 2^n.

Problems: 1977

1. Prove that $3^n + 1$ is not divisible by 8 for any positive integer n.

2. Let k be any nonzero real number and let a, b and c be the roots of $4x^3 - 32x^2 - k^2x + 8k^2 = 0$. Prove that $x^2 + 2kx - abc = 0$ cannot have real roots.

3. Sketch accurately the set of all points which satisfy $[|x|] + |y| = 2$.

4. For any three consecutive positive integers, prove that the cube of the largest cannot be the sum of the cubes of the other two.

5. Prove that $\sin\left(\frac{\pi}{24}\right) = \frac{1}{4}\left(\sqrt{2+\sqrt{2}} - \sqrt{3}\sqrt{2-\sqrt{2}}\right)$.

6. Prove that a triangle with sides of lengths 3, 4 and 5 is a right triangle.

7. If a, b and c denote the lengths of the sides of a triangle, prove that $\sqrt{a(b+c-a)}$, $\sqrt{b(c+a-b)}$ and $\sqrt{c(a+b-c)}$ are also lengths of the sides of a triangle.

8. Prove that 5 is the maximum value of $x + y + z$ subject to $x \geq 0$, $y \geq 0$, $z \geq 0$, $x + y \leq 2$ and $3x - y + z \leq 1$.

9. Three lines on a plane Π meet in a point O. A fourth line ℓ passing through O makes equal angles with each of these lines. Prove that ℓ is perpendicular to Π.

10. If the real numbers a, b, c, x, y and z satisfy $az - 2by + cx = 0$ and $ac - b^2 > 0$, prove that $xz - y^2 \leq 0$.

Problems: 1978

1. Determine all angles θ with $0 \leq \theta \leq 2\pi$ such that $\sin^6\theta + \cos^3\theta = 1$.

2. (a) A corner reflector consists of two straight lines, perpendicular to each other, which are assumed capable of reflecting a ray of light which is in the same plane as the lines. If a ray of light reflects successively off each of the lines, prove that the exit ray is parallel in the opposite sense to the entering ray.

   (b) This time the corner reflector consists of three plane mirrors which are mutually perpendicular. If a ray of light reflects successively off each of the three mirrors, in any order, prove that the exit ray is parallel in the opposite sense to the entering ray.

3. S is a finite set of positive integers, not necessarily different from one another, such that for any three members a, b and c of S, $a+b$ is divisible by c. Determine all such sets S.

4. Let i be a square root of -1 and for real numbers x and y, write the complex number $\frac{x+iy+1}{x+iy-1}$ in the form $a + ib$, where a and b are real numbers. Find the set of points (x, y) for which $a \leq 0$.

5. Prove that for any integers m and n, $mn(m^4 - n^4)$ is divisible by 30.

6. For which values of k do the polynomial equations $x^6+x^5+x^4+x^3+x^2+x-3k = 0$ and $x^6 - x^5 + x^4 - x^3 + x^2 - x - k = 0$ have a common root?

7. Inside a square of side-length 3 are placed 36 points. Prove that there are 3 points which determine a triangle of area no greater than $\frac{1}{2}$.

8. You are given a set of 21 dominoes $\boxed{a|b}$ where a and b are integers from 1 to 6, each pair occurring once. Note that $\boxed{a|b}$ is to be considered the same as $\boxed{b|a}$. Any number of dominoes can be joined to form a chain if they have matching numbers at each join. For example, a 3-chain is given by $\boxed{1|1}\,\boxed{1|6}\,\boxed{6|4}$. Prove that it is not possible to form a 21-chain.

Problems: 1979

1. A hexagon is inscribed in a circle. If two pairs of opposite sides are parallel, prove that the third pair of sides are also parallel.

2. Determine all real numbers A for which the curve by $x^3+y^3-2x^2+Ay^2+x+y = 0$ intersects the line $x - y = 1$ in exactly two distinct points.

3. Prove that for any integer n, $n^5 - n$ is divisible by 5.

4. If a single die is thrown, each of the numbers 1 through 6 has an equal probability of occurring, namely $\frac{1}{6}$. Now suppose that four such dice are thrown together.

   (a) Find the most probable value of the total number obtained.

   (b) Find the probability of obtaining this number.

5. Starting with a given positive integer n, the following procedure is used for obtaining a sequence of integers: Suppose the previous number obtained is m. If m is even, the next number is $\frac{m}{2}$; if m is odd, the next number is the smallest square number larger than m. If 1 is obtained, the sequence terminates. For example, if we started with 20, the sequence would proceed 20, 10, 5, 9, 16, 8, 4, 2, 1. Prove that for any starting number n, the number 1 will be obtained after a finite number of steps.

6. In a round-robin tennis tournament, 16 men played each other once with no tie games allowed. Prove that it is possible to label the men $M_1, M_2, \dots, M_{16}$ so that M_i beat M_{i+1} for $i = 1, 2, \dots, 15$.

Problems: 1980

1. Suppose we start with 7 sheets of paper and then some number of them are each cut into 7 smaller pieces. Then some of the smaller pieces are each cut into 7 still smaller pieces and so on. Finally the process is stopped and it turns out that the total number of pieces of paper is some number between 1976 and 1986. Can one determine the exact final number of pieces of paper?

2. A rectangle is divided into four triangles by joining an interior point to the four vertices with line segments. Three of the triangles have areas of 5, 9 and 11. Find all possible areas for the fourth triangle.

3. A certain sequence of real numbers is defined as follows: $x_1 = 2$ and $x_{n+1} = \frac{2x_n}{3} + \frac{1}{3x_n}$ for $n = 1, 2, 3, \dots$. Prove that for all values of $n > 2$, $1 < x_n < 2$.

4. If all cross-sections of a bounded solid figure are circles, prove that the figure is a sphere.

5. Consider a line segment AB of length $2a$. A point P is chosen at random on this line segment, all points being equally likely. Prove that the probability that $AP \cdot PB > \frac{a^2}{2}$ is $\frac{1}{\sqrt{2}}$.

6. (a) Prove that one root of the equation $x^4 + 5x^2 + 5 = 0$ is $x = \omega - \omega^4$, where ω is a complex fifth root of unity, that is, $\omega^5 = 1$.

   (b) Determine the other three roots as polynomials in ω.

Problems: 1981

1. Prove that the pair of equations $x^4 - x^3 + x^2 + 2x - 6 = 0$ and $x^4 + x^3 + 3x^2 + 4x + 6 = 0$ have a pair of complex roots in common.

2. Trefor wrote down a four-digit number x, transferred the right most digit to the extreme left to obtain a smaller four-digit number y, and then added the two numbers together to obtain a four-digit number z. The next day he was unable to find his calculations but remembered that the last three digits of z were 179. What was x, given that the first digit of y is not 0?

3. A baseball league is made up of 20 teams. Each team plays at least once and there are no tie games. A team's average is defined to be its number of wins divided by its total number of games played.

   (a) If each team played the same number of games, prove that the sum of the averages of all the teams is 10.

   (b) If each team did not play the same number of games, prove that the sum of the averages of all the teams must be at least 1 and at most 19.

4. A farmer owns a fenced yard in the shape of a square, 30 meters by 30 meters. He wishes to divide the yard into three parts of equal area, using 50 meters length of fencing. Find two non-congruent ways that he can do this.

5. If $P'Q'R'$ is the parallel projection of a triangle PQR on to any plane, prove that the volumes of the two tetrahedra $P'Q'R'P$ and $PQRP'$ are the same.

6. Prove that if $x > 0$, then for $n = 1, 2, 3, \ldots$, $\left(\frac{x+1}{n+1}\right)^{n+1} \geq \left(\frac{x}{n}\right)^n$.

Problems: 1982

1. A 9×12 rectangular piece of paper is folded so that a pair of diagonally opposite corners coincide. What is the length of the crease?

2. Let $x = \sin A$, $y = \sin B$ and $z = \sin(A + B)$. Determine $\cos(A + B)$ as a quotient of two polynomials in x, y and z with integral coefficients.

3. A cylindrical tank with diameter 4 meters and open top is partially filled with water. A cone 2 meters in diameter and 3 meters in height is suspended vertex up above the water, so that the bottom of the cone just touches the surface of the water. The cone is then lowered at a constant rate of 10 meters per minute into the water. How long does it take until the cone is completely submerged, given that the water does not overflow?

4. John added the squares of two positive integers and found that his answer was the square of an integer. He subtracted the squares of the same two positive integers and again found that his answer was the square of a positive integers. Prove that John must have made an error in his calculation.

5. Twenty-five Knights gather at the Round Table for a jolly evening. They belong to various Orders, with every two Orders having at least one common member. Members of the same Order occupy consecutive seats at the Round Table.

   (a) If each Order has at most nine members, prove that there is a Knight who belongs to no Order and there is a Knight who belongs to every Order.

   (b) Without the restriction on the sizes of the Orders, prove that there are two Knights such that between them they hold membership to every Order.

Problems: 1983

1. From the top of a hill the angles of depression of two marked points on a level plane from which the hill rises are found to be $45°$ and $30°$ respectively. These points lie due north and due east of the hill top respectively, and are 2 kilometers apart. Determine the height of the hill.

2. Clyde has five pairs of red socks and several pairs of blue ones. He puts them all mixed up in a box. Each morning, Clyde chooses two socks at random and replaces them each night. He has found that the probability that he chooses two socks of the same colour is exactly $\frac{1}{2}$. How many pairs of blue socks does Clyde have?

3. The numbers 1, 2, 3, 4, 5, 6, 7, 8 and 9 are divided in any way into three groups with three numbers in each group. Let a, b and c be the respective sums of the numbers in the groups.

   (a) Prove that $2160 \leq abc \leq 3375$.

   (b) Can abc be equal to 2160?

   (c) Can abc be equal to 3375?

4. A, B, C and D are four points not all on a plane. If $\angle ABC = \angle BCD = \angle CDA = 90°$, prove that $\angle DAB < 90°$.

5. The equation $(x-a)(x-b)(x-c)(x-d) = 1$, where a, b, c and d are real numbers, has four real roots k, ℓ, m and n. Prove that the equation $(x-k)(x-\ell)(x-m)(x-n)+1 = 0$ has four real roots.

Modern Period: 1983–2006

First Round Questions

November 15, 1983

1. The sum of the first two prime numbers greater than 50 is

   (a) 104 (b) 108 (c) 110 (d) 112 (e) 116

2. A rectangular piece of paper is folded in half four times in succession. The final folded piece of paper has area 30 square centimeters. The area in square centimeters of the unfolded piece of paper is

   (a) 120 (b) 480 (c) 900 (d) 14400 (e) none of these

3. The equation of the circle centered at (3,4) and with radius 5 is

   (a) $x^2 + y^2 - 6x - 8y = 0$ (b) $x^2 + y^2 - 6x - 8y = 25$

   (c) $x^2 + y^2 - 3x - 4y = 0$ (d) $x^2 + y^2 - 3x - 4y = 25$ (e) none of these

4. Let a, b and c be real numbers. If $a(x-1)^2 + b(x-1) + c$ has the same value for all real numbers x, then we must have

   (a) $a \neq b$ (b) $a = b \neq 0$ (c) $a = b = 0$

   (d) $a = b = c \neq 0$ (e) $a = b = c = 0$

5. The first, second and fifth terms of an arithmetic progression form a geometric progression. The common ratio of the geometric progression is

   (a) 2 (b) 3 (c) 4 (d) 7 (e) none of these

6. A fair die has the numbers 1, 2, 2, 3, 3 and 4 on its faces. Another fair die has the numbers 1, 3, 4, 5, 6 and 8 on its faces. when these two dice are rolled, the probability that 7 is the total of the numbers on the top faces is

   (a) $\frac{1}{11}$ (b) $\frac{1}{7}$ (c) $\frac{1}{6}$ (d) $\frac{1}{4}$ (e) none of these

7. The product of the complex numbers $-3+4i$ and $3-4i$ is

(a) 7 (b) $7+24i$ (c) -25 (d) $-25+24i$ (e) none of these

8. Rose grows several rows of roses, with the same number of roses in each row. If she has one more row of roses but one less rose in each row, the total number of her roses will decrease by one. The number of Rose's roses must be

(a) a prime number (b) a power of 2 (c) a perfect square

(d) a perfect cube (e) none of these

9. Two sides of an acute triangle have lengths 3 and 7. The number of possible integral lengths for the third side is

(a) 1 (b) 2 (c) 3 (d) 4 (e) more than 4

10. The number of different rectangles formed by the lines $x=1$, $x=2$, $y=1$, $y=2$ and $y=3$ is 3. The number of different rectangles formed by the lines $x=1$, $x=2$, $x=3$, $x=4$, $y=1$, $y=2$, $y=3$, $y=4$ and $y=5$ is

(a) 12 (b) 29 (c) 39 (d) 48 (e) none of these

11. Two circles, of respective radii 3 and 48, touch each other externally. A common tangent touches one of them at the point P and the other at Q. Then PQ is equal to

(a) 12 (b) 24 (c) 36 (d) 48 (e) none of these

12. If $3561x+6439y=55756$ and $6439x+3561y=44244$, then y is equal to

(a) 3 (b) 4 (c) 5 (d) 6 (e) 7

13. The solution set of the inequality $x-2>\frac{1}{x-2}$ is

(a) $\{x : x<1 or x>3\}$ (b) $\{x : x<1\}$ (c) $\{x : 1<x<2$ or $2<x<3\}$

(d) $\{x : x>3\}$ (e) none of these

14. A mini Pacman board consists of 3 horizontal lanes and 3 vertical lanes, enclosing four unit squares in a 2 by 2 array. The lanes, of negligible width, are completely lined with dots. Miss Pacman starts from the center of the board and must eat up all the dots. The minimum distance she must travel is

(a) 12 (b) 13 (c) 14 (d) 15 (e) 16

15. The sequence $\{x_n\}$ is defined by $x_1=1$ and $x_{n+1}=\sqrt{1+x_n}$ for $n\geq 1$. Given that it converges to a finite limit, this limit must be

(a) 1 (b) $\frac{1+\sqrt{5}}{2}$ (c) $\frac{1-\sqrt{5}}{2}$ (d) $\sqrt{2}$ (e) none of these

16. The point P moves in the plane so that its distances from two fixed points in the plane are in a constant ratio of 1 to 2. The locus of P is

(a) a straight line (b) a circle (c) a parabola

(d) an ellipse whose axes are in the ratio 1 to 2

(e) none of these

17. If $1 + \sin\theta < \cos\theta$, then

(a) θ must lie in the first quadrant

(b) θ must lie in the second quadrant

(c) θ must lie in the third quadrant

(d) θ must lie in the fourth quadrant

(e) none of these

18. Suppose $(p-3)x^2 - 2px + 6p = 0$ is a quadratic equation with distinct positive roots. The set of all possible values of p is

(a) $\{p : 0 < p < 3.6\}$ (b) $\{p : p > 3\}$ (c) $\{p : p > 0\}$

(d) $\{p : 3 < p < 3.6\}$ (e) none of these

19. Let a be a positive real number and x be a real number greater than 1. When a is raised to the power $\dfrac{\log_x(\log_x a)}{\log_x a}$, the result is

(a) $\log_x a$ (b) $\frac{\log_x a}{a}$ (c) $x(\log_x a)$ (d) ax (e) none of these

20. Each letter in a 4 by 4 array is either M, S or K. The eight "words" spelt across and down were: MSKK, MMSK, MKSM, MKMS, SKMK, SKSM, KSMK and KSKS, in some order. The "word" spelt down the diagonal from top-left to bottom-right is

(a) MSMK (b) MSSK (c) MKMK (d) MKSK (e) none of these

November 20, 1984

1. The expression $0.049 \div (0.0011 \div 0.005)$ is closest in value to

(a) 0.01 (b) 0.025 (c) 0.25 (d) 1 (e) 1000

2. Bus A leaves the University at noon and every 15 minutes afterwards. Bus B leaves the University at noon and every 25 minutes afterwards. The next time after noon that both buses leave the University together is

(a) 12:25 pm (b) 12:40 pm (c) 1:15 pm (d) 5:00 pm (e) none of these

3. The number of ways of choosing two positive integers neither of which is a prime and with sum equal to 21 is

(a) 2 (b) 3 (c) 4 (d) 5 (e) none of these

4. If $8^m = 16^n$ and $mn \neq 0$, then $\frac{m}{n}$ is equal to

(a) $\frac{1}{2}$ (b) $\frac{3}{4}$ (c) $\frac{4}{3}$ (d) 2 (e) none of these

5. The interior angle at vertex A of triangle ABC is $35°$ and the exterior angle at vertex B is $108°$. The exterior angle at vertex C is

(a) $37°$ (b) $73°$ (c) $107°$ (d) $143°$ (e) none of these

6. If $a = 15$ and $ar^4 = 1215$, then ar^2 is equal to

(a) 81 (b) 135 (c) 225 (d) 615 (e) none of these

7. A father is three times as old as his son in 1984. In 1996, he will be twice as old as his son. He will be one-and-a-half times as old as his son in

(a) 2002 (b) 2008 (c) 2020

(d) the situation is impossible (e) none of these

8. Two flies are in a room 6 meters by 6 meters by 3 meters. The farthest distance, in meters, that they can be apart is

(a) 9 (b) $3 + 6\sqrt{2}$ (c) 15 (d) 81 (e) none of these

9. Suppose a, b, c and d are numbers such that $a + b + c = 4$, $a + b + d = 8$, $a + c + d = -2$ and $b + c + d = 5$. Then $a + b + c + d$ is equal to

(a) 3.75 (b) 5 (c) 7.5 (d) 15 (e) none of these

10. The number of ways of writing down the integers 1, 2, 3 and 4 in a row so that 1 is not in the first place, 2 not in the second place, 3 not in the third place and 4 not in the fourth place is

(a) 6 (b) 7 (c) 8 (d) 9 (e) none of these

11. A girl has three marbles in a bag. They are of the same size but in different colors. She takes one out, looks at it, puts it back and shakes the bag. She does this three times. The probability that she looks at all three marbles is

(a) $\frac{1}{27}$ (b) $\frac{2}{9}$ (c) $\frac{1}{3}$ (d) $\frac{1}{2}$ (e) none of these

12. The sum of two numbers is 18 and the difference of their squares is 144. The difference of the two numbers is

(a) 8 (b) 10 (c) 12 (d) 14 (e) none of these

13. The roots of a cubic polynomial are $\frac{1}{2}$, 1 and -2. The coefficient of x^3 is 4. The coefficient of x is

(a) -10 (b) -5 (c) $-\frac{5}{2}$ (d) $-\frac{5}{4}$ (e) none of these

14. The graph of $|x| + |y| = 1$ is

(a) a straight line (b) two straight lines (c) a circle

(d) a square with sides parallel to the x-axis and y-axis

(e) a square with vertices on the x-axis and y-axis

15. A point X moves along the positive x-axis and a second point Y moves along the positive y-axis so that the distance between X and Y is constant. During this motion, the curve traced by the midpoint of XY is part of a

(a) straight line (b) circle (c) parabola

(d) hyperbola (e) none of these

16. If $x = \dfrac{a-b}{a+b}$, $y = \dfrac{b-c}{b+c}$ and $z = \dfrac{c-a}{c+a}$, where a, b and c are positive, then the expression $\dfrac{(1-x)(1-y)(1-z)}{(1+x)(1+y)(1+z)}$ is equal to

(a) 0 (b) 1 (c) abc (d) xyz (e) none of these

17. If 120 identical spheres are tightly stacked in the form of a regular triangular pyramid, then the number of spheres that lie at the base of the pyramid is

(a) 15 (b) 36 (c) 60

(d) the situation is impossible (e) none of these.

18. The diagonals of a parallelogram are 8 centimeters and 10 centimeters long, and one of the sides is 5 centimeters long. The length, in centimeters, of the longer side is

(a) 6 (b) 7 (c) 8 (d) 9 (e) none of these

19. Checkers are to be placed on a 3 by 8 checkerboard so that the centers of the squares occupied by the checkers do not include the vertices of any rectangle with sides parallel to the sides of the board. The maximum number of checkers that can be so placed is

(a) 9 (b) 10 (c) 11 (d) 12 (e) none of these

20. Each side of a convex polygon has length 1 centimeter and the polygon encloses an area of 1 square centimeter. The number of non-congruent polygons of this type is

(a) 1 (b) 2 (c) 3 (d) 4 (e) none of these

November 19, 1985

1. An ant crawls 1 meter north, then 2 meters east, then 3 meters south and then 4 meters west. It then continues this routine by crawling 1 meter north, then 2 meters east, and so on. The number of meters it has crawled when it crosses its own path for the first time is

   (a) 2 (b) 7 (c) 14 (d) 16 (e) none of these

2. If the sum of 15 consecutive integers is 0, the largest one among them is

   (a) 5 (b) 6 (c) 7 (d) 8 (e) none of these

3. Let k be a positive real number and let $m = k^4$ and $n = k^2$. If $2^m = 4^n$, then the value of k is

   (a) $\sqrt{2}$ (b) 2 (c) 3 (d) 4 (e) none of these

4. O is the center of a circular pond of radius 5 meters. The point A is on the edge of the pond due east of O. A boy swims westward from A to the point B 2 meters from A. Then he turns north and swims to shore, whereupon he swims westward again until he reaches the point C due north of O. The distance in meters between B and C, in meters, is

   (a) 2 (b) 3 (c) 4 (d) 5 (e) none of these

5. A football league has two divisions, with 5 teams in the Western Division and 4 teams in the Eastern Division. All 9 teams are of equal strength, and every team in one division plays every team in the other division exactly twice. On the average, the fraction of these games that will be won by teams from the Western Division is

   (a) $\frac{25}{81}$ (b) $\frac{1}{2}$ (c) $\frac{5}{9}$ (d) $\frac{4}{5}$ (e) none of these

6. When the variable t is eliminated from the equations $x = 2t + 1$ and $y = t(t + 1)$, the resulting equation is

   (a) $4y = x^2 - 2x + 5$ (b) $4y = x^2 - 1$ (c) $4y = x^2 + 2x + 5$

   (d) $4y = x^2 + 4x + 3$ (e) none of these

7. Each of five children at a Halloween party claims that exactly two of the other four are girls. Not all of them are lying. The number of them telling the truth is

   (a) impossible to determine (b) 1 (c) 2

   (d) 3 (e) 4

8. $ABCD$ is a square of side 2 centimeters. Two circular arcs are drawn from B to D, one with A as center and the other with C as center. The area, in square centimeters, of the region enclosed by the two circular arcs is closest to

   (a) 1.5 (b) 1.7 (c) 1.9 (d) 2.1 (e) 2.3

9. Let $\dfrac{x}{a-b} = \dfrac{y}{b-c} = \dfrac{z}{c-a} = 1985$, where $a,\ b,\ c,\ x,\ y$ and z are real numbers. Then $x + y + z$ is equal to

(a) 1 (b) 1985 (c) $a + b + c$ (d) abc (e) none of these

10. Buses travel east along a highway at 40 kilometers per hour and they are spaced exactly 36 minutes apart. A pedestrian is walking east along the highway at 8 kilometers per hour. A bus traveling east passes the pedestrian. The number of minutes that will elapse until the next bus traveling east passes the pedestrian is

(a) 30 (b) 36 (c) 40 (d) 45 (e) none of these

11. An integer consists of 1985 digits each of which is a 2. When this integer is divided by 9, the remainder is

(a) 1 (b) 3 (c) 5 (d) 7 (e) none of these

12. The roots of the equation $2x^2 + 18x + k = 0$ are integers. The largest possible value of k is

(a) 16 (b) 20 (c) 32 (d) 40 (e) none of these

13. Two boys and four girls are to form a line. If the two boys must not be next to each other, the number of ways of forming the line is

(a) 96 (b) 480 (c) 600 (d) 720 (e) none of these

14. A swimming pool 4 meters by 6 meters is inside a rectangular garden such that each corner of the pool touches a different side of the garden. The largest possible area, in square meters, of the garden is

(a) 48 (b) 49 (c) 50 (d) 52 (e) none of these

15. A cabin is built on a flat, circular island. The ocean is 12 kilometers due north and 6 kilometers due south, and the center of the island is 7 kilometers away. The radius, in kilometers, of the island is

(a) $4\sqrt{5}$ (b) $\sqrt{95}$ (c) 11 (d) 13 (e) none of these

16. Let α and β be the nonzero real roots of the equation $x^2 + px + q = 0$. An equation with $\alpha^2 + \alpha\beta$ and $\alpha\beta + \beta^2$ as roots is

(a) $x^2 + p^2x + p2q = 0$ (b) $x^2 - p^2x - p2q = 0$ (c) $x^2 + p^2x - p^2q = 0$

(d) $x^2 - p^2x + p^2q = 0$ (e) none of these

17. Let $a,\ b,\ c,\ d$ and e be real numbers such that $a - b + c = 1,\ b - c + d = 2,\ c - d + e = 3,\ d - e + a = 4$ and $e - a + b = 5$. The largest of $a,\ b,\ c,\ d$ and e is

(a) a (b) b (c) c (d) d (e) e

18. The sum $\frac{1}{\sqrt{0}+\sqrt{2}} + \frac{1}{\sqrt{2}+\sqrt{4}} + \cdots + \frac{1}{\sqrt{398}+\sqrt{400}}$ lies in the closed interval

(a) [5,6] (b) [7,8] (c) [9,10] (d) [11,12] (e) none of these

19. The sum of $k > 1$ integers is equal to their product and this common value is odd. The smallest possible value of k is

(a) 3 (b) 5 (c) 6 (d) 9 (e) none of these

20. A plane is drawn perpendicular to a space diagonal of a cube of side 2 centimeters at its midpoint. The area, in square centimeters, of the plane cross-section is

(a) 4 (b) $\frac{3\sqrt{3}}{2}$ (c) $2\sqrt{2}$ (d) $3\sqrt{3}$ (e) none of these

November 18, 1986

1. A square lawn 10 meters by 10 meters contains 10 blades of grass per square centimeter. The total number of blades of grass on the lawn is

(a) 1000 (b) 10000 (c) 100000 (d) 1000000 (e) 10000000

2. The symbol $*$ in the inequality $n < n * (n+1)$ is to be replaced by +, −, × or ÷. The number of replacements so that the inequality holds for at least one integer n is

(a) 0 (b) 1 (c) 2 (d) 3 (e) 4

3. If $x^2 + Ax + B = (x - A)(x - B)$ for all numbers x and $A \neq B$, then B is equal to

(a) −2 (b) −1 (c) 0 (d) 1 (e) 2

4. An n-sided polygon has sides of lengths $1, 2, 4, \ldots, 2^n$. The smallest possible value of n (at least 3) is

(a) 3 (b) 4 (c) 5 (d) more than 5 (e) non-existent

5. The quadratic polynomials $x^2 + Ax + B$ and $x^2 + Bx + A$, where $A \neq B$, have a common root. The value of $A + B$ is

(a) 1 (b) 2 (c) −1 (d) −2 (e) not determinable

6. Let $T(P)$ be the temperature at a point P on the circumference of a circle C of radius 100 kilometers centered at Red Deer at some given time. $T(P)$ varies continuously with P. If A and B are the end points of a diameter of C, the statement among the following which is always true for at least one choice of A and B is

(a) $T(A) > T(B)$ (b) $T(A) < T(B)$ (c) $T(A) = -T(B)$

(d) $T(A) = T(B)$ (e) none of these

7. The number of digits in the base ten representation of the number $2^{1986}-1$ is closest to

(a) 500 (b) 550 (c) 600 (d) 650 (e) 700

8. The number of planes of symmetry of a rectangular box of dimensions 1 by 1 by 12 is

(a) 2 (b) 3 (c) 5 (d) 9 (e) none of these

9. If three dice are rolled, the probability of getting a sum of six is

(a) $\frac{3}{216}$ (b) $\frac{7}{216}$ (c) $\frac{10}{216}$ (d) $\frac{13}{216}$ (e) none of these

10. A 3 by 3 by 3 cube consists of 27 unit cubes. Two sub-cubes of this cube are different if at least one of them contains a unit cube not contained in the other. The number of different subcubes (excluding the 3 by 3 by 3 cube) is

(a) 8 (b) 29 (c) 31 (d) 35 (e) none of these

11. If $x^y = z$ and $y^z = x$, then z^x is always equal to

(a) y (b) x^{xyz} (c) y^{xyz} (d) x^{y^z} (e) y^{x^z}

12. The closest approximation to $\sqrt[3]{1.0009} - \sqrt[3]{1.00009}$ is

(a) 0.3 (b) 0.03 (c) 0.003 (d) 0.0003 (e) 0.00003

13. If x and y are real numbers, the minimum value of $x^2+2xy+3y^2+4y+3$ is

(a) -1 (b) 0 (c) 1 (d) 2 (e) none of these

14. A sum of three non-zero numbers is eight times the first number, three times the second number and k times the third number. The value of k is

(a) $\frac{13}{24}$ (b) $\frac{24}{35}$ (c) $\frac{24}{19}$ (d) $\frac{24}{13}$ (e) not determinable

15. The largest integer N having the property that N divides k^2-1 for all odd integers k is

(a) 1 (b) 2 (c) 4 (d) 8 (e) 16

16. Let $x,\ y$ and z be real numbers. If exactly one of the following numbers is rational, it must be

(a) $\frac{x+y+z}{x}$ (b) $\frac{x+y+z}{y}$ (c) $\frac{y+z}{x}$ (d) $\frac{z+x}{y}$ (e) $\frac{x+y}{z}$

17. The sum of four positive real numbers is 8. The maximum value of their product is

(a) 4 (b) 8 (c) 16 (d) 18 (e) 32

18. Consider the binomial expansion $(1+x)^n = a_0 + a_1 x + \cdots + a_n x^n$, where n is a positive integer and $x > 0$. The maximum possible number of equal consecutive terms of the sequence $a_0,\ a_1 x, \ldots, a_n x^n$ is

(a) 1 (b) 2 (c) 3 (d) 4 (e) more than 4

19. A two-digit number is divided by the sum of its digits. The remainder cannot possibly be

(a) 11 (b) 12 (c) 13 (d) 14 (e) 15

20. If one root of the equation $x^4 + Ax^3 + Bx^2 + Cx + D = 0$ is $\sqrt{3}+\sqrt{2}$, where $A,\ B,\ C$ and D are integers, then B is

(a) -10 (b) 6 (c) $(\sqrt{3}+\sqrt{2})^2$ (d) $(\sqrt{3}+\sqrt{2})^4$ (e) not determinable

November 17, 1987

1. Suppose $f(1) = 1$ and for all integers $n > 1$, $f(n) = f(n-1) + 1$ if n is even and $f(n) = 2f(n-1)$ if n is odd. Then $f(5)$ is

(a) 7 (b) 10 (c) 13 (d) 16 (e) none of these

2. The sum of the prime factors of 1988 is

(a) 26 (b) 82 (c) 501 (d) 996 (e) 1988

3. The number 2^x is entered in a calculator. When the square-root button is pressed n times in succession, the final answer is 2. The value of x is

(a) n (b) $2n$ (c) n^2 (d) 2^n (e) none of these

4. If the square root of a number is between 6 and 7, then the cube root of the number will be between

(a) 1 and 2 (b) 2 and 3 (c) 3 and 4 (d) 4 and 5 (e) 5 and 6

5. Knights numbered from 1 to 10 inclusive are seated at a round table, evenly spaced. The numbers of two neighboring knights differ by at most two. The number of the knight sitting diametrically opposite knight number 6 is

(a) 3 (b) 5 (c) 7 (d) 9 (e) none of these

6. The equation $\frac{1}{3} + \frac{1}{4} + \frac{1}{x} = \frac{1}{3+4+x}$ has two roots. Their sum is

(a) -12 (b) -7 (c) 7 (d) 12 (e) none of these

7. The function $(\frac{x}{x^2 1})$ is positive if and only if

(a) $x < -1$ or $-1 < x < 1$ or $1 < x$

(b) $-1 < x < 0$ or $1 < x$ (c) $x < -1$ or $1 < x$ (d) $0 < x < 1$ or $1 < x$

(e) $x < -1$ or $0 < x < 1$ or $1 < x$

8. $ABCD$ is a square of side 1. E, F, G and H are points on AB, BC, CD and DA respectively such that $\angle AEH = \angle BFE = \angle CGF = \angle DHG = 30°$. The area of $EFGH$ is

(a) $\frac{2}{2+\sqrt{3}}$ (b) $\frac{1}{2}$ (c) $\sqrt{3}-1$ (d) $\frac{\sqrt{3}}{2}$ (e) $\frac{3}{4}$

9. Two of the terms of an arithmetic progression are 19 and 87. They are not necessarily consecutive. The common difference d of the arithmetic progression is a positive integer. The number of possible values d could have is

(a) 2 (b) 3 (c) 4 (d) 5 (e) 6

10. A rectangle and a triangle are drawn on a piece of paper so that three of the rectangle's four corners are inside the triangle. Then the number of the triangle's corners that are inside the rectangle is

(a) 0 (b) 1 (c) 2 (d) 3 (e) impossible to determine

11. At a school party, every student shook hands once with every other student. There were 55 handshakes between 2 girls, and 66 handshakes between 2 boys. The number of handshakes between a boy and a girl is

(a) 23 (b) 110 (c) 132 (d) 253 (e) none of these

12. The smallest positive multiple of 9 which does not contain any odd digits is

(a) 18 (b) 228 (c) 288 (d) 468 (e) none of these

13. The number of planes of symmetry of an equilateral tetrahedron is

(a) 0 (b) 2 (c) 4 (d) 8 (e) none of these

14. If $\frac{1}{a+b}$, $\frac{1}{b+c}$ and $\frac{1}{c+a}$ form an arithmetic progression of length 3, then so do

(a) a^2, b^2 and c^2 (b) b^2, c^2 and a^2 (c) c^2, a^2 and b^2

(d) b^2+c^2, c^2+a^2 and a^2+b^2 (e) none of these

15. The number of real roots of the equation $|x^2+2x-3| = x-1$ is

(a) 1 (b) 2 (c) 3 (d) 4 (e) none of these

16. Three coins are "loaded" so that each of them is more likely to come up heads than tails. They are then tossed. The most likely outcome is

(a) 3 heads (b) 2 heads and 1 tail (c) 2 tails and 1 head

(d) 3 tails (e) impossible to determine

17. $ABCD$ is a square of side length 1. P is any point on the circle passing through A, B, C and D. The maximum value of $PA^2 + PB^2 + PC^2 + PD^2$ is

(a) 4 (b) $4\sqrt{2}$ (c) 6 (d) 8 (e) none of these

18. If $1 < x < 2$, then $\dfrac{1}{\sqrt{x + 2\sqrt{x-1}}} + \dfrac{1}{\sqrt{x - 2\sqrt{x-1}}}$ is equal to

(a) $\frac{2}{\sqrt{x}}$ (b) $2\sqrt{x}$ (c) $\frac{2}{1+\sqrt{x-1}}$

(d) $2(1 + \sqrt{x-1})$ (e) $\frac{2}{2-x}$

19. APB and CPD are two intersecting perpendicular chords of a given circle. If $AP = 4$, $CP = 6$ and $PD = 8$, the square of the radius of the circle is

(a) 60 (b) 65 (c) 70 (d) 75 (e) impossible to determine

20. If the digits of 1987 are rearranged to form a number divisible by 7, then the two digits which must be adjacent are

(a) 1 and 8 (b) 8 and 9 (c) 9 and 7 (d) nonexistent (e) none of these

November 15, 1988

1. The number of integers between $\sqrt[3]{99}$ and $\sqrt{99}$ is

(a) 3 (b) 4 (c) 5 (d) 6 (e) 7

2. A box of apples costs \$4, a box of oranges costs \$3 and a box of lemons costs \$2. Eight boxes of fruit cost a total of \$23. Among them, the largest possible number of boxes of apples is

(a) 1 (b) 2 (c) 3 (d) 4 (e) 5

3. S is a set of non-congruent right triangles such that each one has area 1 and all the angles of each are a positive whole number of degrees. The maximum number of triangles in S is

(a) 0 (b) 2 (c) 45 (d) 46 (e) unlimited

4. Let x and y be real numbers. John computed $\frac{x^2}{y^2}$ and got 2 as an answer. Mary computed $\frac{x^3}{y^3}$ and got 3 as an answer. The value of $\frac{x^5}{y^5}$ must be

(a) 5 (b) $4\sqrt{2}$ (c) 6 (d) $3\sqrt[3]{9}$

(e) either John or Mary has made a mistake

5. One solution of $\frac{1}{x^2} + \frac{1}{(2-x)^2} = \frac{40}{9}$ is $x = \frac{1}{2}$. A different solution is

(a) $-\frac{1}{2}$ (b) $\frac{1}{4}$ (c) $\frac{3}{2}$ (d) 2 (e) none of these

6. Of the following five numbers, the largest is

(a) 3^{210} (b) 7^{140} (c) 17^{105} (d) 31^{84} (e) 127^{50}

7. O is the center of a regular hexagon of side 1. An equilateral triangle of side 3 is placed so that one of its vertices is at O, but arbitrarily otherwise. The area of the part common to the hexagon and the triangle is

(a) dependent on the actual position of the triangle

(b) $\frac{1}{6}$ (c) $\frac{\sqrt{3}}{2}$ (d) 1 (e) none of these

8. The constant term in the expansion of $(2x^3 - \frac{1}{x})^{12}$ is

(a) -1760 (b) -220 (c) 220 (d) 1760 (e) none of these

9. Arthur rolls a fair cubical die with faces numbered 1, 3, 4, 5, 6 and 8. Betty rolls a fair octahedral die with faces numbered 1, 2, 3, 4, 5, 6, 7 and 8. The probability that Arthur's die-roll is higher than Betty's is

(a) $\frac{5}{12}$ (b) $\frac{7}{16}$ (c) $\frac{11}{24}$ (d) $\frac{1}{2}$ (e) none of these

10. $ABCD$ is a 6×8 rectangle. Starting from the midpoint of AB, a bug crawls first to some point on AD, then to some point on CD, and finally arrives at the midpoint of BC. The minimum distance the bug must have crawled is

(a) 14 (b) 15 (c) $\sqrt{52} + \sqrt{73}$ (d) 17 (e) none of these

11. The number of integral solutions of the equation $7(x + \frac{1}{x}) - 2(x^2 + \frac{1}{x^2}) = 9$ is

(a) 0 (b) 1 (c) 2 (d) 3 (e) 4

12. The integer closest to 1988 that is the product of three integers in arithmetic progression is

(a) 1875 (b) 1891 (c) 1989 (d) 1995 (e) 2016

13. A is the center of a sphere of radius 4 and B is the center of a sphere of radius 3. The spheres rest on the floor and touch each other. The extension of AB meets the floor at the point C. The length of AC is

(a) 7 (b) 14 (c) 21 (d) 24 (e) 28

14. Of the following five numbers, $\sqrt[3]{1.001001001} - \sqrt[3]{1.001001}$ is closest to

(a) 0 (b) 10^{-6} (c) $5 \cdot 10^{-9}$ (d) 10^{-9} (e) 10^{-10}

15. Let $a_1, a_2, \ldots, a_{19}$ be the first 19 terms of an arithmetic progression where $a_1 + a_8 + a_{12} + a_{19} = 224$. The sum of these 19 terms is

(a) 969 (b) 1064 (c) impossible to determine

(d) the situation cannot occur (e) none of these

16. A tetrahedron has vertices with rectangular coordinates (1,0,0), (0,1,0), (0,0,1) and (1,1,1) respectively. Its volume is

(a) $\frac{1}{6}$ (b) $\frac{1}{4}$ (c) $\frac{1}{3}$ (d) $\frac{1}{2}$ (e) none of these

November 21, 1989

1. Let a be the smallest integer greater than 1 which is both a cube and a fourth power. The number of digits of a in base-ten is

(a) 1 (b) 2 (c) 3 (d) 4 (e) 5

2. A square island 1 kilometer on a side is centered in a circular lake 2 kilometers in diameter. The shortest distance, in kilometers, from the island to the circular shore of the lake is

(a) $1 - \frac{1}{\sqrt{2}}$ (b) $\frac{1}{2}$ (c) 1 (d) $2 - \frac{1}{\sqrt{2}}$ (e) none of these

3. A hotel charges \$48 per day for stays of less than 7 days and \$36 per day for stays of 7 days or more. Alice and Brian check into the hotel on the same day but leave on different days, Alice leaving first. Their rooms cost them exactly the same amount. The number of days Alice stays in the hotel is

(a) 3 (b) 4 (c) 5 (d) 6

(e) impossible to determine

4. Four numbers are written in a row. The average of the first two numbers is 7, the average of the middle two numbers is 2.3 and the average of the last two numbers is 8.4. The average of the first number and the last number is

(a) 5.9 (b) 10.7 (c) 13.1

(d) impossible to determine (e) none of these

5. Five dice are in the form of a regular tetrahedron, a cube, a regular octahedron, a regular dodecahedron and a regular icosahedron. The number of red faces of these dice are 1, 2, 3, 4 and 5 respectively. The total number of faces on the die with the highest probability, when rolled, of landing on a red face is

(a) 4 (b) 6 (c) 8 (d) 12 (e) 20

6. For a pair of integers x and y, $axy + 4x^2 + 625y^2 = 9$. A possible value for a is

(a) 28 (b) 60 (c) 80 (d) 96 (e) 100

7. If $f(x) = 2^x$, then 4^8 is equal to

(a) $f(f(2))$ (b) $f(f(f(2)))$ (c) $f(f(f(f(2))))$

(d) $f(f(f(f(f(2)))))$ (e) none of these

8. Consider the statement: "If a circle is tangent to n sides (not their extensions) of a regular polygon, then it is tangent to all sides." The smallest value of n for which this statement is true is

(a) 1 (b) 2 (c) 3 (d) 4 (e) 5

9. Mary has 13 coins adding up to 70 cents. The maximum number of dimes she could have had is

(a) 1 (b) 4 (c) 5 (d) 6 (e) 7

10. Consider the statement: "The sum of n consecutive integers is always odd." This is true if $n = 2$. The smallest value of n greater than 2 for which this statement is true is

(a) 3 (b) 4 (c) 5 (d) 6 (e) none of these

11. A group of students write a competition with 6 "true or false" questions. For each pair of students, there is at least one question to which the two give opposite answers. The maximum number of students in such a group is

(a) 4 (b) 6 (c) 15 (d) 36 (e) 64

12. In a certain mathematics examination, the average grade of the students passing is $X\%$, while the average of those failing is $Y\%$. The average of all the students taking the examination is Zpercentage of students who failed the examination is

(a) $\frac{10OXY}{Z^2}$ (b) $\frac{100YZ}{X^2}$ (C) $\frac{10OZX}{Y^2}$ (d) $\frac{100(X-Y)}{X-Z}$ (e) $\frac{100(X-Z)}{X-Y}$

13. If a, b, c, d and e are real numbers such that $a+b < c+d$, $b+c < d+e$, $c+d < e+a$ and $d+e < a+b$, then

    (a) the largest of the five numbers is a and the smallest is b

    (b) the largest of the five numbers is e and the smallest is c

    (c) the largest of the five numbers is a and the smallest is c

    (d) the largest of the five numbers is e and the smallest is b

    (e) none of these

14. $ABCD$ is a convex quadrilateral with AB and AD both of length $\sqrt{1989}$ centimeters. E is a point inside $ABCD$ such that $BCDE$ is a square and EA, EB and ED divide $ABCD$ into three parts of equal area. In square centimeters, the area of $BCDE$ is

    (a) 117 (b) 153 (c) 221 (d) 1989 (e) none of these

15. The sum of the first n terms of a geometric progression is $2n$ whereas the sum of the first $2n$ terms of it is n. The sum of the first $3n$ terms of the progression is

    (a) 0 (b) n (c) $\frac{3n}{2}$ (d) $2n$ (e) $\frac{5n}{2}$

16. A circle is inscribed in a triangle whose sides are 40, 40 and 48 centimeters respectively. A smaller circle is tangent to the two equal sides of the triangle and to the first circle. The radius, in centimeters, of the smaller circle is

    (a) 3 (b) $3\sqrt{3}$ (c) 4 (d) $\frac{29}{8}$ (e) none of these

November 20, 1990

1. A publisher puts out a new edition of a problem book by moving a block of problems at the very end of the book to the very beginning. If the new #7 is the old #33 and the new #18 is the old #8, the total number of problem in this book is

    (a) 26 (b) 33 (c) 36 (d) 46 (e) 62

2. A circular piece of maximum size is cut out of a square piece of metal. Then a square piece of maximum size is cut out of this circular piece. The ratio of the areas of the squares is

    (a) 2:1 (b) $2\sqrt{2} : 1$ (c) 3:1 (d) $\pi : 1$ (e) 4:1

3. A number of $2 \times 3 \times 5$ bricks are arranged in three stacks. In the first stack, the horizontal side of each brick is 3×5. In the second stack, each horizontal side is 2×5. In the third stack, each horizontal side is 2×3. The three stacks have exactly the same height. The smallest number of bricks in the three stacks combined is

    (a) 10 (b) 15 (c) 29 (d) 30 (e) 31

4. The numbers $a_1,\ a_2,\ a_3,\ldots$ are are defined by $a_1 = 10,\ a_2 = 20$ and for $n \geq 2,\ a_{n+1} = \frac{1}{n}(a_1 + a_2 + \cdots + a_n)$. Then a_{1990} is equal to

(a) 10 (b) 15 (c) 20 (d) 19900 (e) none of these

5. Some of the squares of a 2×9 board contain markers. Every square either contains a marker or shares an edge with a square that contains a marker. The smallest number of markers on the board is

(a) 5 (b) 6 (c) 7 (d) 8 (e) 9

6. The number of pairs of positive integers (x, y) such that $x^2 - y^2 = 99$ is

(a) 1 (b) 2 (c) 3 (d) 6 (e) infinite

7. $ABCD$ is a quadrilateral with AB parallel to DC, $AB = 5$ and $DC = 8$. E is a point on CD such that the area of triangle AED is equal to the area of the quadrilateral $ABCE$. The ratio $DE : CE$ is

(a) 7:5 (b) 8:5 (c) 13:7 (d) 13:5 (e) 13:3

8. The real numbers $x,\ y$ and z satisfy $2^{x+y} = 10,\ 2^{y+z} = 20$ and $2^{z+x} = 30$. Then 2^x is

(a) $\frac{\sqrt{6}}{2}$ (b) $\frac{3}{2}$ (c) $\sqrt{15}$ (d) $10\sqrt{6} - 20$ (e) none of these

9. Each face of a cube is to be coloured red, yellow or green, such that there are two faces of each colour. The maximum number of distinctly coloured cubes, independent of how they are held, is

(a) 3 (b) 4 (c) 5 (d) 6 (e) none of these

10. Andrea has three short straws each 10 centimeters in length, and three long straws each 30 centimeters in length. She draws at random three of the straws and tries to form a triangle by joining them at the ends. The probability that the three straws can form a triangle is

(a) $\frac{1}{10}$ (b) $\frac{9}{20}$ (c) $\frac{1}{2}$ (d) $\frac{11}{20}$ (e) $\frac{3}{4}$

11. If x and y are positive numbers, the largest of the following is

(a) $\sqrt{3xy}$ (b) $x + y$ (c) $\frac{x^2+y^2}{x+y}$ (d) $\frac{x^4+x^2y^2+y^4}{x^3+y^3}$

(e) dependent on x and y

12. For $n = 1, 2, 3, \ldots,\ A_nB_nC_nD_nE_n$ are regular pentagons such that $B_1,\ B_2,\ B_3, \ldots,$ $C_1,\ C_2, C_3, \ldots$ all lie on one straight line while $D_1,\ D_2,\ D_3, \ldots,\ E_1,\ E_2,\ E_3, \ldots$

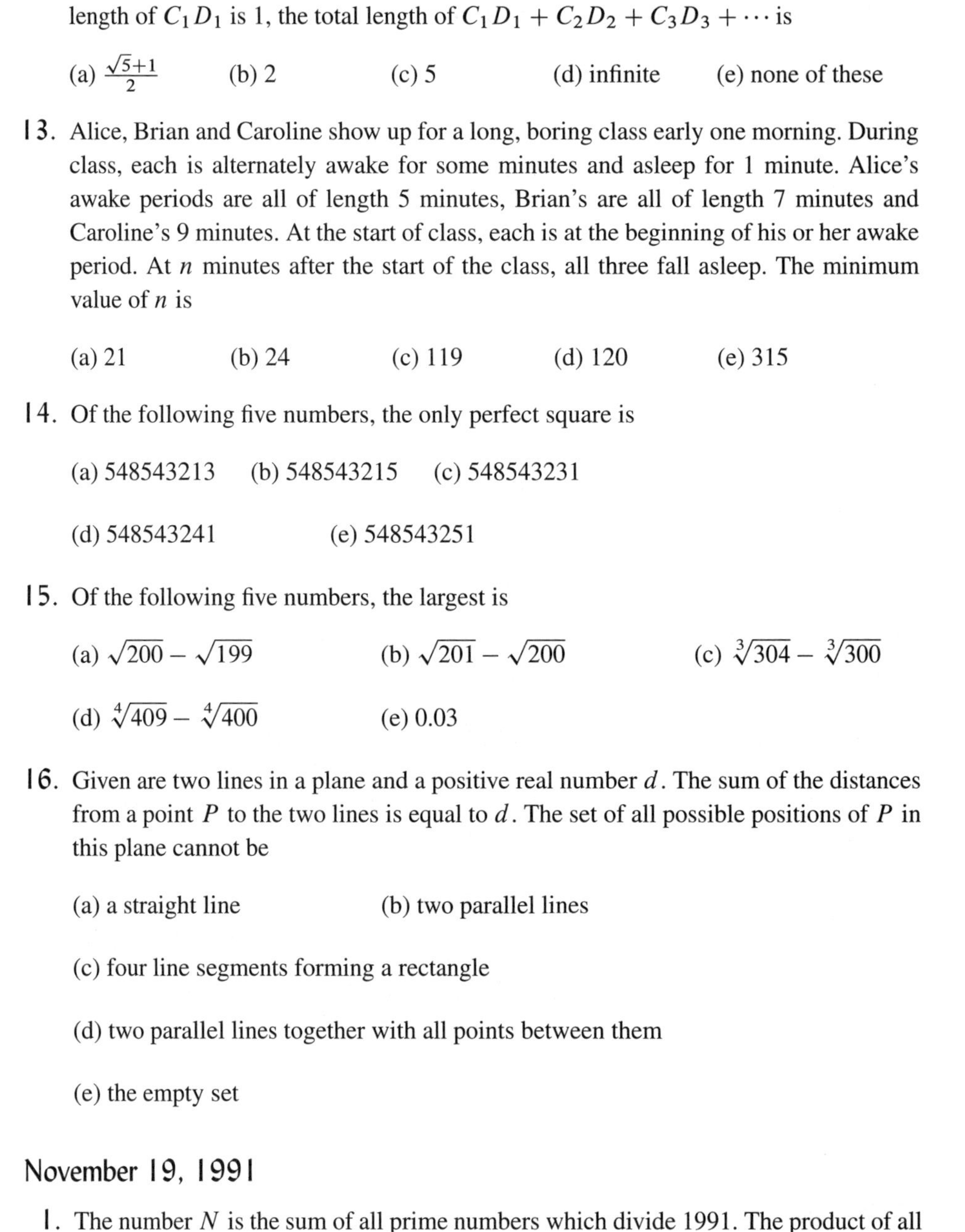

all lie on another straight line. For $n \geq 2$, A_n is the midpoint of $C_{n-1}D_{n-1}$. If the length of C_1D_1 is 1, the total length of $C_1D_1 + C_2D_2 + C_3D_3 + \cdots$ is

(a) $\frac{\sqrt{5}+1}{2}$ (b) 2 (c) 5 (d) infinite (e) none of these

13. Alice, Brian and Caroline show up for a long, boring class early one morning. During class, each is alternately awake for some minutes and asleep for 1 minute. Alice's awake periods are all of length 5 minutes, Brian's are all of length 7 minutes and Caroline's 9 minutes. At the start of class, each is at the beginning of his or her awake period. At n minutes after the start of the class, all three fall asleep. The minimum value of n is

(a) 21 (b) 24 (c) 119 (d) 120 (e) 315

14. Of the following five numbers, the only perfect square is

(a) 548543213 (b) 548543215 (c) 548543231

(d) 548543241 (e) 548543251

15. Of the following five numbers, the largest is

(a) $\sqrt{200} - \sqrt{199}$ (b) $\sqrt{201} - \sqrt{200}$ (c) $\sqrt[3]{304} - \sqrt[3]{300}$

(d) $\sqrt[4]{409} - \sqrt[4]{400}$ (e) 0.03

16. Given are two lines in a plane and a positive real number d. The sum of the distances from a point P to the two lines is equal to d. The set of all possible positions of P in this plane cannot be

(a) a straight line (b) two parallel lines

(c) four line segments forming a rectangle

(d) two parallel lines together with all points between them

(e) the empty set

November 19, 1991

1. The number N is the sum of all prime numbers which divide 1991. The product of all prime numbers which divide N is

(a) 5 (b) 6 (c) 15 (d) 192 (e) 1991

2. If $x - (x - (x - (x - (x - 1)))) = 1$, then x is

(a) -2 (b) -1 (c) 0 (d) 1 (e) 2

3. The sum $\frac{1}{2}+\frac{3}{4}+\frac{7}{8}+\frac{15}{16}+\cdots+\frac{1023}{1024}$ lies between

(a) 5 and 6 (b) 6 and 7 (c) 7 and 8 (d) 8 and 9 (e) 9 and 10

4. A piece of cardboard is 40 centimeters by 60 centimeters. It is cut into 500 pieces to make a jigsaw puzzle. The piece of cardboard is not moved while it is being cut. The total length of cutting done is 20 meters. The average perimeter, in centimeters, of each of the 500 pieces is

(a) 4 (b) 4.2 (c) 4.4 (d) 8 (e) 8.4

5. Let x be a positive number. If $x^x = y$ and $x^y = 100000^{100000}$, then x lies between

(a) 1 and 10 (b) 10 and 100 (c) 100 and 1000

(d) 1000 and 10000 (e) 10000 and 100000

6. $ABCD$ is a square. Three parallel lines ℓ_1, ℓ_2 and ℓ_3 pass through A, B and C respectively. The distance between ℓ_1 and ℓ_2 is 5 centimeters, and the distance between ℓ_2 and ℓ_3 is 7 centimeters. The area, in square centimeters, of $ABCD$ is

(a) 35 (b) 70 (c) 74 (d) 144 (e) none of these

7. If $(a+b+c+x)(a-b-c+x)=(a-b+c-x)(a+b-c-x)$, where a, b and c are real numbers not equal to 0, then x is equal to

(a) $\frac{a+b+c}{3}$ (b) $\frac{bc}{a}$ (c) $\frac{ca}{b}$ (d) $\frac{ab}{c}$ (e) none of these

8. The book *Mathematical Challenges* contains 213 problems from 12 years of competitions in Scotland. At first there were 25 problems per year, then 16 problems per year, and finally 20 per year. The number of years when there were 25 problems was

(a) 1 (b) 2 (c) 3 (d) 4 (e) 5

9. The number of terms of the fifth degree which can be formed from products of the variables a, b, c, d and e, with repetition allowed, is

(a) 49 (b) 50 (c) 51 (d) 125 (e) 126

10. The King collects tax in multiples of \$100. For each \$100 he collects, he celebrates by letting off 4 firecrackers which costs \$10 each. On a particularly good day, the King collected a lot of tax and bought many firecrackers. At the end of the day, he found that he had \$1000 and 200 firecrackers more than he did when the day began. The number of firecrackers he bought on that day was

(a) 240 (b) 320 (c) 336 (d) 368 (e) 400

11. The number of real roots of $1+x+x^2+\cdots+x^{1991}$ is

(a) 0 (b) 1 (c) 2 (d) 3 (e) more than 3

12. The diagonals AC and BD of the quadrilateral $ABCD$ intersect at a point E inside $ABCD$. The areas of the triangles EAB, EBC, ECD and EDA are w, x, y and z respectively. Of the two equations $w + y = x + z$ and $wy = xz$.

(a) both are always true. (b) both are sometimes true.

(c) one is always true and the other is never true.

(d) one is always true and the other is sometimes true.

(e) one is sometimes true and the other is never true.

13. The four regions of a mini-dartboard are labelled 1, 2, 3 and 4. On each throw, the player gets a score equal to the label of the region hit, and 0 otherwise. The object of the game is to get a total of exactly 4 points in two or fewer throws. When Brenda throws a dart at one of the regions, she hits it 40% of the time, hits some other region 40% of the time, and misses the board altogether 20% of the time. The probability, expressed as a percentage, that she will win using the best strategy is

(a) 16 (b) 20 (c) 40 (d) 64 (e) 80

14. The value of $\sqrt[3]{\frac{4}{9}} - \sqrt[3]{\frac{2}{9}} + \sqrt[3]{\frac{1}{9}}$ is

(a) $\dfrac{1}{\sqrt[3]{9}(\sqrt[3]{2}+1)}$ (b) $\dfrac{\sqrt[3]{3}}{\sqrt[3]{2}+1}$ (c) $\dfrac{1}{\sqrt[3]{3}}$

(d) $\dfrac{1}{\sqrt[3]{9}(\sqrt[3]{2}-1)}$ (e) $\dfrac{\sqrt[3]{3}}{\sqrt[3]{2}-1}$

15. If PA, PB, PC and PD are four rays in space such that

$$\angle APB = \angle BPC = \angle CPA = \angle DPB = \angle DPC = 120^\circ,$$

then $\angle DPA$ is equal to

(a) 30° (b) 60° (c) 90° (d) 120° (e) none of these

16. Debbie writes down two positive integers m and n. It turns out that exactly one of the following statements about m and n is false. The false statement is

(a) $m + n$ is divisible by 3 (b) $2m + n$ is divisible by 8

(c) $m + 1$ is divisible by n (d) $m + 7n$ is a prime number (e) $m = 2n + 5$

November 17, 1992

1. If 2^{x+1} is between 8 and 10, then 4^{x+2} is between

   (a) 16 and 20 (b) 32 and 40 (c) 64 and 100 (d) 128 and 200 (e) none of these

2. $ABCDE$ is a pentagon. F is a point on AE such that AF is of length 1 centimeter. FAB, FBC, FCD and FDE are all isosceles right triangles, with the right angles at A, B, C and D respectively. The area, in square centimeters, of $ABCDE$ is

   (a) 5 (b) $\frac{13}{2}$ (c) $5\sqrt{2}$ (d) $\frac{15}{2}$ (e) none of these

3. Seven children ate 24 peanuts altogether. Andy ate one peanut, Candy two and Randy three. Sandy ate more peanuts than any of the others. The minimum number of peanuts Sandy must have eaten is

   (a) 4 (b) 5 (c) 6 (d) 7 (e) 8

4. Consider all compound fractions of the form $\frac{a}{b}/\frac{c}{d}$, where a, b, c and d are 2, 3, 5 and 7 in some order. The total number of different values these compound fractions have is

   (a) 6 (b) 12 (c) 16 (d) 24 (e) none of these

5. In triangle ABC, $AB = 5$, $BC = 4$ and $CA = 3$. D is the foot of perpendicular from C to AB. The length of AD is

   (a) $\frac{9}{5}$ (b) 2 (c) $\frac{12}{5}$ (d) 3 (e) $\frac{16}{5}$

6. Two statements are made about four points in space. P: "All four points lie in a plane." Q: "Three of the four points lie on a line."

   (a) P and Q are equivalent.

   (b) P implies Q but not vice versa.

   (c) Q implies P but not vice versa.

   (d) Neither P nor Q implies the other and both can be true.

   (e) Neither P nor Q implies the other and both can be false.

7. The smallest square determined by four points of a square lattice is 1. No four points of the lattice determine a square of area

   (a) 25 (b) 26 (c) 34 (d) 35 (e) 37

8. A positive integer is called a fiver if the sum of its digits is divisible by 5. In any block of 7 consecutive positive integers, the maximum number of fivers is

   (a) 1 (b) 2 (c) 3 (d) 4 (e) 5

9. A circle of radius 3 centimeters is tangent to both sides of a given angle. A circle of radius 7 centimeters is tangent to these two sides and the first circle. A third circle is tangent to the same two sides and the second circle. The radius, in centimeters, of the third circle is

(a) 11 (b) 15 (c) $\frac{49}{3}$ (d) 21 (e) none of these

10. If one of the following inequalities has solution $0 < x \leq 1$, it is

(a) $\frac{\sqrt[3]{1-x}}{x} \geq 0$ (b) $\frac{\sqrt[3]{1-x^2}}{x} \geq 0$ (c) $\frac{\sqrt[3]{1-x}}{x^2} \geq 0$

(d) $\frac{\sqrt[3]{1-x^2}}{x^2} \geq 0$ (e) none of these

11. An object is known to have integral weight, but this cannot be determined using a balance and three weights of values 1, 4 and 11. The minimum weight of this object is

(a) 2 (b) 3 (c) 9 (d) 13 (e) 17

12. The product of three prime numbers is 47 times their sum. One of them is

(a) 2 (b) 3 (c) 7 (d) 13 (e) 17

13. A square carpet 99 centimeters by 99 centimeters is in the middle of a large room. The room is filled with spheres having diameter 1 meter. The largest number of these spheres that can touch the carpet is

(a) 1 (b) 2 (c) 3 (d) 4 (e) more than 4

14. The value of $\sqrt[3]{1+\frac{2}{3}\sqrt{\frac{7}{3}}}+\sqrt[3]{1-\frac{2}{3}\sqrt{\frac{7}{3}}}$ is

(a) 1 (b) strictly between 1 and 2

(c) 2 (d) irrational, but not between 1 and 2

(e) none of these

15. In a hockey tournament, each of six teams plays every other team once. Two points are awarded for a win, one point for a tie and zero points for a loss. At the end of the tournament, each team has a different point total and the third place team has more points than the bottom three teams combined. It is not possible to have a tied game between

(a) the first and second place teams

(b) the second and third place teams

(c) the third and fourth place teams

(d) the fourth and fifth place teams

(e) the fifth and sixth place teams

16. The sum $\frac{1}{1+1^2+1^4}+\frac{2}{1+2^2+2^4}+\frac{3}{1+3^2+3^4}+\cdots+\frac{99}{1+99^2+99^4}$ lies in the open interval

(a) (0.46,0.47) (b) (0.47,0.48) (c) (0.48,0.49)

(d) (0.49,0.50) (e) (0.50,0.51)

November 16, 1993

1. On each question in this competition, you will receive 5 marks for a correct answer, 2 marks for no answer and 0 marks for an incorrect answer. Suppose you know that two of the alternative answers are wrong and you randomly guess one of the remaining three. If you answer many questions this way, your average score on each such question will be

   (a) $\frac{1}{2}$ (b) $\frac{2}{3}$ (c) $\frac{5}{3}$ (d) $\frac{7}{3}$ (e) $\frac{5}{2}$

2. Assume that the average volume of a raindrop is 10 cubic millimeters. A square city 10 kilometers by 10 kilometers received 1 centimeter of rain. The number of raindrops that fell on the city is

   (a) 10^{10} (b) 10^{11} (c) 10^{12} (d) 10^{13} (e) 10^{14}

3. The product of the repeating decimals $0.\overline{3}$ and $0.\overline{6}$ is

   (a) $0.\overline{18}$ (b) $0.1\overline{9}$ (c) $0.\overline{2}$ (d) $0.\overline{9}$ (e) none of these

4. A circular table is 2 meters across and 1 meter high. A square tablecloth laid on the table just touches the floor at its four corners. The length in meters of one side of the tablecloth is

   (a) $2\sqrt{2}$ (b) 4 (c) $4\sqrt{2}$ (d) 4π (e) none of these

5. Danny has the same number of brothers as sisters. His sister has twice as many brothers as sisters. The number of children Danny's parents have is

   (a) 3 (b) 4 (c) 5 (d) 6 (e) 7

6. Each of A, B, C and D either always tells the truth or always lies. They make the following statements about themselves.

   A: None of us tells the truth.
   B: At most one of us tells the truth.
   C: At most two of us tell the truth.
   D: At most three of us tell the truth.

   The number of them who tell the truth is

   (a) 0 (b) 1 (c) 2 (d) 3

   (e) impossible to determine

7. Seven 3 by 4 rectangles are to be assembled, without overlap, into one large rectangle. The minimum perimeter of this rectangle is

(a) 25 (b) 31 (c) 50 (d) 62 (e) none of these

8. On an island, 99% of the population are natives. Some natives emigrate, so that only 98% of the population are natives. If the initial population of the island was 1000, the number of natives who emigrated is

(a) 10 (b) 20 (c) 100 (d) 200 (e) 500

9. Samira runs the first 2 kilometers of a 6-kilometer race at 6 kilometers per hour, the second 2 kilometers at 4 kilometers per hour, and the final 2 kilometers at 3 kilometers per hour. In kilometers per hour, her average speed over the whole race is

(a) $\frac{13}{4}$ (b) 4 (c) $\frac{13}{3}$ (d) 5 (e) none of these

10. A certain government backbencher's age is equal to his IQ. If he were to join the cabinet, he would have raised its average age from 50 to 51 and lowered its average IQ from 114 to 111. The number of cabinet members now is

(a) 12 (b) 15 (c) 20 (d) 30 (e) none of these

11. The product of a three-digit number and the number with the same digits in reverse order is 433755. The tens-digit of this number is

(a) 1 (b) 5 (c) 6 (d) 7 (e) none of these

12. Starting with the number a, suppose you perform the following four operations: add 2, subtract 2, multiply by 2 and divide by 2, one after another, each exactly once, but not necessarily in that order. The largest number you can end up with is

(a) a (b) $a + 1$ (c) $a + 2$ (d) $a + 4$

(e) dependent on whether a-is even or odd

13. $ABCD$ and $DEFG$ are rectangles with $AB = DE = 15$ and $BC = EF = 9$. E lies on AB, and EF intersects BC at H. The area of $DEHC$ is

(a) 60 (b) 65 (c) 70 (d) 75 (e) none of these

14. Starting with the number 3, repeat 30 times the operation of doubling and subtracting 1. You get

(a) $3 \cdot 2^{30}$ (b) $3 \cdot 2^{31}$ (c) $2^{31} + 1$ (d) $2^{31} - 1$ (e) none of these

15. A 10 meters by $10\sqrt{2}$ meters rectangular building is surrounded by a large grassy field. A peg is pounded into the grass at a distance $5\sqrt{2}$ meters straight out from

the midpoint of one of the long walls of the building. A goat is tied to the peg by a 20-meter rope. In square meters, the area of grass available to be eaten by the goat is

(a) $\frac{325\pi}{2} + 25$ (b) $250\pi + 25$ (c) $300\pi + 50$ (d) $325\pi + 50$

(e) none of these

16. If $F(x) = \dfrac{1}{\sqrt{x + 2\sqrt{x-1}}} + \dfrac{1}{\sqrt{x - 2\sqrt{x-1}}}$, then $F(\frac{3}{2})$ is equal to

(a) $2\sqrt{3}$ (b) $\frac{7}{2}$ (c) 4 (d) $3\sqrt{2}$ (e) none of these

November 15, 1994

1. You have 5 sticks, of lengths 10, 20, 30, 40 and 50 cm. The number of non-congruent triangles that can be formed by choosing three of the sticks to make the sides is

   (a) 3 (b) 5 (c) 8 (d) 9 (e) 10

2. A glass box 7 cm by 12 cm by 18 cm, closed on all six sides, is partly filled with coloured water. When the box is placed on one of its 7 by 12 sides, the water level is 15 cm above the table. When the box is placed on one of its 7 by 18 sides the water level above the table, in cm, will be

   (a) 7.5 (b) 9 (c) 10 (d) 12.5 (e) none of these

3. There are four cottages on a straight road. The distance between Lil's and Ted's cottages is 3 kilometers. Both Jack's and Jill's cottages are twice as far from Lil's as from Ted's. In kilometers, the distance between Jack's and Jill's cottages is

   (a) 1 (b) 2 (c) 3 (d) 4 (e) 6

4. Some playing cards from an ordinary deck are arranged in a row. To the right of some King is at least one Queen. To the left of some Queen is at least one other Queen. To the left of some Heart is at least one Spade. To the right of some Spade is at least one other Spade. The minimum number of cards in this row is

   (a) 2 (b) 3 (c) 4 (d) 5 (e) 6

5. Ace, Bea, Cec and Dee have \$16, \$24, \$32 and \$48 respectively. Their father proposed that Ace and Bea share their wealth equally, then Bea and Cec do likewise, and then Cec and Dee. Their mother's plan is the same except that Dee and Cec begin by sharing equally, then Cec and Bea and then Bea and Ace. The number of children who end up with more money under their father's plan than under their mother's is

   (a) 0 (b) 1 (c) 2 (d) 3 (e) 4

6. If n metal spheres 3 mm in diameter are to be melted down to form one metal sphere of diameter 10 mm, then the minimum value of n satisfies

   (a) $n < 5$ (b) $5 \leq n < 10$ (c) $10 \leq n < 15$

   (d) $15 \leq n < 20$ (e) $n \geq 20$

7. The sum of all positive integers which are less than 600 and are not multiples of 3 is

(a) 60,000 (b) 90,000 (c) 120,000 (d) 150,000 (e) 180,000

8. Five infinite straight lines are to be drawn in the plane. No three lines pass through the same point. The number of non-negative integers which can serve as the total number of points of intersection of the five lines is

(a) less than 7 (b) 7 (c) 8 (d) 9 (e) 11

9. In a warehouse, a stack of 6 mattresses was piled up. Each mattress was originally 12 cm thick. Each compressed by a third each time an additional mattress was piled on top. The height h of the pile, in cm, satisfies

(a) $h < 24$ (b) $24 \leq h < 30$ (c) $30 \leq h < 32$

(d) $32 \leq h < 40$ (e) $h \geq 40$

10. If $r^2 - r - 10 = 0$, then $(r+1)(r+2)(r-4)$ is

(a) integral (b) positive and irrational (c) negative and irrational

(d) rational but non-integral (e) non-real

11. I give Sarah N dollars for getting a good mark in school. Then, since Tim got a better mark, I give him just enough 2 dollar bills so that he gets **more** money than Sarah. Finally, since Ursula got the best mark, I give her just enough 5 dollar bills so that she gets **more** money than Tim. The largest amount of money, in dollars, that Ursula could get is

(a) $N+2$ (b) $N+5$ (c) $N+6$ (d) $N+7$ (e) none of these

12. The number of ordered pairs of (positive or negative) integers (x, y) which satisfy

$$8x^3 + y^3 - 12x^2 + 6y^2 + 6x + 12y = 21$$

is

(a) 0 (b) 1 (c) 2 (d) 3 (e) more than 3

13. Two integers are declared equivalent if both are divisible by the same prime numbers. The maximum number of positive integers less than 25 which are pairwise non-equivalent is

(a) 9 (b) 10 (c) 16 (d) 17 (e) 24

14. The points (x, y) on the coordinate plane satisfying $|x| \leq 2$, $|y| \leq 2$ and $||x| - |y|| \leq 1$, where $|a|$ is the absolute value of a, define a region with area

(a) 8 (b) 10 (c) 12 (d) 14 (e) 16

15. The center of each of two circular discs of radius 1 lies on the circumference of the other. The area of the intersection of the discs is

(a) $\frac{\pi}{3} - \frac{\sqrt{3}}{3}$ (b) $\frac{\pi}{3}$ (c) $\frac{2\pi}{3} - \frac{\sqrt{3}}{2}$ (d) $\frac{2\pi}{3} - \frac{\sqrt{3}}{3}$ (e) $\frac{\pi}{3}$

16. Wilma and Roberta have five coins between them, a dollar, a quarter, a dime, a nickel and a penny. One of them belongs to Wilma and the other four to Roberta. Each tosses her coin or coins. Whoever has more heads wins all the coins. If it is a tie, they toss again. If this game is as fair as possible moneywise, then Wilma must have the

(a) dollar (b) quarter (c) dime (d) nickel (e) penny

November 21, 1995

1. A circle and a parabola are drawn in the plane. The number of regions they divide the plane into is at most

(a) 3 (b) 4 (c) 5 (d) 6 (e) 7

2. The number of different primes $p > 2$ such that p divides $71^2 - 37^2 - 51$ is

(a) 0 (b) 1 (c) 2 (d) 3 (e) 4

3. Suppose that your height this year is 10% more than it was last year, and last year your height was 20% more than it was the year before. During the last two years, your height has increased by

(a) 30% (b) 31% (c) 32% (d) 33% (e) none of these

4. Multiply the consecutive even positive integers together until the product $2 \cdot 4 \cdot 6 \cdot 8 \cdots$ becomes divisible by 1995. The last even integer you use is

(a) between 1 and 21 (b) between 21 and 31 (c) between 31 and 41

(d) bigger than 41 (e) non-existent

5. A rectangle contains three circles as in the diagram, all tangent to the rectangle and to each other. If the height of the rectangle is 4, then the width of the rectangle is

(a) $3 + 2\sqrt{2}$ (b) $4 + \frac{4\sqrt{2}}{3}$ (c) $5 + \frac{2\sqrt{2}}{3}$ (d) 6 (e) $5 + \sqrt{10}$

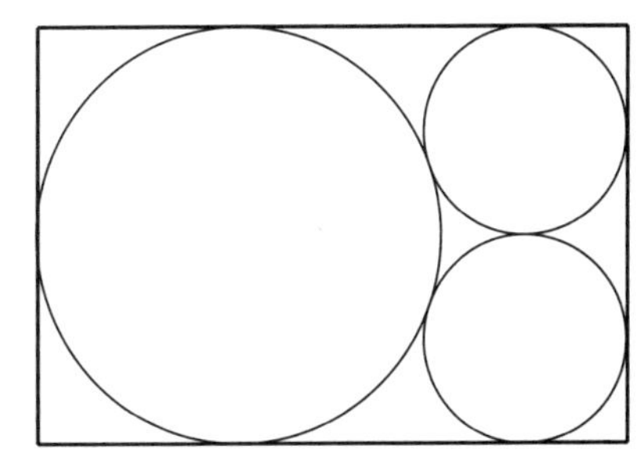

6. Mary Lou works a full day and gets her usual pay. Then she works some overtime hours, each at 150% of her usual hourly salary. Her total pay that day is equivalent to 12 hours at her usual hourly salary. The number of hours that she usually works each day is

(a) 6 (b) 7.5 (c) 8 (d) 9

(e) not uniquely determined by the given information

7. A fair coin is tossed 10000 times. The probability p of obtaining at least three heads in a row satisfies

(a) $0 \le p < \frac{1}{4}$ (b) $\frac{1}{4} \le p < \frac{1}{2}$ (c) $\frac{1}{2} \le p < \frac{3}{4}$
(d) $\frac{3}{4} \le p < 1$ (e) $p = 1$

8. In the plane, the angles of a regular polygon with n sides add up to less than n^2 degrees. The smallest possible value of n satisfies

(a) $n < 40$ (b) $40 \le n < 80$ (c) $80 \le n < 120$
(d) $120 \le n < 160$ (e) $n \ge 160$

9. A polynomial $P(x)$ of degree 3 has $P(1) = 1$, $P(2) = 2$, $P(3) = 3$ and $P(4) = 5$. The value of $P(6)$ is

(a) 7 (b) 10 (c) 13 (d) 16

(e) not uniquely determined by the given information

10. The positive numbers x and y satisfy $xy = 1$. The minimum value of $\dfrac{1}{x^4} + \dfrac{1}{4y^4}$ is

(a) $\frac{1}{2}$ (b) $\frac{5}{8}$ (c) 1 (d) $\frac{5}{4}$ (e) nonexistent

11. Of the points (0,0), (2,0), (3,1), (1,2), (3,3), (4,3) and (2,4), at most how many can lie on a circle?

(a) 3 (b) 4 (c) 5 (d) 6 (e) 7

12. The number of different positive integer triples (x, y, z) satisfying simultaneously the equations $x^2 + y - z = 100$ and $x + y^2 - z = 124$ is

(a) 0 (b) 1 (c) 2 (d) 3 (e) none of these

13. Of the following conditions, the one which does not guarantee that the convex quadrilateral $ABCD$ is a parallelogram is

(a) $AB = CD$ and $AD = BC$ (b) $\angle A = \angle C$ and $\angle B = \angle D$

(c) $AB = CD$ and $\angle A = \angle C$ (d) $AB = CD$ and AB is parallel to CD
(e) All of these do guarantee that $ABCD$ is a parallelogram.

14. For all possible numbers x and y for which $x > y$, the number of the expressions $x^3 + y^4, x^4 + y^3, x^3 + y^3$ and $x^4 + y^4$ which are positive is

(a) 0 (b) 1 (c) 2 (d) 3 (e) 4

15. In triangle ABC, the altitude from A to BC meets BC at D, and the altitude from B to CA meets AD at H. If $AD = 4$, $BD = 3$ and $CD = 2$, then the length of HD is

(a) $\frac{\sqrt{5}}{2}$ (b) $\frac{3}{2}$ (c) $\sqrt{5}$ (d) $\frac{5}{2}$ (e) $\frac{3\sqrt{5}}{2}$

16. The value of $\dfrac{(2^3-1)(3^3-1)(4^3-1)\cdots(100^3-1)}{(2^3+1)(3^3+1)(4^3+1)\cdots(100^3+1)}$ is closest to

(a) 0 (b) $\frac{1}{2}$ (c) $\frac{2}{3}$ (d) $\frac{7}{10}$ (e) $\frac{3}{4}$

November 19, 1996

1. An eight-inch pizza is cut into 3 equal slices. A ten-inch pizza is cut into 4 equal slices. A twelve-inch pizza is cut into 6 equal slices. A fourteen inch pizza is cut into 8 equal slices. From which pizza would you take a slice if you want as much pizza as possible?

(a) 8-inch (b) 10-inch (c) 12-inch (d) 14-inch (e) does not matter

2. One store sold red plums at four for a dollar and yellow plums at three for a dollar. A second store sold red plums at four for a dollar and yellow plums at six for a dollar. You bought m red plums and n yellow plums from each store, spending a total of ten dollars. The total number of plums you bought was

(a) 10 (b) 20 (c) 30 (d) 40

(e) not enough information

3. Six identical cardboard pieces are piled on top of one another, and the result is shown in the diagram. The third piece to be placed is

(a) A (b) B (c) C (d) D (e) E

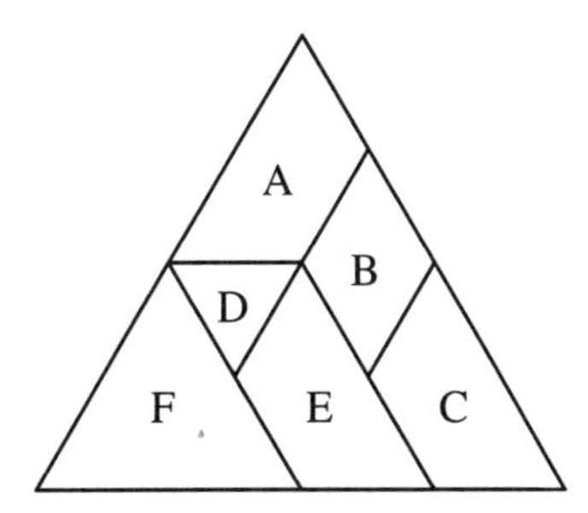

4. A store offered triple the GST in savings. A salesclerk calculated the selling price by first reducing the original price by 21% and then adding the 7% GST based on the reduced price. A customer protested, saying that the store should first add the 7% GST and then reduce that total by 21%. They agreed on a compromise: the clerk just reduced the original price by the 14% difference. From the customer's point of view,

   (a) the clerk's way is the best. (b) the customer's way is the best.

   (c) the compromise is the best. (d) all three ways are the same.

   (e) the compromise is the worst while the other two are the same.

5. If m and n are integers such that $2m - n = 3$, then $m - 2n$ is equal to

   (a) -3 only (b) 0 only (c) only multiples of 3

   (d) any integer (e) none of these

6. If x is $x\%$ of y, and y is $y\%$ of z, where x, y and z are positive real numbers, then z is

   (a) 100 (b) 200 (c) 10000

   (d) non-exist (e) not uniquely determined

7. A regular hexagon is rotated about a line through some angle θ, $0° < \theta < 360°$, so that the hexagon again occupies its original position. The number of such lines is

   (a) 1 (b) 3 (c) 4 (d) 6 (e) 7

8. AB is a diameter of a circle of radius 1 unit. CD is a chord perpendicular to AB that cuts AB at E. If the arc CAD is $\frac{2}{3}$ of the circumference of the circle, the length of the segment AE is

   (a) $\dfrac{2}{3}$ (b) $\dfrac{3}{2}$ (c) $\dfrac{\pi}{2}$ (d) $\dfrac{\sqrt{3}}{2}$ (e) none of these

9. One of Kerry and Kelly lies on Mondays, Tuesdays and Wednesdays, and tells the truth on the other days of the week. The other lies on Thursdays, Fridays and Saturdays, and tells the truth on the other days of the week. At noon, the two had the following conversation:

   Kerry: I lie on Saturdays.
   Kelly: I will lie tomorrow.
   Kerry: I lie on Sundays.

   This conversation takes place on a

   (a) Monday (b) Wednesday (c) Thursday (d) Saturday (e) Sunday

10. The number of integer pairs (m, n) whichi satisfy the equation $m(m+1) = 2^n$ is

(a) 0 (b) 1 (c) 2 (d) 3 (e) more than 3

11. Of the following triangles given by the lengths of their sides, the one which has the greatest area is

(a) 5,12,12 (b) 5,12,13 (c) 5,12,14 (d) 5,12,15 (e) 5,12,16

12. Suppose that $x < y$ and $x < 0$. Of the following numbers, the one that is never greater than any of the others is

(a) $x + y$ (b) $x - y$ (c) $x + |y|$ (d) $x - |y|$ (e) $-|x + y|$

13. An x by y flag, with $x < y$, consists of two perpendicular white stripes of equal width and four congruent blue rectangles at the corners. If the total area of the blue rectangles is half that of the flag, the length of the shorter side of each blue rectangle is

(a) $\dfrac{x - y + \sqrt{x^2 + y^2}}{4}$ (b) $\dfrac{x - y + \sqrt{x^2 + y^2}}{2}$ (c) $\dfrac{3x + y + \sqrt{x^2 + y^2}}{4}$

(d) $\dfrac{3x + y + \sqrt{x^2 + y^2}}{2}$ (e) none of these

14. A game is played with a deck of ten cards numbered from 1 to 10. Shuffle the deck thoroughly.

(i) Take the top card. If it is numbered 1, you win. If it is numbered k, where $k > 1$, go to (ii).

(ii) If this is the third time you have taken a card, you lose. Otherwise, put the card back into the deck at the kth position from the top and go to (i).

The probability of winning is

(a) $\frac{1}{5}$ (b) $\frac{5}{18}$ (c) $\frac{13}{45}$ (d) $\frac{3}{10}$ (e) none of these

15. Five of the angles of a convex polygon are each equal to $108°$. The maximum angle of all such polygons lies in the interval

(a) $(105°, 120°)$ (b) $(120°, 135°)$ (c) $(135°, 150°)$

(d) $(150°, 165°)$ (e) $(165°, 180°)$

16. Of the following numbers, the one which cannot be expressed as the difference of the squares of two integers is

(a) 314159265 (b) 314159266 (c) 314159267

(d) 314159268 (e) 314159269

November 18, 1997

1. The positive integer 199619971998 has exactly five different digits. If a and b are also numbers with exactly five different digits, the smallest possible number of different digits in $a + b$ is

   (a) 1 (b) 2 (c) 5 (d) 9 (e) 10

2. Some tarts are missing and one of three children ate them. The children made the following statements.
   Ace: Bea did not eat them.
   Bea: Either Ace or Cec ate them.
   Cec: Ace ate them.
   At least one child told the truth and at least one lied. The one who ate the tarts was

   (a) Ace (b) Bea (c) Cec

   (d) not enough information (e) The situation is impossible.

3. One half of a flock of birds flew south, one quarter flew east and one seventh flew west. The rest flew north. The minimum number of birds which flew north is

   (a) 1 (b) 2 (c) 3 (d) 4 (e) 7

4. Let a and b be real numbers such that $2^a = b^3$. Of the following expressions, the one which is always equal to 4^a is

   (a) b^5 (b) b^6 (c) b^8 (d) b^9 (e) $2b^3$

5. The number of odd positive integers n such that $11 \cdot 14^n + 1$ is a prime number is

   (a) 0 (b) 1 (c) 2 (d) 3 (e) more than 3

6. If a and b are positive integers such that $a^2 + 4b^2 = 25$, the value of $a + b$ is

   (a) 4 (b) 5 (c) 6 (d) 7

   (e) not uniquely determined

7. One thousand equal circles are arranged end to end; that is, their centers lie on the same straight line and each circle except the two at the ends is tangent to both of its neighbors.

   ← 1 meter →

   The total length of their diameters is 1 meter and the total length of their circumferences is c meters. Of the following integers, the one which is closest to c is

   (a) 0 (b) 1 (c) 2 (d) 3 (e) 4

8. The value of $100 - 1 + 98 - 3 + 96 - 5 + \cdots + 4 - 97 + 2 - 99$ is

(a) 0 (b) 50 (c) 100 (d) 147 (e) none of these

9. when written out in base 10, the number of zeros in $(10^{50})^2$ is exactly 100, and the number of zeros in $(10^{50} - 1)^2$ is

(a) 0 (b) 49 (c) 50 (d) 98 (e) 99

10. Young MacDonald has a farm on a 3×3 plot. The farmhouse is in the central lot, and pigs are kept in the other eight lots, with at least one pig in each. The total number of pigs in the three lots on any of the four sides of the farm is 9. An example is shown in the diagram below.

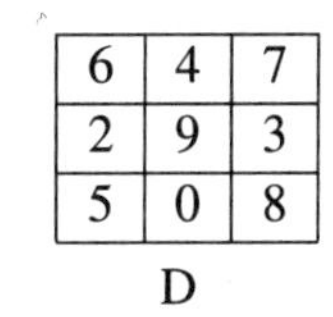

Of the following numbers, the one which cannot be the total number of pigs is

(a) 20 (b) 24 (c) 27 (d) 32 (e) 36

11. In each of the following mazes, you start from the square marked 0 and move either horizontally or vertically, one square at a time. You may not visit the same square twice. You score the sum of the points on all the squares you have visited.

4	8	4
8	4	8
4	0	4

A

9	1	9
1	9	1
9	0	9

B

3	9	3
9	3	9
3	0	3

C

6	4	7
2	9	3
5	0	8

D

To get the highest possible score, the maze you should choose is

(a) A only (b) B only (c) C only (d) D only

(e) At least two of them can yield the highest score.

12. The diagram below shows four figures each consisting of eight unit squares, and a V-shaped piece consisting of three unit squares.

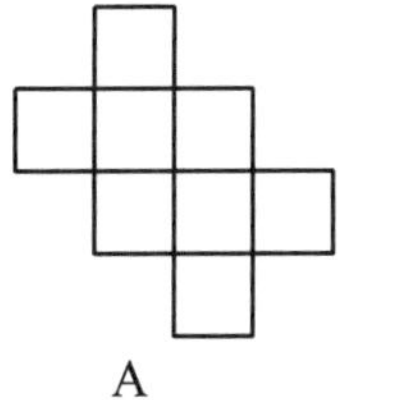
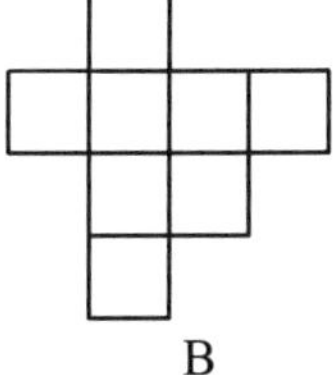
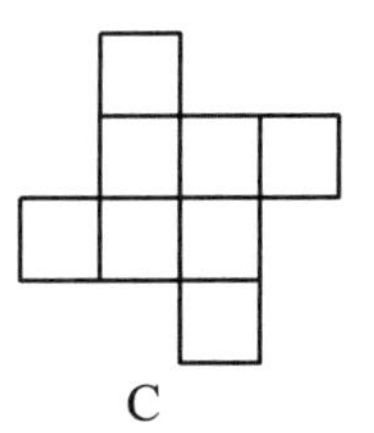
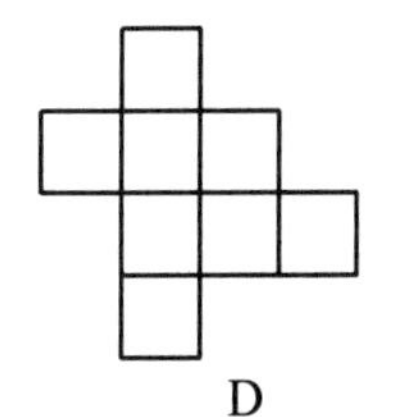
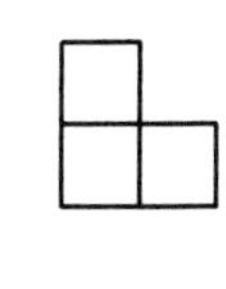

A B C D

We use three copies of the V-shaped piece to try to cover each of the figures. The pieces may be turned or flipped, and they are allowed to overlap. The figure which cannot be covered is

(a) A only (b) B only (c) C only (d) D only

(e) All of them can be covered.

13. The number of real numbers x which satisfy $|x-1|+|x-2|+|x-3|=|x-4|$ is

(a) 1 (b) 2 (c) 3 (d) 4 (e) infinitely many

14. If $x+\frac{1}{x}=3$, the value of $x^3+\frac{1}{x^3}$ is

(a) 7 (b) 9 (c) 18 (d) 21 (e) 27

15. D is a point on side BC and F is a point on side AB of triangle ABC. AD and CF intersect at the point G. It is given that $\angle ABC = 30° = \angle CAD$ and $\angle ACF = 15° = \angle BCF$. Of the following equalities, the one which is incorrect is

(a) $AB = AC$ (b) $AD = CD$ (c) $AF = AG$ (d) $AF = BF$

(e) All four are correct.

16. In a geometric progression, the p-th term is equal to q and the q-th term is equal to p. The value of the $(2p-q)$-th term is

(a) $\frac{p^2}{q}$ (b) $\frac{q^2}{p}$ (c) p^2q (d) q^2p (e) $2q-p$

November 17, 1998

1. A restaurant usually sells its bottles of wine at a 100% profit. Recently it managed to buy some bottles of its most popular wine for half of what it usually pays for them, but still charged its customers what it would normally charge. For these bottles of wine, the profit percentage is

(a) 50 (b) 200 (c) 300 (d) 400

(e) not uniquely determined

2. The number of integers n which satisfy the inequality $34n \geq n^2+289$ is

(a) 0 (b) 1 (c) 2 (d) 3 (e) more than 3

3. A university evaluates five magazines. Last year, the rankings were MacLuck with a rating of 150, followed by MacLock with 120, MacLick with 100, MacLeck with 90,

and MacLack with 80. This year, the ratings of these magazines are down 50%, 40%, 20%, 10% and 5% respectively. The ranking change for MacLeck is

(a) up three places (b) up two places (c) up one place (d) unchanged

(e) down one place

4. Parallel lines are drawn on a rectangular piece of paper. The paper is then cut along each of the lines, forming n identical rectangular strips. If the strips have the same length to width ratio as the original, this ratio is

(a) $\sqrt{n} : 1$ (b) $n : 1$ (c) $n : \dfrac{\sqrt{5}+1}{2}$ (d) $n : 2$ (e) $n^2 : 1$

5. "The smallest integer which is at least $a\%$ of 20 is 10." The number of integers a for which this statement is true is

(a) 1 (b) 2 (c) 3 (d) 4 (e) 5

6. Let $S = 1 + 2 + 3 + \cdots + 10^n$. The number of 2s in the prime factorization of S is

(a) 0 (b) 1 (c) $n - 1$ (d) n (e) $n + 1$

7. When $1 + x + x^2 + x^3 + x^4 + x^5$ is factored as far as possible into polynomials with integral coefficients (not counting trivial factors consisting of the constant polynomial 1), the number of such factors is

(a) 1 (b) 2 (c) 3 (d) 4 (e) 5

8. In triangle ABC, $AB = AC$. The perpendicular bisector of AB passes through the midpoint of BC. If the length of AC is $10\sqrt{2}$ centimeters, the area in square centimeters of ABC is

(a) 25 (b) $25\sqrt{2}$ (c) 50 (d) $50\sqrt{2}$ (e) none of these.

9. If $f(x) = x^x$, $f(f(x))$ is equal to

(a) $x^{(x^{(x^x)})}$ (b) $x^{(x^x)}$ (c) $x^{(x^2)}$ (d) $x^{(x^{(x+1)})}$ (e) $x^{((x+1)^x)}$

10. A certain TV station has a logo which is a rotating cube in which one face has an A on it and the other five faces are blank. Originally the A-face is at the front of the cube as shown in (a) below. Then you perform the following sequence of three moves over and over: rotate the cube 90° around the vertical axis, so that the front face moves to the left; then rotate the cube 90° around a horizontal axis, so that the front face moves down; then rotate the cube 90° around the vertical axis again, so that the front face moves to the left. After you have performed this sequence of three moves a total of 1998 times, the front face look like

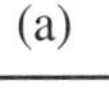
(a) (b) (c) (d) (e)

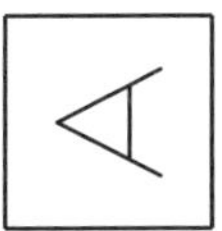

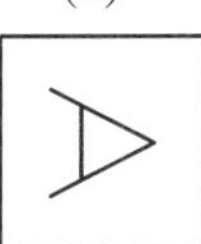

11. The number of triples (x, y, z) of real numbers satisfying the simultaneous equations $x + y = 2$ and $xy - z^2 = 1$ is

(a) 1 (b) 2 (c) 3 (d) 4 (e) infinitely many

12. A line ℓ through A is parallel to BC, where ABC is an equilateral triangle of side length 3. A line cuts ℓ, the side AB, the side AC and the extension of BC at P, Q, R and S respectively. If $PQ = QR = RS$, the length of BR is

(a) 2 (b) $\sqrt{6}$ (c) $\frac{3\sqrt{3}}{2}$ (d) $\sqrt{7}$ (e) none of these

13. Let a, b, c and d be the roots of the equation $x^4 - 8x^3 - 21x^2 + 148x - 160 = 0$. The value of $\frac{1}{abc} + \frac{1}{abd} + \frac{1}{acd} + \frac{1}{bcd}$ is

(a) $-\frac{4}{37}$ (b) $-\frac{1}{20}$ (c) $\frac{1}{20}$ (d) $\frac{4}{37}$ (e) none of these

14. Wei writes down, in order of size, all positive integers b with the property that b and 2^b end in the same digit when they are written in base 10. The 1998th number in Wei's list is

(a) 19974 (b) 19976 (c) 19994 (d) 19996 (e) none of these

15. Suppose $x = 3^{1998}$. The number of integers are between $\sqrt{x^2 + 2x + 4}$ and $\sqrt{4x^2 + 2x + 1}$ is

(a) $3^{1998} - 2$ (b) $3^{1998} - 1$ (c) 3^{1998} (d) $3^{1998} + 1$ (e) $3^{1998} + 2$

16. The lengths of all three sides of a right triangle are positive integers. The area of the triangle is 480. The length of the hypotenuse is

(a) 13 (b) 17 (c) 34 (d) 52 (e) 68

November 16, 1999

1. Subtracting 99% of 19 from 19% of 99, the difference d satisfies

(a) $d < -1$ (b) $d = -1$ (c) $-1 < d < 1$ (d) $d = 1$ (e) $d > 1$

2. Suppose you multiply three different positive prime numbers together and get a product which is greater than 1999. The smallest possible size of the largest of your primes is

(a) 11 (b) 13 (c) 17 (d) 19 (e) undefined

3. Suppose you multiply three different positive prime numbers together and get a product which is greater than 1999. The largest possible size of the smallest of your primes is

(a) 3 (b) 5 (c) 7 (d) 11 (e) undefined

4. The number of two-digit positive integers such that the difference between the integer and the product of its digits is 12 is

(a) 1 (b) 2 (c) 3 (d) 4 (e) none of these

5. The non-zero slope of a certain straight line is equal to its y-intercept if and only if the x-intercept a satisfies

(a) $a = 1$ (b) $a = -1$ (c) $a > 0$ (d) $a < 0$ (e) none of these

6. A and B are positive integers. The sum of the digits of A is 19. The sum of the digits of B is 99. The smallest possible sum of the digits of the number $A + B$ is

(a) 1 (b) 19 (c) 20 (d) 118 (e) none of these

7. O is the origin of the coordinate plane. A, B and C are points on the x-axis such that $OA = AB = BC = 1$. D, E and F are points on the y-axis such that $OD = DE = EF \geq 1$. If $CD \cdot AF = BE^2$, then OD is

(a) 1 (b) $\sqrt{7}$ (c) 7 (d) 49 (e) none of these

8. The integer closest to $100(12 - \sqrt{143})$ is

(a) 2 (b) 3 (c) 4 (d) 5 (d) 6

9. A bag contains four balls numbered -2, -1, 1 and 2. Two balls are drawn at random from the bag, and the numbers on them are multiplied together. The probability that this product is either odd or negative (or both) is

(a) $\frac{1}{6}$ (b) $\frac{1}{2}$ (c) $\frac{9}{16}$ (d) $\frac{2}{3}$ (e) $\frac{3}{4}$

10. The number of positive perfect cubes which divide 9^9 is

(a) 6 (b) 9 (c) 18 (d) 27 (e) none of these.

11. In the quadratic equation $x^2 - 14x + k = 0$, k is a positive integer. The roots of the equation are two different prime numbers p and q. The value of $\frac{p}{q} + \frac{q}{p}$ is

(a) 2 (b) $\frac{106}{45}$ (c) $\frac{130}{33}$ (d) $\frac{170}{13}$ (e) none of these

12. In the quadrilateral $ABCD$, AB is parallel to CD, $AB = 4$ and $BC = CD = 9$. X is on BC and Y is on DA such that XY is parallel to AB. If the quadrilaterals $ABXY$ and $YXCD$ are similar, distance BX is

(a) 3 (b) 3.6 (c) 5.4 (d) 6 (e) none of these

13. The country of Magyaria has three kinds of coins, each worth a different integral number of dollars. Matthew collected four Magyarian coins with a total worth of 28 dollars, while Daniel collected five with a total worth of 21 dollars. Each had at least one Magyarian coin of each kind. In dollars, the total worth of the three kinds of Magyarian coins is

(a) 16 (b) 17 (c) 18 (d) 19 (e) none of these

14. Colin wants a function f which satisfies $f(f(x)) = f(x+2) - 3$ for all integers x. If he chooses $f(1)$ to be 4 and $f(4)$ to be 3, then he must choose $f(5)$ to be

(a) 3 (b) 6 (c) 9 (d) 12 (e) 15

15. Lindsay summed all the integers from a to b, including a and b. She chose these numbers so that $1 \leq a \leq 10$ and $11 \leq b \leq 20$. This sum cannot be equal to

(a) 91 (b) 92 (c) 95 (d) 98 (e) 99

16. A set of points in the plane is such that each of the numbers 1, 2, 4, 8, 16 and 32 is a distance between two of the points in the set. The minimum number of points in this set is

(a) 4 (b) 5 (c) 6 (d) 7 (e) more than 7

November 21, 2000

1. Amy has 58 coins totaling one dollar. They are all pennies, nickels and dimes. The number of nickels she has is

(a) 2 (b) 3 (c) 4 (d) 5 (e) 6

2. Nima runs for 5 kilometers at 10 kilometers per hour, followed by 10 kilometers at 5 kilometers per hour. Her average speed in kilometers per hour for the whole trip is

(a) 6 (b) 6.5 (c) 7 (d) 7.5 (e) 8

3. The sum of the digits of the number $2^{2000}5^{2004}$ is

(a) 8 (b) 9 (c) 11 (d) 13 (e) 14

4. A bug crawls 1 centimeter north, 2 centimeters west, 3 centimeters south, 4 centimeters east, 5 centimeters north and so on, at 1 centimeter per second. Each segment is 1 centimeter longer than the preceding one, and at the end of a segment, the bug makes a left turn. The direction in which the bug is facing 1 minute after the start is

(a) north (b) west (c) south (d) east (e) changing

5. The average of fifteen different positive integers is 13. The maximum value for the second largest of these integers is

   (a) 28 (b) 46 (c) 52 (d) 90 (e) none of these

6. A polynomial $p(x)$ has four terms, so that it is of the form $p(x) = ax^k + bx^\ell + cx^m + dx^n$, where $k,\ \ell,\ m$ and n are four different non-negative integers and $a,\ b,\ c$ and d are non-zero real numbers. The largest number of terms the polynomial $(p(x))^2$ can have is

   (a) 6 (b) 8 (c) 10 (d) 12 (e) 16

7. If $x^2 - x - 1 = 0$, then the value of $x^3 - 2x + 1$ is

   (a) $\frac{1-\sqrt{5}}{2}$ (b) 0 (c) $\frac{1+\sqrt{5}}{2}$ (d) 2

   (e) not uniquely determined

8. A thin rod 6 meters long has one end on the level ground and is leaning against a sphere on the ground, 4 meters in diameter. The rod touches the sphere at a point 1 meter from the top end of the rod. The distance, in meters, between the bottom of the sphere and the bottom of the rod is

   (a) 3 (b) $2\sqrt{5}$ (c) 5 (d) 6 (e) none of these

9. David went to a Casino with $64 and played a certain game six times. Each time, he bet half the amount of money he had at the time. The chance of winning this game was $\frac{1}{2}$, and David won three of the six games, though it was not recorded which three they were. In each of those three games, he won an amount of money equal to his bet. After these six games, the amount of money David had was

   (a) $27 (b) $32 (c) $64 (d) $128

   (e) dependent on which games David lost

10. The number of integers n which satisfy $(n^2 - n - 1)^{n+2} = 1$ is

    (a) 1 (b) 2 (c) 3 (d) 4 (e) 5

11. In triangle ABC, $\angle BCA = 90°$. Points E and F lie on the hypotenuse AB such that $AE = AC$ and $BF = BC$. If $\angle ECF = x$, then

    (a) $x = 30°$ (b) $30° < x < 45°$ (c) $x = 45°$

    (d) $45° < x < 60°$ (e) $x = 60°$

12. The integer part of a number is the largest integer not exceeding the number, and the fractional part of a number is the difference between it and its integer part. For instance, the integer part of $\frac{5}{3}$ is 1 and the fractional part is $\frac{2}{3}$. Suppose that the product

of the integer parts of two positive rational numbers is 5, and the product of their fractional parts is $\frac{1}{4}$. Their product is

(a) $\frac{5}{4}$ (b) $\frac{21}{4}$ (c) $\frac{13}{2}$ (d) $\frac{33}{4}$

(e) not uniquely detcrmined

13. The sum of the positive divisors of 120 is 360. The sum of the reciprocals of the positive divisors of 120 is

(a) 2 (b) 3 (c) 4 (d) 6

(e) not uniquely determined

14. The expression $\dfrac{a-b}{a+b}+\dfrac{b-c}{b+c}+\dfrac{c-a}{c+a}+\dfrac{(a-b)(b-c)(c-a)}{(a+b)(b+c)(c+a)}$ simplifies to

(a) 1 (b) $a+b+c$ (c) $\dfrac{1}{a+b+c}$

(d) $\dfrac{2(a-b)(b-c)(c-a)}{(a+b)(b+c)(c+a)}$ (e) none of these

15. A certain positive integer has m digits when written in base 3 and $m+1$ digits when written in base 2. The largest possible value of m is

(a) 1 (b) 2 (c) 3 (d) 4 (e) 5

16. $ABCD$ is a square. A line through B intersects the extension of CD at E, the side AD at F and the diagonal AC at G. If $BG = 3$ and $GF = 1$, then the length of FE is

(a) 4 (b) 6 (c) 8 (d) 10 (e) 12

November 20, 2001

1. The sum of 2001 and the square of a prime number p is another prime number. The number of possible values of p is

(a) 0 (b) 1 (c) 2001 (d) infinite (e) none of these

2. A recent newspaper article reported that among the Canadian tourists who had been arrested abroad, 96.5% of them were later found guilty, while 3% of the male tourists and 0.5% of the female tourists were later found not guilty. The percentage of Canadians arrested abroad who were male is

(a) 50% (b) 75% (c) 87.5%

(d) not uniquely determined (e) The situation is impossible.

3. The number of integers n for which $\frac{2n+1}{n+1}$ is an integer is

(a) 0 (b) 1 (c) 2 (d) 3 (e) infinite

4. The number of pairs (x, y) of integers satisfying $(|x| - 1)^2 + (|y| - 1)^2 < 2$ is

(a) 16 (b) 17 (c) 18 (d) 25 (e) none of these

5. When multiplied out, the last digit of the product $3^{101} \times 7^{102} \times 13^{103}$ is

(a) 1 (b) 3 (c) 7 (d) 9 (e) none of these

6. Between Dima and Sunera, they bought at least one dog, at least one cat, at least one hamster but no other pets. Neither Dima nor Sunera bought two pets of the same kind. A possible combination of their purchases is for Dima to buy a dog and a cat and Sunera to buy a dog and a hamster. The number of possible combinations of their purchases is

(a) 15 (b) 16 (c) 24 (d) 25 (e) 27

7. The number of triples (x, y, z) of positive integers such that $xy + yz = 29$ and $xz + yz = 81$ is

(a) 0 (b) 1 (c) 2 (d) 3 (e) infinite

8. A, B, C, D and E are five points on a line in that order while F, G, H, I, J and K are six points on another line in that order. Triangles AFG, BGH, CHI, DIJ and EJK are all equilateral. If $AF = 1$ and $EJ = 25$, then the value of CH is

(a) 5 (b) 9 (c) 13 (d) 17 (e) none of these

9. Two thousand and one claims are made about a real number x. If n is odd, the n-th claim is that $x > n$. If n is even, the n-th claim is that $x < n$. In other words, the claims are $x > 1$, $x < 2$, $x > 3, \ldots, x > 2001$. The largest number of these claims that can be true is

(a) 1 (b) 1000 (c) 1001 (d) 2000 (e) none of these

10. Monika and Andy are paying \$750 a month each for their one-bedroom apartments. Starting in 2002, Monika's rent will go up \$50 every January and July, while Andy's will go up \$10 every month. Counting from January 2002, the total accumulated rent paid by Andy will exceed that paid by Monika in

(a) June 2002 (b) December 2003 (c) January 2004

(d) February 2004 (e) This will never happen.

11. If r and s are the solutions of the equation $x^2 - 3x + 1 = 0$, then the value of $\sqrt{r} + \sqrt{s}$ is

(a) $\sqrt{3}$ (b) $\sqrt{5}$ (c) 3 (d) 5 (e) none of these

12. The largest possible area of triangle ABC, with $AB = 5$ and $AC + BC = 7$, is

(a) $\frac{5\sqrt{6}}{4}$ (b) $\frac{5\sqrt{6}}{2}$ (c) $5\sqrt{6}$ (d) $\frac{35}{2}$ (e) none of these

13. Xenia, Yvonne and Zeke were given a positive integer each and the information that the sum of their numbers is 16. Xenia could deduce that all three numbers were different while Yvonne could only deduce that Zeke's number was different from Xenia's. When they told Zeke about their deductions, Zeke could determine all three numbers. The product of the three numbers must be

(a) 44 (b) 52 (c) 54 (d) 66 (e) 90

14. Let x and y be real numbers such that $x + y = x^2 + y^2$. The largest possible value of x is

(a) 1 (b) $\frac{\sqrt{2}+1}{2}$ (c) $\sqrt{5} - 1$ (d) $\frac{\sqrt{5}+1}{2}$ (e) unlimited

15. Suppose for all positive integers n, we have $f(4 + n^2) = an + 2$ and $f(9 - n^2) = 3n - b$ for some numbers a and b. Then the value of $f(13)$ is

(a) -7 (b) -3 (c) 7

(d) not uniquely determined (e) The situation is impossible.

16. Three lines in space are such that no two are parallel and no two intersect. The number of lines which intersect all three is

(a) 0 (b) 1 (c) 2 (d) 3 (e) infinite

November 19, 2002

1. The expression $(\cdots((((1 - 2) + 3) - 4) + 5)\cdots) - 2002$ is equal to

(a) -2001 (b) -1001 (c) -1 (d) 0 (e) none of these

2. There are 10 locks labeled 1 to 10, and 10 keys also labeled 1 to 10. Each lock can be opened by 9 of the keys and each key can open 9 of the locks, but it is not known which key will not open which lock. In each move, you select a lock and a key. The minimum number of moves you must make in order to guarantee that the selected key opens the selected lock is

(a) 2 (b) 3 (c) 9 (d) 10 (e) none of these

3. The number of positive integers less than 500000 with initial digit 5 is

(a) 1000 (b) 5000 (c) 10000 (d) 11000 (e) none of these

4. Let a and b be positive integers such that the units digit of $2a + b$ is 2 and the units digits of $a + 2b$ is 5. The units digit of a is

(a) 1 (b) 3 (c) 7 (d) 9

(e) not uniquely determined

5. The base of a pyramid with positive height is a regular n-sided polygon and all edges of the pyramid have the same length. The largest possible value of n is

(a) 3 (b) 4 (c) 6 (d) 8 (e) none of these

6. Initially, 2002 is written on a blackboard. At any time, we may write a new integer on the blackboard, which is either 1 greater than some existing integer, or half of some existing even integer. The minimum numbers of integers on the blackboard when the integer 1 appears is

(a) 15 (b) 16 (c) 18 (d) 2002 (e) none of these

7. The numbers 1, 2, 3, 4 and 5 are to be placed in the remaining spaces in the H-shaped figure below, one number per space, such that the sum of the three numbers along each of the three lines of the letter H is the same. The number of ways the spaces can be filled is

(a) 0 (b) 1 (c) 2 (d) 3 (e) 4

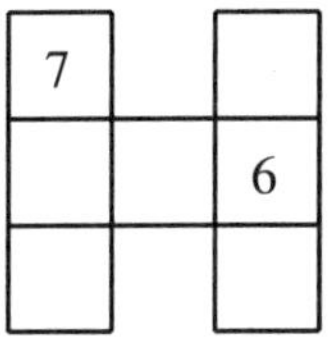

7		
		6

8. The value of $\sqrt[6]{24 - 16\sqrt{2}} \cdot \sqrt[3]{4 + 2\sqrt{2}}$ is

(a) $\sqrt[6]{2}$ (b) $\sqrt[3]{2}$ (c) $\sqrt{2}$ (d) 2 (e) none of these

9. The straight lines $ax + \frac{1}{2}y = 1$ and $(a+1)x + \frac{1}{2}y = 1$ are perpendicular to each other. The value of the constant a is

(a) $-\frac{1}{2}$ (b) $\frac{\sqrt{2}-1}{2}$ (c) $\frac{1}{2}$ (d) $\frac{\sqrt{2}+1}{2}$

(e) not uniquely determined

10. A 22×50 rectangle is divided into unit squares. One diagonal of the rectangle is drawn. The number of unit squares containing a positive length of this diagonal is

(a) 70 (b) 71 (c) 72 (d) 73 (e) none of these

11. In a chess tournament, each participant was to play every other exactly once. However, three participants withdrew after each had played 2 games. As a result, only 40 games were played in the whole tournament. The total number of games played between two of those three participants was

(a) 0 (b) 1 (c) 2 (d) 3

(e) not uniquely determined

12. A function f is such that for any real number x, $f(x)+2f(1-x) = 3x^2$. The value of $f(1)$ is

(a) -1 (b) 0 (c) 1 (d) 2 (e) 3

13. Two circles of unequal sizes are tangent to each other externally. A quadrilateral has two parallel sides, each a diameter of one of the circles. The maximum area of such a quadrilateral is 144 square centimeters. The distance, in centimeters, between the centers of the two circles is

(a) 8 (b) 10 (c) 12 (d) 14

(e) not uniquely determined

14. There are 4 red and 3 white marbles in a bag. A first marble is drawn at random and is replaced with the addition of k more marbles of the same colour. A second marble is then drawn at random. The probability of this marble being red is

(a) $\frac{16}{49}$ (b) $\frac{1}{2}$ (c) $\frac{4}{7}$ (d) $\frac{3}{4}$

(e) dependent on the value of k

15. The minimum value of $|x+1|+|x|-|x-1|$ for any real number x is

(a) -2 (b) -1 (c) 0 (d) 1 (e) 2

16. Consider the positive integers $121 + 19n$ where $n = 0, 1, 2, \ldots, 3000$. The number of these numbers which are the squares of integers is

(a) 12 (b) 13 (c) 19 (d) 25 (e) 2002

November 18, 2003

1. At the beginning of the school year, only 60% of a class liked mathematics. At the end of the school year, 70% of this class liked mathematics. Of the students who did not like mathematics before, the percentage who have changed their opinion was

(a) 10% (b) 12% (c) 15% (d) 20% (e) 25%

2. Daniel walks at a constant speed and runs at a higher constant speed. He goes to school everyday by walking 10 minutes and then running 2 minutes, getting there just in time. One day, he has been walking for 2 minutes when he realizes that his watch is 5 minutes slow. He starts running and gets to school just in time. The ratio of Daniel's running speed to his walking speed is

(a) 8:5 (b) 2:1 (c) 8:3 (d) 10:3 (e) none of these

3. Lily and Lala both start saying "Yakkity-yak" at the same time at different uniform rates, and stop at the same time. Lily has said it 80 times while Lala has said it 72 times. The number of moments when at least one girl is beginning to say "Yakkity-yak" is

(a) 72 (b) 80 (c) 108 (d) 144 (e) 152

4. The straight line through the points $(\frac{1}{a}, b)$ and $(\frac{1}{b}, a)$ has slope

(a) $\frac{1}{ab}$ (b) $-\frac{1}{ab}$ (c) ab (d) $-ab$ (e) none of these

5. Let a, b and c be integers. Among $\frac{b+c}{2}$, $\frac{c+a}{2}$ and $\frac{a+b}{2}$, the possible number of integers is

(a) 3 only (b) 1 or 3 (c) 2 or 3 (d) 1, 2 or 3 (e) none of these

6. The value of $\sqrt{2+\sqrt{3}} + \sqrt{2-\sqrt{3}}$ is

(a) $\sqrt{6}$ (b) $2\sqrt{2}$ (c) $2\sqrt{3}$ (d) 6 (e) none of these

7. The real numbers a, b, c and d satisfy $abc = 1$, $bcd = 2$, $cda = 4$ and $dab = 8$. The value of $a + b + c + d$ is

(a) $\frac{15}{8}$ (b) $\frac{15}{4}$ (c) $\frac{15}{2}$ (d) 15 (e) none of these

8. A 4×6 rectangle is constructed using six pieces, five of which are shown below. The sixth piece is a duplicate of one of them. The piece which appears twice is

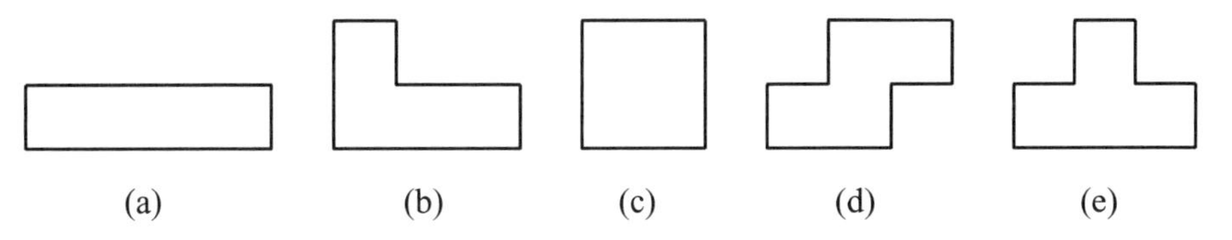

9. A function f is defined by $f(x) = x^a$ where a is some positive real number. If $f(f(f(2))) = 222$, then a satisfies

(a) $0 < a \leq 1$ (b) $a = 1$ (c) $1 < a \leq 2$ (d) $a = 2$ (e) $2 < a$

10. When $x^{100} - 2x^{52} + 2$ is divided by $x^2 + 1$, the remainder is

(a) 0 (b) 1 (c) 2 (d) x (e) none of these

11. In the Stanley Cup Championship, seven games are scheduled. Games 1, 2, 5 and 7 are to be played in the arena of one team, while Games 3, 4 and 6 are to be played in the arena of the other team. However, the Championship is awarded as soon as either team wins four games. If exactly two games are won by the home team, then the total number of games cannot be

(a) 4 (b) 5 (c) 6 (d) 7 (e) all possible

12. Three balls resting on a table are tangent to one another. The pairwise distances between the points of contact of the table with the balls are 2, 3 and 4. The product of the radii of the three balls is

(a) 1 (b) 2 (c) 3 (d) 4 (e) none of these

13. A function f is such that for any real number x, $2f(x) - f(1-x) = 6x$. The value of $f(1) + f(2)$ is

(a) 4 (b) 6 (c) 8 (d) 10 (e) 12

14. Jerry tried to use a calculator to compute $\frac{a+b}{c}$, where a, b and c are positive integers. He pressed $a,\ +,\ b,\ /,\ c$ and $=$ in that order, and got $a + \frac{b}{c} = 11$. Then he pressed $b,\ +,\ a,\ /,\ c$ and $=$ in that order, and got $b + \frac{a}{c} = 14$. Actually, he should have pressed $a,\ +,\ b\ =,\ /,\ c$ and $=$ in that order for the correct answer

(a) 1 (b) 4 (c) 5 (d) 20 (e) 24

15. In triangle ABC, the distance from B to the midpoint of CA is 18. The distance from C to the midpoint of AB is 24. The maximum area of ABC is

(a) 96 (b) 144 (c) 192 (d) 288 (e) none of these

16. The sum of the maximum and the minimum values of $\sin^6 x + \cos^6 x$ is

(a) $\frac{1}{4}$ (b) $\frac{3}{4}$ (c) 1 (d) $\frac{5}{4}$ (e) none of these

November 16, 2004

1. Mount Rundle may be approximated as a triangular prism with its three parallel edges horizontal. Each of them is 10 kilometers in length. Each cross-section perpendicular to these edges is a triangle with base 5 kilometers and height 1800 meters. The volume of Mount Rundle, in cubic meters, is

(a) 9×10^9 (b) 30×10^9 (c) 45×10^9 (d) 9×10^{10} (e) 45×10^{10}

2. In this contest, there are 16 questions. A contestant gets 20 points as the base score, plus 5 points for each correctly answered question and 2 points for each unanswered

question. An incorrectly answered question is worth 0 points. Of the scores from 20 to 100, the number of those which cannot be attained is

(a) 2 (b) 3 (c) 4 (d) 5 (e) 6

3. The remainder when $x^{2004} - 2x + 2$ is divided by $x^2 - 1$ is

(a) $-x + 2$ (b) $-2x + 3$ (c) $2x - 3$ (d) $-3x + 2$ (e) $3x - 2$

4. Consider all four-digit numbers formed from one each of 0, 1, 2 and 3. The number of those not beginning in 0 and not ending in 1 is

(a) 8 (b) 10 (c) 12 (d) 14 (e) 16

5. The positive integer n is such that exactly n of the following five statements are true:

$$\sqrt{n} < 1, \ \sqrt{n} < 2, \ \sqrt{n} < 3, \ \sqrt{n} < 4, \ \sqrt{n} < 5.$$

The number n is

(a) 1 (b) 2 (c) 3 (d) 4 (e) non-existent

6. A prophet makes the following prediction in 2004: "Everyone who is now alive and who will live to at least age n^2 will be alive during at least part of some year which is a perfect square." The smallest positive integral value of n which makes the prediction true is

(a) 5 (b) 8 (c) 9 (d) 10 (e) none of these

7. The number of pairs (x, y) of non-zero real numbers such that three of $x + y$, $x - y$, xy and $\frac{x}{y}$ have the same value is

(a) 0 (b) 1 (c) 2 (d) 4 (e) none of these

8. A sequence of eight numbers is such that starting from the second, each is obtained from the preceding one by adding a fixed positive number. The sum of the second and the fourth numbers is -4 and the product of the first and the fifth numbers is -12. The sum of all eight numbers is

(a) -16 (b) -8 (c) 0 (d) 8 (e) 16

9. A regular polygon has 100 sides each of length 1. Another regular polygon has 200 sides each of length 2. When the area of the larger polygon is divided by the area of the smaller polygon, the quotient is closest to the integer

(a) 2 (b) 4 (c) 8 (d) 12 (e) 16

10. The sum of the digits of the smallest multiple of 999 which does not contain any digit 9 is

(a) 18 (b) 27 (c) 36 (d) 45 (e) 54

11. The absolute value of a real number x, denoted by $|x|$, is defined by $|x| = x$ if $x \geq 0$ and $|x| = -x$ if $x < 0$. The set of all solutions to the equation $x^2 - |x| - 1 = 0$ is

(a) $\{\frac{1\pm\sqrt{5}}{2}\}$ (b) $\{\frac{-1\pm\sqrt{5}}{2}\}$ (c) $\{\pm\frac{1+\sqrt{5}}{2}\}$ (d) $\{\pm\frac{1-\sqrt{5}}{2}\}$ (e) $\{\frac{\pm1\pm\sqrt{5}}{2}\}$

12. P and Q are points on BC, R is a point on CA and S is a point on AB such that $PQRS$ is a square inscribed in triangle ABC. The areas of triangles ARS, BSP and CQR are 1, 3 and 1 respectively. The area of $PQRS$ is

(a) 2 (b) 3 (c) 4 (d) 6 (e) none of these

13. If $x^2 - 13x + 1 = 0$, then the units-digit of $x^4 + x^{-4}$ is

(a) 1 (b) 3 (c) 7 (d) 9 (e) none of these

14. The largest integer n such that 6125^n divides $1 \times 2 \times 3 \times \cdots \times 2003 \times 2004$ is

(a) 165 (b) 166 (c) 331 (d) 499 (e) none of these

15. In the quadrilateral $ABCD$, $AB = \sqrt{6}$, $\angle ABC = 135°$, $BC = 5-\sqrt{3}$, $\angle BCD = 120°$ and $CD = 6$. The length of AD is

(a) 8 (b) $\sqrt{73}$ (c) $2\sqrt{19}$ (d) $\sqrt{91}$ (e) none of these

16. If x and y are positive real numbers such that $x + y = 1$, the maximum value of $x^4y + xy^4$ is

(a) $\frac{1}{16}$ (b) $\frac{1}{12}$ (c) $\frac{1}{8}$ (d) $\frac{1}{4}$ (e) none of these

November 15, 2005

1. The value of $2005 \times 20042004 - 2004 \times 20052005$ is

(a) 0 (b) 10000 (c) 2003000 (d) 2005000 (e) none of these

2. An ice cream store has 20 kinds of ice cream. A customer may get one scoop or two scoops of ice cream, and if she gets two scoops they can be the same or different, but the order of the scoops does not matter. The number of different cones possible is

(a) 210 (b) 220 (c) 230 (d) 420 (e) none of these

3. In triangle ABC, let D be the midpoint of BC and let E be on AD such that $ED = 2AE$. If the area of triangle ABC is 150, then the area of triangle ABE is

(a) 25 (b) 32.5 (c) 50 (d) 75 (e) none of these

4. Let a, b and c be real numbers and $x = 11c - a - b$. If $b - a - 3c \leq -2$ and $b - 2a + c \geq 3$, then

(a) $x \in [0, 1]$ (b) $x \in [2, 5]$ (c) $x \in [6, 9]$

(d) $x \in [10, 11]$ (e) $x \in [12, \infty)$

5. Penelope has 3 red socks, 3 yellow socks and 3 blue socks. If she picks them without looking, the smallest number she must take in order to guarantee that she has four socks that form two pairs of socks of matching colours is

(a) 4 (b) 5 (c) 6 (d) 7 (e) 8

6. The base of a gazebo is 3 meters square and the walls are 2 meters high. The roof-braces from the top of the roof to the corners on top of the walls are 3 meters. The height, in meters, of the top of the roof from the ground is

(a) $2 + \frac{\sqrt{2}}{2}$ (b) $\sqrt{\frac{17}{2}}$ (c) $2 + \frac{3\sqrt{2}}{2}$ (d) $\sqrt{22}$ (e) none of these

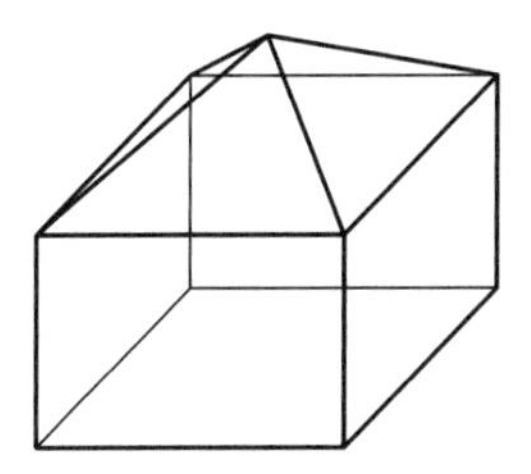

7. If f is a function defined on the positive real axis and $f(x) + 3f(\frac{1}{x}) = x - \frac{5}{x} + 4$ for all $x > 0$, then $f(\frac{1}{2})$ is

(a) $\frac{1}{2}$ (b) 1 (c) $\frac{3}{2}$ (d) 2 (e) 4

8. In the sequence obtained by omitting the squares and the cubes from the sequence of positive integers, 2005 sits on the position

(a) 1949 (b) 1950 (c) 1951 (d) 1952 (e) 1953

9. The positive integer a is such that the inequality $2a + 3x \leq 101$ has exactly six solutions in positive integers x. The number of possible values of a is

(a) 1 (b) 2 (c) 3 (d) 4 (e) 5

10. P is a point inside a parallelogram $ABCD$. If the area of triangle PAD is one-third that of $ABCD$, and the area of triangle PCB is 6 square centimeters, then the area, in square centimeters, of the parallelogram is

(a) 24 (b) 36 (c) 48 (d) 60 (e) 72

11. There are three problems in a contest. Students win bronze, silver or gold medals if they solve 1, 2 or 3 problems respectively. Each problem is solved by exactly 60 students, and there are exactly 100 medalists. The difference between the number of bronze medalists nad the number of gold medalists is

(a) 0 (b) 10 (c) 20 (d) 30

(e) not uniquely determined

12. The figure in the diagram below is placed on an infinite piece of graph paper so that it covers exactly 5 squares. It may be turned over or rotated. We wish to paint the squares of the graph paper in such a way that no matter where the tile is placed, it never covers 2 squares of the same color. The smallest number of colors needed is

(a) 5 (b) 6 (c) 7 (d) 8 (e) 9

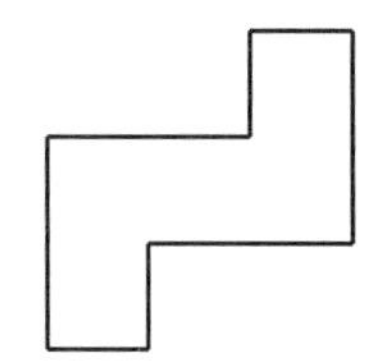

13. The positive numbers $a,\ b$ and c are such that $a+b+c=7$ and $\frac{1}{b+c}+\frac{1}{c+a}+\frac{1}{a+b}=\frac{10}{7}$. The value of $\frac{a}{b+c}+\frac{b}{c+a}+\frac{c}{a+b}$ is

(a) 1 (b) 3 (c) 7 (d) 10

(e) dependent on $a,\ b$ and c

14. The number of ordered pairs of integers (x, y) such that $xy = x^2 + y^2 + x + y$ is

(a) 3 (b) 4 (c) 6 (d) 8 (e) none of these

15. The positive integer that satisfies the equation $\log_2 3 \log_3 \cdots \log_n(n+1) = 2005$ should be a multiple of

(a) 2 (b) 3 (c) 5 (d) 31 (e) none of these

16. Three spheres of unit radius sit on a plane and they are tangent to one another. A large sphere with its center in the plane contains all three unit spheres and is tangent to them. The radius of the large sphere is

(a) $\sqrt{2}+1$ (b) $\sqrt{\frac{5}{3}}+1$ (c) $\sqrt{\frac{7}{3}}+1$ (d) $\sqrt{3}+1$ (e) none of these

Second Round Problems

February 14, 1984

1. How many seven-digit numbers are there in base 10 which are multiples of 11 and whose digit-sums are 61?

2. A cylindrical glass is partly filled with water and is placed on a level table. The glass is then tilted slightly so that the bottom remains in contact with the table at one point. Has the distance from the water level to the table increased, decreased or remained the same? Justify your answer.

3. (a) Prove that for all real numbers x, $3(16^x + 1) \geq 4(1 - x)(16^x + 4x)$.

   (b) Find all values of x for which equality holds.

4. A and B are opposite vertices of a unit cube. C and D are opposite vertices of another unit cube. The two cubes are placed side by side so that A and C are opposite vertices of a 1 by 2 rectangle. Find the shortest distance between the lines AB and CD.

5. (a) A first player changes some *'s in the polynomial $x^3 * a_1x^2 * a_2x * a_3$ into +'s and the rest into −'s, after which a second player changes the a's into positive numbers. Prove that regardless of the first player's choices, the second player can ensure that the completed polynomial will have 3 distinct real roots.

   (b) If instead the starting polynomial is of the form $x^n * a_1x^{n-1} * \cdots * a_{n-1}x * a_n$, can the second player still ensure that it will have n distinct real roots?

February 12, 1985

1. The General has 5 subordinates, each either a soldier or an officer. A soldier has no subordinates. An officer has 5 subordinates, each either a soldier or another officer. Each soldier or officer is subordinate directly to only one officer (or the General). If there are 20 officers other than the General, how many soldiers are there?

2. Let t be a positive number. Define $r_0 = 0$ and $r_n = \sqrt{t + r_{n-1}}$ for $n = 1, 2, 3, \ldots$. Prove that $r_n < \frac{1}{2}(1 + \sqrt{1 + 4t})$ for all n.

3. Let a, b and c be integers. Consider the following three infinite sequences of integers:

$$\{a, a+1, a+4, \ldots, a+n^2, \ldots\},$$
$$\{b, b+1, b+4, \cdots, b+n^2, \ldots\},$$
and
$$\{c, c+1, c+4, \ldots, c+n^2, \ldots\}.$$

   Prove that there is an integer which belongs to at least two of these three sequences.

4. Let AB and CD be two line segments, not necessarily on the same plane. Let E be the midpoint of AB and F be the midpoint of CD. Prove that $2EF \leq AD + BC$.

5. A straight line of telephone poles, equally spaced apart, stands on level ground and stretches forever in both directions. Each telephone pole is a vertical cylinder of radius 1 meter. Prove that at most 100 telephone poles are completely visible to an observer standing 99 meters away from the line joining the centers of the telephone poles.

February 11, 1986

1. The number of boys writing this mathematics contest is 20% higher this year than in 1980. The number of girls writing this contest has increased by 50% over the same period, while the total number of contestants has increased by only 30%. What fraction of the students writing the contest this year are girls?

2. An island has a number of lakes and rivers. Each river flows either from one lake to another or from a lake to the sea. If it is possible to walk between any two dry points on the island without crossing water, prove that there must be at least as many lakes as rivers.

3. The altitude to the hypotenuse of a right triangle has length 1.

   (a) Prove that the radius of the incircle of the triangle is less than $\frac{1}{2}$.

   (b) Find the minimum possible value of the inradius of the triangle.

4. (a) For each number k which can be expressed in the form $4t + 1$, where t is a positive integer, find k positive integers (not necessarily distinct) such that their sum is equal to their product and such that this common value is odd.

   (b) Let $k > 1$ be an integer. Suppose there are k positive integers (not necessarily distinct) such that their sum is equal to their product and such that this common value is odd. Prove that k can be expressed in the form $4t+1$, where t is a positive integer.

5. If two circles not in the same plane intersect in two points, prove that they lie on a common sphere.

February 10, 1987

1. If a and b are two positive integers that have no common factor greater than 1 and such that $a + b$ is even, prove that the product $ab(a - b)(a + b)$ is divisible by 24.

2. A badly designed table is constructed by nailing through its center a circular disc of diameter two meters to a sphere of diameter one meter. The table tips over to bring the edge of the disc into contact with the floor. As the table rolls, the two points of contact with the floor trace out a pair of concentric circles. What are the radii of these circles?

3. Find all pairs of real numbers (x, y) that are solutions to the simultaneous equations $x^{x+y} = y^3$ and $y^{x+y} = x^6y^3$.

4. Adrian thinks of an integer A between 1 and 12 inclusive and Bernice tries to guess it. To each guess B that Bernice makes, Adrian responds according to the following rules:

   - If $B = A$, Adrian replies, "That's my number."
   - If $0 < |B - A| \leq 3$, Adrian replies, "You're close."
   - If $|B - A'| > 3$, Adrian replies, "You're not close."

Prove that with four or fewer guesses, Bernice can always force Adrian to say, "That's my number."

5. $ABCD$ is a rectangular sheet of paper. E is the point on the side CD such that if the paper is folded along BE, C will coincide with a point F on the side AD. G is the point of intersection of the lines AB and EF. Prove that CG is perpendicular to BD.

February 9, 1988

1. A girl wishes to videotape and save 7 hockey games, 11 movies and 20 episodes of Star Trek. She knows that one videotape can hold 2 hockey games plus 1 episode of Star Trek, or 3 episodes of Star Trek plus 2 movies. There may be some room left over in either case. Prove that 9 videotapes will be enough for the girl to record all her programs, but that 8 may not be.
2. Mark has an old ruler which has 11 grooves on it. The distance between every pair of adjacent grooves is 1 centimeter. Originally, the grooves were colored, but the color has worn off. Mark decides to re-color some of the grooves so that for each integer value of n between 1 and 10 inclusive, there will be at least two colored grooves n centimeters apart. What is the minimum number of grooves Mark must re-color?
3. Find all triples (a, b, c) of real numbers such that $a + b + c = ab + bc + ca = abc$.
4. Determine all integer pairs (x, y) such that $3^4 2^3 (x^2 + y^2) = x^3 y^3$.
5. Prove that the tetrahedron of largest volume which can be inscribed in a given sphere is the equilateral one.

February 14, 1989

1. A number of students are in the library, some wearing glasses and some not. If one of the students wearing glasses leaves the library, the fraction of students in the library who wear glasses will become $\frac{1}{5}$. If instead, another student wearing glasses comes into the library, the fraction will become $\frac{1}{4}$. How many students in the library do not wear glasses?
2. Let b be a real number.

   (a) Prove that if $b^3 = b + 1$, then $b^5 = b^4 + 1$.

   (b) Prove or disprove that if $b^5 = b^4 + 1$, then $b^3 = b + 1$.
3. Three circles γ_1, γ_2 and γ_3 all pass through a point O. In addition, γ_1 meets γ_2 at A, γ_2 meets γ_3 at B and $\gamma_3 3$ meets γ_1 at C, where A, B and C are distinct points. P is a point on the arc BC of γ_3 which does not contain O. PC meets γ_1 again at Q, and PB meets γ_2 again at R. Prove that Q, A and R lie on a straight line.
4. The positive integer X has $2n$ digits, all of which are 4's. The positive integer Y has n digits, all of which are 8's. Determine $\sqrt{X - Y}$.
5. P is a point in space. PA_1, $PA_2, \ldots, PA_n$ are distinct rays such that $\angle A_i PA_j$ is constant for all distinct i and j. Prove that $n \leq 4$.

February 13, 1990

1. For which positive integers n is the sum of n consecutive integers

   (a) always odd;

   (b) sometimes odd and sometimes even;

   (c) always even?

2. A square 10 centimeters on a side is given.

   (a) Show how to cut the square into two pieces of perimeters 30 centimeters and 60 centimeters respectively.

   (b) Prove or disprove that the square can be cut into two pieces of perimeters 50 centimeters and 100 centimeters respectively.

3. $ABCD$ is a convex quadrilateral such that $\angle ACD = 40°$, $\angle BAC = 50°$, $\angle ACB = 70°$ and $\angle CAD = 80°$ respectively. Determine $\angle ABD$.

4. A number of Knights gathered at the Round Table for a jolly evening. During the day, each Knight had jousted with exactly two of the others, and no two Knights had jousted with the same two opponents. At the Round Table, each Knight notices that his opponents were seated next to each other.

   (a) Prove that the number of Knights cannot be even.

   (b) Can the number of Knights be any odd number? Justify your answer.

5. A spherical ball of radius 2 meters is touching two walls and the floor of a rectangular room. A smaller spherical ball is touching the same two walls and the floor and also the first ball. Determine the radius of the smaller ball.

February 12, 1991

1. Call a positive integer "special" if it is equal to the sum of its digits plus the product of its digits. For instance, 19 is special because $19 = 1 + 9 + 1 \cdot 9$. Find all special numbers.

2. E is a point on AB and F is a point on AC such that $AE = AF$. D is a point on CB extended and G is a point on BC extended such that $BD = BE$ and $CF = CG$. Prove that D, E, F and G lie on a circle.

3. The arithmetic progression $a_1, a_2, \ldots, a_{16}$ with $a_1 \neq 0$ is such that $a_7 + a_9 = a_{16}$. Find all three-term geometric progressions contained in this arithmetic progression.

4. In a certain board game, each creature has a rating from 1 to 6. In each round of a battle, a six-sided die is rolled for each creature. If the die-roll is less than or equal to its rating, the creature scores a hit and eliminates an opposing creature. Both sides roll simultaneously, and hits are applied at the end of each round. If a side with two or more creatures suffers losses, the creature with the lowest rating is eliminated first. The battle continues until at least one side is completely eliminated. If both sides are

wiped out, the battle is a draw. Otherwise, the surviving side wins. One team has a single creature with rating 6. A second team has a single creature with rating 1. The latter is reinforced by another creature to make the battle as even as possible. What is the rating of this new creature?

5. If $x + y + z = ax + by + cz = a^2x + b^2y + c^2z = 1$, determine the value of $a^3x + b^3y + c^3z$ in terms of $a,\ b$ and c.

February 11, 1992

1. The Committee to Halt Excessive Amount of Photocopying (CHEAP) is itself accused of over-expenditure in photocopying, even though it never makes more than one copy of anything. The new committee set up to investigate this accusation makes, for each of its 13 members, a photocopy of everything CHEAP has photocopied, so that it can study whether the expenditure has been justified. Each committee is charged 7 cents per page for the first 2000 pages and 5 cents per page thereafter. It turns out that the photocopying expenditure of the new committee is 10 times that of CHEAP. How many photocopies did CHEAP make? Find all possible solutions.

2. The base of a tub is a square with sides of length 1 meter. It contains water 3 centimeters deep. A heavy rectangular block is placed in the tub three times. Each time, the face that rests on the bottom of the tub has a different area. When this is done, the water in the tub ends up being 4 centimeter, 5 centimeters and 6 centimeters deep. Find the dimensions of the block.

3. A positive integer is said to have the 32-property if the sum of the five remainders obtained when it is divided by some five consecutive positive integers is 32. For example, 24 has the 32-property since when 24 is divided by 11, 12, 13, 14 and 15, the respective reainders 2, 0, 11, 10 and 9 add up to 32.

   (a) Verify that 26 has the 32-property.

   (b) Determine the smallest positive integer with the 32-property.

   (c) Prove that there are infinitely many positive integers with the 32-property.

4. Suppose $x,\ y$ and z are real numbers which satisfy the equation $ax + by + cz = 0$, where $a,\ b$ and c are given positive numbers.

   (a) Prove that $x^2 + y^2 + z^2 \geq 2xy + 2yz + 2xz.$

   (b) Determine when equality holds in (a).

5. $ABCD$ is a square piece of paper with sides of length 1 meter. A quarter-circle is drawn from B to D with center A. The piece of paper is folded along EF, with E on AB and F on AD, so that A falls on the quarter-circle. Determine the maximum and minimum areas that triangle AEF could have.

February 9, 1993

1. Prove that $1992^{1993} - 1992$ is divisible by $1992^2 + 1993$.

2. Mary tosses a fair coin repeatedly and records a sequence of H's and T's according to whether the coin lands heads or tails. She is looking out for the subsequences HTT, TTH and THT. What is the probability that

   (a) HTT appears before TTH;

   (b) TTH appears before THT?

3. Three spheres of different sizes are tangent to the plane of a triangle at the vertices of the triangle. They are also tangent to one another. Determine the radii of the spheres in terms of the sides a, b and c of the triangle.

4. Todd and Steven play a game by alternating choosing a previously unselected number from $\{1, 2, \ldots, n\}$, where n is a positive integer. Steven chooses first, and the game continues until all numbers have been selected. Steven wins if and only if the sum of the numbers he has chosen is even. For which n can Steven force a win?

5. If $a_1, a_2, \ldots, a_n$ are the lengths of the sides of an n-sided polygon, prove that $\frac{a_1}{1+a_1}$, $\frac{a_2}{1+a_2}, \ldots, \frac{a_n}{1+a_n}$ are possible lengths of the sides of another n-sided polygon.

February 8, 1994

1. Find all polynomials $P(x)$ that satisfy the equation

$$P(x^2) + 2x^2 + 10x = 2xP(x+1) + 3.$$

2. An isosceles triangle is called an **amoeba** if it can be divided into two isosceles triangles by a straight cut. How many different (that is, not similar) amoebas are there?

3. (a) Show that there is a positive integer n so that the interval $((n + \frac{1}{1994})^2, (n + \frac{1}{1993})^2)$ contains an integer N.

   (b) Find the smallest integer N which is contained in such an interval for some n.

4. $ABCDE$ is a convex pentagon in the plane. Through each vertex draw a straight line which cuts the pentagon into two parts of the same area. Prove that for some vertex, the line through it must intersect the **opposite side** of the pentagon, where the opposite side to vertex A is the side CD, the opposite side to B is DE, and so on.

5. Let a, b and c be real numbers. Their pairwise sums $a+b, b+c$ and $c+a$ are written on three round cards, one sum on each card, and their pairwise products ab, bc and ca are written on three square cards. We call (a, b, c) a **tadpole** if we can divide the six cards into three pairs, each pair consisting of one round card and one square card with the same number on both. An example of a tadpole is (0,0,0).

   (a) Find all possible tadpoles of the form (a, a, a).

   (b) Prove that there is a tadpole that is not of the form (a, a, a). You do **not** have to find the actual values of a, b and c.

February 14, 1995

1. Five friends attend the same school. When walking alone, the speed of each is 2.0 meters per second. The speed of any pair is 1.5 m/s, that of any trio is 1.0 m/s and that of any quartet is 0.5 m/s. Two of them leave school together at noon one day, a third at 12:01, a fourth at 12:02 and the fifth at 12:03, all walking in the same direction along a straight road. Any of them that catches up with some others joins the group. How far are they from school when they are all together?

2. Find all positive real numbers a, b and c satisfying $a = bc - 1$, $b = ca - 2$ and $c = ab + 1$.

3. Find all positive integers a, b and c satisfying $\frac{a}{a+1} = \frac{b}{b+2} + \frac{c}{c+3}$.

4. A, B, C, D, E and F are six points in space such that $AB = DE$, $BC = EF$ and $CD = AF$. Moreover, AB is parallel to ED, BC to FE and CD to AF. Prove that the mid-points of these six line segments all lie on the same plane.

5. Let $P(x)$ and $Q(x)$ be two polynomials with integer coefficients. If $\frac{P(n)}{Q(x)}$ is an integer for every integer n, prove that $P(x) = Q(x)R(x)$, where $R(x)$ is a polynomial with rational coefficients.

February 13, 1996

1. Several different positive integers are given in base 10. All the digits of each are 4's. Must the sum of these positive integers contain the digit 4 at least once? Either give a proof or find a counterexample.

2. Find all non-congruent pentagons with the following properties:
   - each of two adjacent angles is equal to 135°;
   - the side between these two angles has length 3;
   - each of the other three angles is less than 180°;
   - the lengths of the other four sides are chosen from 2, 3 and 4.

3. (a) Prove that every integer n can be written in the form

$$n = a_1b_1 + a_2b_2 + a_3b_3,$$

where $a_1 < a_2 < a_3$ and either $b_1 < b_2 < b_3$ or $b_1 > b_2 > b_3$ are arithmetic progressions of integers.

(b) Determine whether every integer n can be written in the form

$$n = a_1b_1 + a_2b_2 + a_3b_3 + a_4b_4,$$

where $a_1 < a_2 < a_3 < a_4$ and either $b_1 < b_2 < b_3 < b_4$ or $b_1 > b_2 > b_3 > b_4$ are arithmetic progressions of integers.

(c) Find the smallest integer $k > 4$ such that every integer n can be written in the form

$$n = a_1b_1 + a_2b_2 + \cdots + a_kb_k,$$

where $a_1 < a_2 < \ldots < a_k$ and either $b_1 < b_2 < \ldots < b_k$ or $b_1 > b_2 > \cdots > b_k$ are arithmetic progressions of integers.

4. Let a, b, c and d be non-negative real numbers not exceeding 1. Prove that

$$4(a^4 + b^4 + c^4 + d^4) - (a^2bc + b^2cd + c^2da + d^2ab) \leq 12.$$

5. $ABCDE$ is a pentagon inscribed in a circle, with $AB = AD$ and $\frac{AB}{AC} = \frac{AE}{AD}$. BD intersects AC at F and CE at G.

(a) Prove that $AF = AE$.

(b) Prove that AG bisects $\angle CAE$.

February 11, 1997

1. Find all real numbers x satisfying $|x - 7| > |x + 2| + |x - 2|$.

Remark: Note that $|a|$ is called the absolute value of the real number a. It has the same numerical value as a but is never negative. For example, $|3.5| = 3.5$ while $|-2| = 2$. Of course, $|0| = 0$.

2. Two lines b and c form a $60°$ angle at the point A, and B_1 is a point on b. From B_1, draw a line perpendicular to the line b meeting the line c at the point C_1. From C_1 draw a line perpendicular to c meeting the line b at B_2. Continue in this way obtaining points C_2, B_3, C_3, and so on. These points are the vertices of right triangles AB_1C_1, AB_2C_2, AB_3C_3, ... Denote by $[P]$ the area of the polygon P. If $[AB_1C_1] = 1$, determine

$$[AB_1C_1] + [AB_2C_2] + [AB_3C_3] + \cdots + [AB_{1997}C_{1997}].$$

3. A and B are two points on the diameter MN of a semicircle. C, D, E and F are points on the semicircle such that $\angle CAM = \angle EAN = \angle DBM = \angle FBN$. Prove that $CE = DF$.

4. (a) Suppose that p is an odd prime number and a and b are positive integers such that p^4 divides $a^2 + b^2$ and p^4 also divides $a(a + b)^2$. Prove that p^4 also divides $a(a + b)$.

(b) Suppose that p is an odd prime number and a and b are positive integers such that p^5 divides $a^2 + b^2$ and p^5 also divides $a(a + b)^2$. Show by an example that p^5 does not necessarily divide $a(a + b)$.

5. The picture shows seven houses represented by the dots, connected by six roads represented by the lines. Each road is exactly 1 kilometer long. You live in the house marked B. For each positive integer n, how many ways are there for you to run n kilometers if you start at B and you never run along only part of a road and turn around

between houses? You have to use the roads, but you may use any road more than once, and you do not have to finish at B. For example, if $n = 4$, then three of the possibilities are: B to C to F to G to F; B to A to B to C to B; and B to C to B to A to B.

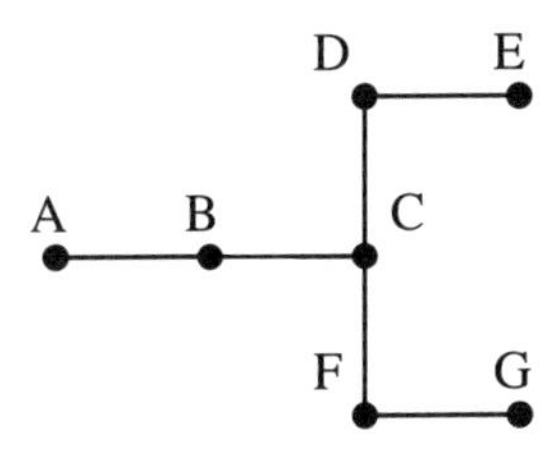

February 10, 1998

1. The sum of the fractions $\frac{1}{2}$ and $\frac{8}{5}$ is $\frac{21}{10}$. If we switch their denominators, the new sum $\frac{1}{5} + \frac{8}{2} = \frac{21}{5}$ is twice the earlier sum. Find two positive fractions such that if we switch their denominators, the sum after the switch is 100 times the sum before.

2. Twenty dots are arranged in a 2 by 10 array. We wish to label them $P_1, P_2, \ldots, P_{20}$ such that no two of the segments $P_1P_2, P_2P_3, \ldots, P_{19}P_{20}, P_{20}P_1$ are parallel to each other. If this is possible, show how to do it. If this is impossible, explain why.

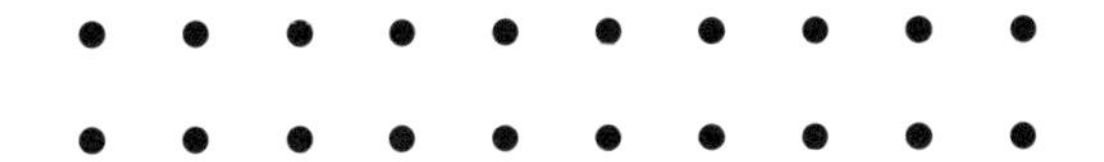

3. D and E are the respective midpoints of the sides BC and CA of triangle ABC which has area 36. If $AD = 6$ and $BE = 9$, prove that AD is perpendicular to BE.

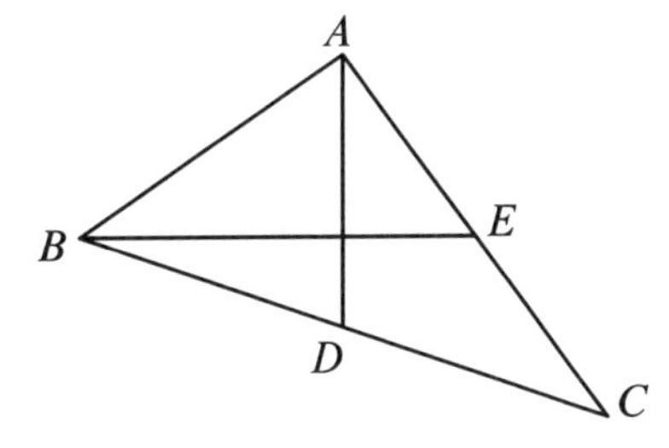

4. At some point during a tournament, each team has won twice and lost twice. It turns out that for any two teams X and Y, either X has beaten Y or has beaten some team Z which has beaten Y.

   (a) Prove that there are less than eight teams in this tournament.

   (b) The following diagram shows that this tournament may have five teams. An arrow going from X to Y indicates that X has beaten Y. Give a similar construction to show that this tournament may have six teams. Teams which have not yet played each other will not be connected by an arrow.

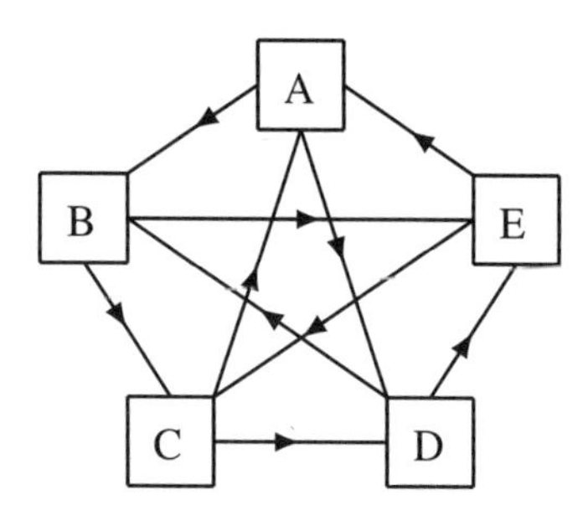

(c) Can this tournament have exactly seven teams?

5. If a, b, c and d are real numbers such that $a + b + c + d > 0$, $ab + ac + ad + bc + bd + cd > 0$, $abc + abd + acd + bcd > 0$ and $abcd > 0$, prove that a, b, c and d are all positive.

February 3, 1999

1. How many integers between 10000 and 99999 have the property that all five digits are distinct and the last digit is the sum of the other four?

2. Let A, B and C be angles, all between $0°$ and $90°$ inclusive, such that $\sin A = \cos B = \tan C$. Determine the maximum and minimum values of $A + B + C$.

3. Let m and k be integers such that $0 < m < k$. A chicken from Minsk lays an egg a day for m consecutive days, takes a day off, and then repeats the cycle. A chicken from Kiev lays an egg a day for k consecutive days, takes a day off, and then repeats the cycle. The chickens do not have to take the same days off. Cathy has 18 chickens, each from either Minsk or Kiev, and they lay exactly 11 eggs each day. Determine all possible values of m and k. *Be sure to check that they are indeed possible.*

4. Show that there exist real numbers x and y such that $(x^4 + y^4)^2 < 2xy(x^6 + y^6)$.

5. A, B and C are points on a circle with center O, such that $AB = AC$. The extension of CO intersects the line AB at D. If $AD = BC$, determine all possible values of $\angle BAC$.

February 2, 2000

1. A goblin with strength 8 is fighting two elves with respective strengths 6 and 10, but one at a time. When a character with strength x fights a character with strength y, the first beats the second with a probability of $\frac{x}{x+y}$, and the victor's new strength afterwards becomes the maximum of x and y. The goblin wins if it beats both elves, and the elves win otherwise.

   (a) Which elf should fight first if the elves want to maximize their chances of winning?

   (b) What is the answer to the question in (a) if the new strength of the victor is $x + y$ instead?

2. Suppose that a and b are any positive integers. Prove that there must exist positive integers A and B so that the sum of the digits of A is a, the sum of the digits of B is b, **and** the sum of the digits of the number $A + B$ is at most 9.

3. Determine the angles of a right triangle if the length of its hypotenuse is $2\sqrt{bc}$ and the lengths of its other two sides are b and c.

4. In triangle ABC, $AB > AC$. The altitude from A meets the line BC at K. The perpendicular bisector of BC meets BC at D and AB at F, while the line through F parallel to BC meets AC at E. Prove that the line DE passes through the midpoint of AK.

5. Determine the maximum possible value of

$$\sqrt{4x + (y - z)^2} + \sqrt{4y + (z - x)^2} + \sqrt{4z + (x - y)^2}$$

where x, y and z are nonnegative real numbers such that $x + y + z = 1$.

February 7, 2001

1. Every day, Fred cycles either a or b kilometers, where a and b are positive integers with $a > b$. It is possible for Fred to cycle exactly 100 kilometers in 9 days or 11 days, but not in 10 days. Determine all possible values of a and b.

2. Prove that among any five integers, there are always two whose sum or difference is divisible by 7.

3. If $x^3 - 4x^2 = y$, $y^2 - 4x = 1$ and $x \neq y$, prove that $x^3(y + 4) = 1$.

4. Let ABC be a triangle. Let M be a point on the side AB and N be a point on the side AC such that $\angle ABN = \angle ACM$. Prove that $MN < BC$.

5. Ace prepares a list of questions about two statements P and Q for Bea, who knows whether P and Q are true or false. Each question must be one of the following:

   (a) Is P true?
   (b) Is Q True?
   (c) Is one of P and Q true and the other false?

   The same question may appear several times on the list. After receiving the list, Bea answers each question "Yes" or "No", but may lie at most twice. Determine the minimum number of questions on the list in order for Ace to find out for sure whether P and Q are true or false.

February 6, 2002

1. A certain kind of molecule is formed from four types of atoms, A, B, C and D, arranged into a circle of more than one atom. As many of each type of atom as we like can be used, provided that the following rules are obeyed:

   (i) No two atoms of the same type can be next to each other.

(ii) Every A atom must have a D atom next to it on each side.

(iii) Every D atom with a B atom next to it on one side must have a B atom next to it on the other side too.

Prove that every molecule formed this way must have an even number of atoms.

2. A point E inside a square $ABCD$ is such that $AE = 5$, $BE = 2\sqrt{2}$ and $CE = 3$. Determine the area of $ABCD$.

3. Prove that for any positive real numbers x and y, at least one of $\frac{2x+y}{2x+2y+1}$, $\frac{2y}{2x+2y+1}$ and $\frac{x+1}{2x+2y+1}$ must be less than or equal to $\frac{4}{9}$.

4. All digits of the positive integer M are odd and no two are the same. All digits of the positive integer N are even and no two are the same. If N is a multiple of M, determine the largest possible value of M.

5. (a) Let $a \neq 0$, b and c be real constants such that $|ax^2+bx+c| \leq 1$ for $-1 \leq x \leq 1$. Prove that $|ax^2 + bx + c| \leq 17$ for $1 \leq x \leq 3$.

   (b) Find real constants $a \neq 0$, b and c such that $|ax^2 + bx + c| \leq 1$ for $-1 \leq x \leq 1$ and the maximum value of $|ax^2 + bx + c|$ for $1 \leq x \leq 3$ is 17.

February 5, 2003

1. Prove that if a positive integer is not divisible by 11, then it is possible to make it divisible by 11 by either adding a digit at the end or inserting a digit somewhere in the number.

2. Two birds 90 meters apart walk towards each other, both at 45 meters per minute. A bee starts from the penguin and flies towards the pelican. As soon as it reaches one bird, it turns around immediately and flies back towards the other bird, until the two birds meet. The bee always flies at 80 meters per minute toward the pelican, but at 60 meters per minute towards the penguin. What is the total distance covered by the bee?

3. Two of the integers a, b and c are odd and one is even. Prove that the polynomial

$$x^3 + ax^2 + bx + c$$

cannot have three integer roots.

4. E is the midpoint of the side CA and F is the midpoint of the side AB of triangle ABC. BE and CF intersect at G. Prove that no two lines passing through G divide the triangle into four parts of equal area.

5. Find all finite sets of real numbers such that whenever the number x is in the set, the number $3|x| - 4x^2 - 1$ is also in the set.

February 4, 2004

1. Bill had a supply of identical rectangular bricks with which he asked a workman to build a decorative jagged walk from the northwest corner to the southeast corner of his yard, which measured 10 meters by 20 meters. What Bill had in mind was shown in the first diagram below, where each brick stuck out from the previous one a distance equal to $\frac{1}{13}$ of its length. What the workman built was shown in the second diagram below, where each brick stuck out from the previous one a distance equal to $\frac{1}{3}$ of its length. Determine the length and width of each brick.

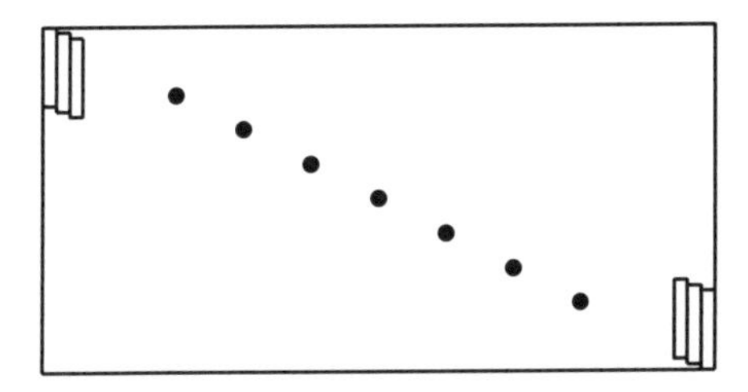
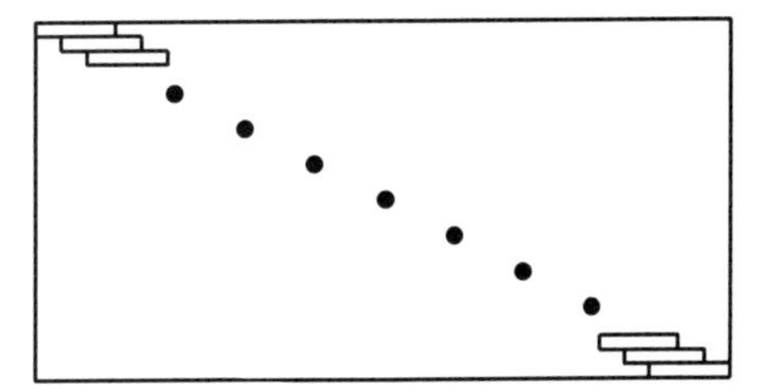

2. The government consists of an Upper House with 6 members and a Lower House with 13 members. The Upper house is investigating the Lower House for corruption while the Lower House is investigating the Upper House for extravagance. Each of the 19 members bills the government \$600 per day up to the first 30 days in which the respective House is holding a hearing, and \$900 per day thereafter. The Upper House hearing lasts twice as many days as the Lower House hearing, and the total bill from the members of the Upper House is equal to that from the members of the Lower House. Based on this information, the Auditor General calculates and announces to the public what this fiasco may have cost the government. However, the actual figure is λ times as large, where λ is some number greater than 1. Determine λ.

3. How many positive integers $n \leq 447560$ are there such that n is divisible by $\lfloor\sqrt{n}\rfloor$, which is obtained by rounding $\sqrt{n}$ down to the nearest integer?

4. In triangle ABC, D is the midpoint of BC. If $\angle BAD = 30°$ and $\angle DAC = 15°$, determine $\angle ABC$.

5. (a) The lengths of the sides of a triangle are a, b and c. Prove that $\frac{a}{a+1}+\frac{b}{b+1} \geq \frac{c}{c+1}$.

   (b) The lengths of the n sides of a convex polygon are $a_1, a_2, \dots, a_n$. Prove that

$$\frac{a_1}{a_1+1}+\frac{a_2}{a_2+1}+\cdots+\frac{a_{n-1}}{a_{n-1}+1} \geq \frac{a_n}{a_n+1}.$$

February 2, 2005

1. In a mathematics examination at the university, each score is an integer. The average score of one class is exactly 70.6 and the average of another class is exactly $70.\overline{7} = 70.777\ldots$. The combined average of the two classes is exactly $70.\overline{67} = 70.676767\ldots$. If the total number of students in both classes is less than 500, find the number of students in each class.

2. A son and a daughter inherited six pearls strung together by five silk threads as shown below.

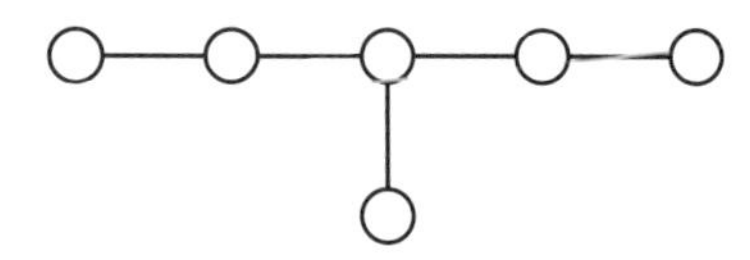

The inheritors had decided to share the pearls by alternating cutting off a silk thread so as to free one pearl from the remaining part, the freed pearl going to the cutter. The daughter would go first, and when she made her third and final cut, she got to choose one of the two remaining pearls while the son had to take the one that was left. The value of each pearl was known. Prove that the daughter could always get three pearls whose total value would be at least the total value of the other three, regardless of the values of the individual pearls.

3. Let S be a set of real numbers with the following properties:

   (i) The number 1 belongs to S.

   (ii) If the number x belongs to S, then the number $x^2 + 4x + 5$ also belongs to S.

   (iii) If the number $x^2 + 1$ belongs to S, then the number x also belongs to S.

   Prove that for any integers m and n, the number $m + \sqrt{2}n$ belongs to S.

4. $ABCDEF$ is a hexagon such that $AB = BC = CD = 1$ and $DE = EF = FA = 2$. If all six vertices lie on the same circle, determine the radius.

5. $ABCD$ is a quadrilateral inscribed in a circle, with $AB = AD$. The diagonals intersect at E. F is a point on AC such that $\angle BFC = \angle BAD$. If $\angle BAD = 2\angle DFC$, determine $\frac{BE}{DE}$.

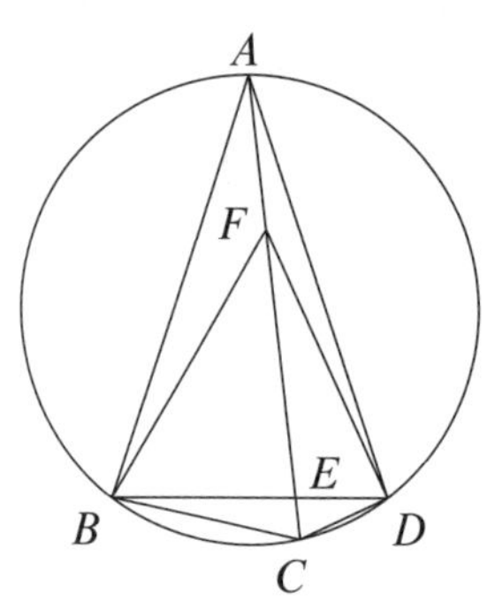

February 1, 2006

1. The diagram below shows a square with one corner on a side of a $2 \times \ell$ rectangle, two corners on the extensions of the diagonals of that rectangle, and the fourth corner on the extension of the line joining its opposite corner to the center of the rectangle. If the square and the rectangle have the same area, determine ℓ.

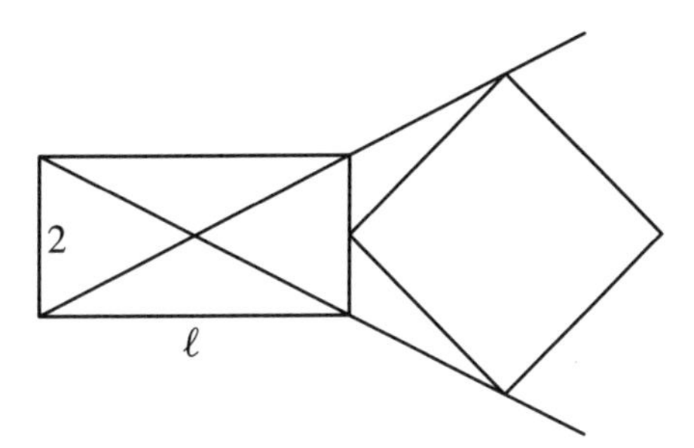

2. A jacket with an original price of $100 was put on sale and the price was reduced by an integer percentage. After a while, the price was reduced again by a greater integer percentage. The new price was now $79.17. Determine the percentages of the two price reductions.

3. Determine all positive integers such that each is 13 times the sum of its digits in base ten.

4. A rectangle $ABCD$ has three of its corners lying on the parabola $y = x^2$: $A(0,0)$, $B(b, b^2)$ for some $b > 0$ and $D(d, d^2)$ for some $d < 0$. The fourth corner C lies somewhere above the parabola.

   (a) Prove that C lies on the parabola $y = x^2 + 2$.

   (b) Express the ratio $\frac{AB}{AD}$ in terms of b only.

5. Determine the maximum number of elements in a set of positive integers such that

$$|m - n| \geq \frac{(m+1)(n+1)}{20}$$

for any two elements m and n in the set.

Solutions

Ancient Period: 1957–1966

1957 **1.** $\frac{3+x^2}{x(1-x)(1+x)}$. **2.** 5. **3.** $\frac{29}{2}$ metres. **4.** -2. **5.** $\frac{-1\pm\sqrt{3}i}{2}$. **6.** 2, $-1\pm 2i$. **8.** $5-\sqrt{7}\times 5-\sqrt{7}$. **9.** $x=0, y=1, z=2$. **10.** no solutions. **11.** (a) 3 milligrams; (b) $\frac{3}{\sqrt{2}}$ milligrams. **12.** $\frac{4\pi-3\sqrt{3}}{6}$. **13.** 48 kph. **14.** 9. **18.** 30%.

1958 **1.** $\frac{241}{168}$. **2.** $\frac{ac(2c^2+ac+b)}{(ac+b)(ac-b)}$. **3.** 255. **4.** $\frac{5\pm\sqrt{5}}{2}$. **5.** $\frac{7\pm\sqrt{5}}{2}$. **6.** $x=\frac{9}{25}, y=-\frac{14}{25}, z=\frac{26}{25}$. **8.** 17. **9.** (a) 91390; (b) 84645. **11.** 28501. **12.** 2.962. **14.** $-1, \frac{3}{2}, \frac{3}{2}$. **17.** an ellipse and two lines. **18.** $x=\frac{3(-1\pm4\sqrt{17}i}{13}, y=\frac{1\pm4\sqrt{17}i}{13}$, or $x=\frac{27\pm4\sqrt{3}i}{37}, y=\frac{3(27\pm4\sqrt{3}i}{37}$. **19.** $\frac{pr(r^5-1)}{r-1}$ dollars.

1959 **1.** 2. **2.** 2, -2, -3. **3.** $x=4, y=6, z=8$. **4.** (a) $a=c$; (b) $b=0$. **6.** intersection of AC and BD. **7.** 480. **10.** yes. **11.** no. **12.** $\frac{1+3\cos^2\theta}{2}$. **13.** $\frac{635}{2}$. **15.** 44. **16.** (b) $\sqrt{ab}$. **17.** The second last line should have been $x+1-\frac{2x+1}{2}=\frac{2x+1}{2}-x$. **18.** (c) $(2+\frac{a}{2},0), x=2-\frac{a}{2}$.

1960 **1.** 13. **2.** 194. **3.** $\frac{\sqrt{3}}{4}$. **4.** (a) 2; (b) $\pm\frac{3}{2}$. **5.** $3x^3+7x^2-4=0$. **6.** $-1, 2, 3$. **7.** (b) no. **8.** (a) $\frac{x^5-x^3+x)}{x^6-x^4+x^2-1}$. (b) $\frac{1-2x^2-3x^4-x^6}{x^5+3x^3+2x}$. **11.** $12(\sqrt{3}-1)$ km, S45°W. **13.** (a) 45; (b) 9; (c) 120; (d) 36; (e) 8. **14.** 127. **17.** (a) $\frac{682}{27}$ metres; (b) 30 metres. **19.** no.

1961 **1.** $\frac{x^5-x^4+8x^3-23x^2+7x-76}{x(x^2+1)(x^2+7)}$. **2.** $\frac{2a-a^2c+b^2c}{\sqrt{(a+b)(a-b)}}$. **3.** 68. **4.** no solutions. **5.** $-\log_{10}2$. **6.** $x>1$ or $x<0$. **7.** 30. **9.** a line perpendicular to AB. **11.** a circle and its centre. **13.** (b) -2. **14.** 7. **15.** $\frac{49}{333}$. **19.** (a) 1680; (b) 1280. **20.** $\frac{25!}{10!}$. **21.** $0°, 60°, 180°, 300°$. **22.** $30°, 150°, 210°, 330°$.

1962 **1.** $\frac{3}{2}$. **2.** (a) $a=2, b=3$; (b) $A=2, B=3$. **3.** $-4, 5$. **4.** no solutions. **5.** -7. **7.** 0. **9.** $0°, 120°, 240°$. **10.** $15°, 30°, 75°, 90°, 150°, 195°, 210°, 255°, 270°, 330°$. **11.** $90°, 210°, 270°, 330°$. **12.** (a) $\frac{n}{n+1}$; (b) 101. **13.** $ax^2-(4a-b)x+(4a-2b+c)=0$. **15.** (a) 7; (b) 105. **21.** (b) t^2-2; (c) $2\pm\sqrt{3}, \frac{-3\pm\sqrt{5}}{2}$. **22.** (b) $A=B$.

1963 **1.** $-\frac{2x}{x^2+1}$. **2.** $\frac{a}{1+2x}$. **3.** x. **4.** $16x^2+29$. **5.** $\frac{a(17+12\sqrt{2})+b(7-4\sqrt{3})}{(7-4\sqrt{3})(17+12\sqrt{2})}$. **6.** $x=35, y=45$ or $x=45, y=35$. **7.** 1600. **8.** $\frac{4}{N}$. **9.** $\frac{125}{2}$ kph, $\frac{200}{3}$ kph. **10.** 8. **12.** $8\sqrt{13}$ metres. **13.** (c)

$6n^5 + 2n^3$. **15.** (b) $\frac{a+ax(1-x^{n-1})}{1-x} - (n^2+2n-1)ax^n + n^2ax^{n+1}$. **16.** (a) $\frac{15}{2}$ metres; (b) $\frac{35}{6}$ metres. **17.** (c) $2 \pm \sqrt{3}, -2 \pm 2\sqrt{2}$. **18.** $\frac{\sqrt{3}ab}{4}$. **21.** (b) $\frac{8}{\tan 8x}$.

1964 **1.** $\frac{25(3+\sqrt{17})}{2}$. **2.** (a) 6, 8, 10. **3.** (a) 126 cm; (b) 2. **4.** (b) no modification. **5.** $-\log_{10} 3$. **7.** (b) 64. **8.** $-\frac{9}{4}$. **9.** $4x^2+x+1=0$. **11.** $\frac{5}{3}, \frac{65}{6}, 20, \frac{175}{6}, \frac{115}{3}$. **13.** (a) $\pm(3-\sqrt{5}i)$. **16.** 20 km. **17.** 3.9979. **18.** $\frac{3b^8}{8a^5}$. **19.** $\frac{858}{20825}$.

1965 **1.** 2. **4.** $a = \frac{1}{2}, b = \frac{\sqrt{3}}{2}$. **5.** 4. **6.** $2p$. **7.** $-1, -3, 4$. **8.** (b) $n^2 - (\frac{ah}{a-h})^2 = m^2 - a^2$. **12.** 45 kph. **15.** (c) $\frac{1}{2(n-1)} - \frac{1}{n} + \frac{1}{2(n+1)}$; (d) $\frac{(n-1)(n+2)}{4n(n+1)}$. **16.** 16π square metres. **17.** $3\sqrt{3}$ metres.

1966 **1.** $(x+1)(x^2-x+1)(x^2+x+1)$. **5.** $1 < x < 2+\sqrt{3}$ or $-1 < x < 2-\sqrt{3}$. **6.** $x=1, y=2, z=4$ or $x=\frac{16}{13}, y=\frac{22}{13}, z=\frac{53}{13}$. **8.** the square with vertices at (2,0), (0,2), $(-2,0)$ amd $(0,-2)$. **9.** (a) $b > -2$. **11.** (a) $\frac{57}{16}$; (c) 4. **14.** 6 cm, 6 cm, $\frac{15\sqrt{7}}{2}$ cm. **16.** (a) $\frac{2x^2}{(1-x)(1+x)}$.

Medieval Period: 1967–1983

Answers to Questions with Multiple Choices

1967	e	c	e	b	e	a	b	e	d	a	c	c	d	c	d	b	a	e	b	b	a	d	b	e	c
1968	d	b	c	c	a	d	e	d	e	c	e	b	a	d	b	b	e	a	c	a	a	e	d	e	b
1969	c	e	a	e	a	c	e	d	e	b	c	d	a	b	c	c	b	d	c	e	d	b	c	e	d
1970	c	b	a	d	b	d	e	e	b	c	a	c	a	d	d	b	c	b	b	a	e	d	a	c	d
1971	b	a	a	d	d	c	d	c	b	b	d	e	b	c	e	a	b	e	c	d					
1972	d	d	a	c	a	e	c	a	b	b	d	b	c	d	e	c	e	b	a	e					
1973	c	b	a	c	a	c	d	b	c	d	e	a	c	b	c	c	d	b	c	b					
1974	c	c	c	b	d	b	e	d	d	a	b	e	c	b	e	b	c	d	b	a					
1975	d	a	d	c	c	c	a	d	e	a	d	a	b	e	c	b	d	d	c	b					
1976	c	e	d	b	a	b	e	a	e	c	b	c	e	d	c	d	d	d	a	a					
1977	d	d	b	b	a	d	c	c	b	e	d	a	e	d	a	e	c	b	c	a					
1978	c	c	d	a	d	b	a	b	d	b	b	e	e	a	d	d	e	e	c	e					
1979	b	d	c	b	e	e	d	a	e	c	a	c	d	b	e	d	b	a	e	d					
1980	b	b	d	d	b	c	c	a	a	b	a	e	d	a	b	c	a	e	c	d					
1981	a	e	c	b	c	c	c	b	a	a	e	a	d	a	e	c	b	b	e	b					
1982	b	c	c	a	c	b	b	a	e	c	c	c	a	a	d	a	b	d	a	c					
1983	a	b	e	e	e	b	c	c	c	d	c	b	a	b	c	e	c	c	b	a					

Answers to Problems requiring Full Solutions

1967 **1.** (b) 0, 2, 6 or infinitely many. **5.** (c) 1.

1968 **3.** 153. **7.** (a) 11111111100. **8.** 2. **9.** (a) $\frac{1}{2}$; (b) $\frac{5}{12}$. **10.** (a) true; (b) true; (c) false; (d) false; (e) true.

1969 **5.** (a) $7 - 5\sqrt{2}$; (b) $x^2 - 198x + 1 = 0$. **9.** $PQ = 4$, $QR = 3$.

1970 **1.** (a) 6; (b) $6x$. **2.** $H + \sqrt{H^2 + 4A}$. **3.** $2 + 2\sqrt{2}$.

1971 **2.** $\sqrt{2}$. **4.** -9. **6.** The expression containing b is bigger. **8.** (a) $\{1, 2, 3, \dots, m, m - 1, m - 2, \dots, 1\}$. **9.** False.

1972 **1.** 3968, 3969, 3970, 3971, 3972. **2.** $S = \{0\}$. **3.** a rectangle. **4.** $g(x) = \frac{x}{3}$. **5.** $\frac{11}{2}$. **6.** $\frac{\sqrt{3}}{2}$. **7.** 5. **9.** $(r, s) = (7, 5)$ or (10,7).

1973 **4.** (a) $A = 1, B = 36$. **5.** (a) 2; (b) 4; (c) 7; (d) 11; (e) 5051. **7.** $1 + \sqrt{2}$. **8.** $\frac{3-2\sqrt{2}}{2}$. **9.** 7, 2 and 10. **10.** $-3, -2, -1, -\frac{1}{2}$ and $\pm 4i$.

1974 **2.** Yes. **5.** $\sqrt{5}^{\sqrt{3}}$ is greater. **9.** (a) 3^n; (b) $3^n - 3 \cdot 2^n + 3$.

1975 **1.** $\sqrt{2}-1$. **3.** $-7, -4, 8$ and 11. **5.** 10. **6.** false. **9.** $a_1 + a_2 + \cdots + a_k$.

1976 **2.** (a) 10; (b) 4. **3.** $\frac{1}{4}$ or 0. **5.** $\frac{5+\sqrt{13}}{2}$. **6.** 1. **7.** $a = b = 1$ and $c = 0$. **8.** $A = 4$ and $B = -3$. **9.** 200.

1977 **3.** In the first quadrant, the graph consists of the segments from (0,2) to (1,2), from (1, 1) to (2,1) and from (0,2) to (0,3). In each case, the right endpoint is omitted. The whole graph may be obtained by reflection about the coordinate axes.

1978 **1.** $0°$, $90°$ and $270°$. **3.** $\{a, \ldots, a\}$, $\{a, \ldots, a, 2a\}$, or $\{a, 2a, 3a\}$. **4.** $\{(x, y) | x^2 + y^2 \leq 1, (x, y) \neq (1, 0)\}$. **6.** 0 or 42.

1979 **2.** $-1 \pm 2\sqrt{2}$. **4.** (a) 14; (b) $\frac{73}{648}$.

1980 **1.** yes. **2.** 3, 7 or 15. **6.** $\omega^4 - \omega, \omega^3 - \omega^2$ and $\omega^2 - \omega - 3$.

1981 **2.** 1981.

1982 **1.** $\frac{45}{4}$. **2.** $\frac{z^2-x^2-y^2}{2xy}$. **3.** $\frac{11}{40}$ minutes.

1983 **1.** 1. **2.** 3 pairs. **3.** (b) yes; (c) yes.

Modern Period: 1983–2006

First Round Answers

#	1	2	3	4	5	6	7	8	9	10	11	12	13	14	15	16	17	18	19	20
1983/84	d	b	a	c	b	c	b	c	a	e	b	e	e	d	b	b	d	d	a	b
1984/85	c	c	b	c	c	b	c	a	b	d	b	a	a	e	b	b	b	e	c	e
1985/86	c	c	a	d	b	b	d	e	e	d	a	d	b	c	c	d	c	c	b	d
1986/87	e	e	a	e	c	d	c	c	c	d	c	d	c	d	d	e	c	b	d	a
1987/88	b	b	d	c	b	b	b	a	e	e	c	c	e	c	a	e	a	e	b	a
1988/89	c	c	c	e	c	c	e	a	b	b	b	c	e	e	b	c				
1989/90	d	a	d	c	c	e	b	c	c	d	e	e	a	b	c	a				
1990/91	c	a	e	b	a	c	e	c	d	d	b	b	c	d	a	a				
1991/92	b	e	e	e	a	c	b	a	e	e	b	d	d	b	e	a				
1992/93	e	d	c	a	a	c	d	c	c	a	e	d	c	a	c	d				
1993/94	c	e	c	a	e	c	e	e	b	b	c	c	d	c	d	c				
1994/95	a	c	d	b	c	e	c	b	d	a	d	c	c	c	c	d				
1995/96	d	d	c	c	a	e	d	e	d	c	c	b	c	a	b	c				
1996/97	b	d	b	e	c	a	e	b	b	e	b	d	a	c	a	b				
1997/98	a	c	c	b	a	b	d	b	b	e	a	c	e	c	d	b				
1998/99	c	b	a	a	e	c	c	e	d	a	a	d	b	b	b	c				
1999/00	c	c	e	b	b	a	b	c	d	e	c	b	b	d	d	b				
2000/01	e	a	d	c	e	c	d	c	a	d	c	e	b	e	d	c				
2001/02	a	e	c	a	d	e	c	a	c	d	b	b	c	b	a	e				
2002/03	b	a	e	b	e	b	e	d	a	a	c	a	c	c	b	d				
2003/04	e	c	d	c	b	a	c	e	c	b	c	c	d	c	d	d				
2004/05	c	e	b	d	e	d	c	d	e	b	c	c	c	a	e	b				
2005/06	a	d	a	e	c	b	d	c	a	b	c	c	e	c	d	c				

First Round Statistics

The following tables show the numbers of responses of each type for each question, "x" standing for no responses. The second set of numbers to the right are the percentages. The numbers for the correct responses are in boldface.

83/84	a	b	c	d	e	x	a	b	c	d	e	x
1	183	5	97	**236**	1	15	34	1	18	**44**	0	3
2	43	**430**	5	5	48	6	8	**80**	1	1	9	1
3	**166**	38	75	70	64	124	**31**	7	14	13	12	23
4	38	16	**172**	64	156	91	7	3	**32**	12	29	17
5	48	**129**	38	16	91	215	9	**24**	7	3	17	40
6	16	16	**322**	54	75	54	3	3	**60**	10	14	10
7	32	**199**	11	48	166	81	6	**37**	2	9	31	15
8	43	38	**263**	27	97	69	8	7	**49**	5	18	13
9	**70**	43	59	38	220	116	**13**	8	11	7	41	20
10	97	32	16	43	**199**	150	18	6	3	8	**37**	28
11	27	**75**	59	32	150	194	5	**14**	11	6	28	36
12	27	81	54	38	**145**	192	5	15	10	7	**27**	36
13	97	32	43	183	**75**	107	18	6	8	34	**14**	20
14	64	43	70	**134**	102	124	12	8	13	**25**	19	23
15	64	**54**	32	59	43	285	12	**10**	6	11	8	53
16	124	**38**	64	102	59	150	23	**7**	12	19	11	28
17	54	59	43	**140**	38	198	10	11	9	**26**	7	37
18	59	86	48	**54**	48	242	11	16	9	**10**	9	45
19	**48**	59	32	48	91	259	**9**	11	6	9	17	48
20	32	**145**	59	43	43	215	6	**27**	11	8	8	40

84/85	**a**	**b**	**c**	**d**	**e**	**x**	**a**	**b**	**c**	**d**	**e**	**x**
1	35	55	**512**	42	21	27	5	8	**74**	6	3	4
2	1	7	**602**	14	55	13	0	7	**87**	2	8	2
3	318	**118**	42	7	83	124	46	**17**	6	1	12	18
4	48	90	**381**	62	63	48	7	13	**55**	9	9	7
5	76	152	**339**	21	55	49	11	22	**49**	3	8	7
6	14	**567**	7	7	55	42	2	**82**	1	1	8	6
7	83	42	**325**	90	62	90	12	6	**47**	13	9	13
8	**388**	69	28	7	76	124	**56**	10	4	1	11	18
9	7	**235**	35	69	62	284	1	**34**	5	10	9	41
10	62	35	55	**360**	138	42	9	5	8	**52**	20	6
11	263	**118**	145	7	55	104	38	**17**	21	1	8	15
12	**498**	7	55	7	55	70	**72**	1	8	1	8	10
13	**187**	21	48	28	62	346	**27**	3	7	4	9	50
14	207	76	42	35	**215**	115	30	11	6	5	**31**	17
15	208	**36**	97	131	35	145	30	**11**	14	19	5	21
16	55	**235**	42	14	55	291	8	**34**	6	2	8	42
17	125	**131**	35	159	55	187	18	**19**	5	23	8	27
18	55	83	76	69	**90**	319	8	12	11	10	**13**	46
19	28	55	**55**	90	97	367	4	8	**8**	13	14	53
20	131	42	35	14	**48**	422	19	6	5	2	**7**	61

85/86	**a**	**b**	**c**	**d**	**e**	**x**	**a**	**b**	**c**	**d**	**e**	**x**
1	16	8	**626**	40	95	8	2	1	**79**	5	12	1
2	16	32	**619**	40	63	23	2	4	**78**	5	8	3
3	**563**	87	8	24	40	71	**71**	11	1	3	5	9
4	8	79	95	**333**	182	46	1	10	12	**42**	23	12
5	16	**301**	190	79	79	128	2	**38**	24	10	10	16
6	40	**285**	40	32	111	285	5	**36**	5	4	14	36
7	190	32	111	**365**	16	79	24	4	14	**46**	2	10
8	63	63	48	79	**317**	223	8	8	6	10	**40**	28
9	48	79	71	87	**190**	318	6	10	9	11	**24**	40
10	56	71	174	**262**	119	111	7	9	22	**33**	15	14
11	**317**	79	63	32	190	112	**40**	10	8	4	24	14
12	87	56	48	**309**	95	198	11	7	6	**39**	12	25
13	135	**119**	48	32	238	221	17	**15**	6	4	30	28
14	167	79	**103**	127	79	238	21	10	**13**	16	10	30
15	48	87	**174**	119	103	262	6	11	**22**	15	13	33
16	103	56	32	**71**	32	499	13	7	4	**9**	4	63
17	56	143	**87**	48	135	324	7	18	**11**	6	17	41
18	40	48	**87**	32	103	483	5	6	**11**	4	13	61
19	167	**71**	32	32	143	348	21	**9**	4	4	18	44
20	79	63	111	**32**	48	460	10	8	14	**4**	6	58

86/87	a	b	c	d	e	x	a	b	c	d	e	x
1	58	38	167	118	**598**	1	6	4	17	12	**61**	0
2	78	127	382	137	**157**	99	8	13	39	14	**16**	10
3	**196**	78	235	98	78	295	**20**	8	24	10	8	30
4	167	284	69	78	**206**	176	17	29	7	8	**21**	18
5	69	98	**118**	39	245	411	7	10	**12**	4	25	42
6	39	29	157	**274**	245	236	4	3	16	**28**	25	24
7	97	68	**138**	216	138	323	10	7	**14**	22	14	33
8	88	216	**216**	78	186	196	9	22	**22**	8	19	20
9	284	78	**216**	59	225	118	29	8	**22**	6	23	12
10	167	78	78	**196**	176	285	17	8	8	**20**	18	29
11	392	49	**138**	147	49	205	40	5	**14**	15	5	21
12	147	147	108	**235**	108	235	15	15	11	**24**	11	24
13	49	78	**186**	78	353	236	5	8	**19**	8	36	24
14	78	49	59	**216**	225	353	8	5	6	**22**	23	36
15	147	138	59	**274**	88	274	15	14	6	**28**	9	28
16	108	78	88	49	**157**	500	11	8	9	5	**16**	51
17	29	98	**686**	78	49	40	3	10	**70**	8	5	4
18	58	**68**	69	59	167	559	6	**7**	7	6	17	57
19	206	59	118	**157**	157	283	21	6	12	**16**	16	29
20	**49**	69	127	39	176	521	**5**	7	13	4	18	53

87/88	a	b	c	d	e	x	a	b	c	d	e	x
1	58	**268**	35	58	338	410	5	**23**	3	5	29	35
2	82	**595**	140	70	47	233	7	**51**	12	6	4	20
3	222	292	105	**233**	175	140	19	25	9	**20**	15	12
4	47	128	**864**	35	23	70	4	11	**74**	3	2	6
5	117	**595**	58	58	175	162	10	**51**	5	5	15	14
6	35	**373**	105	58	163	433	3	**32**	9	5	14	37
7	128	**292**	222	93	47	385	11	**25**	19	8	4	33
8	**117**	280	47	58	93	572	**10**	24	4	5	8	49
9	128	175	163	82	**198**	421	11	15	14	7	**17**	36
10	712	105	23	70	**187**	70	61	9	2	6	**16**	6
11	35	128	**222**	82	257	445	3	11	**19**	7	22	38
12	105	47	**747**	35	175	58	9	4	**64**	3	15	5
13	58	93	315	128	**163**	410	5	8	27	11	**14**	35
14	70	58	**105**	187	93	654	6	5	**9**	16	8	56
15	**338**	280	128	82	35	304	**29**	24	11	7	3	26
16	257	292	12	12	**455**	139	22	25	1	1	**39**	12
17	**210**	105	70	70	105	607	**18**	9	6	6	9	52
18	47	70	163	82	**47**	758	4	6	14	7	**4**	65
19	47	**70**	70	23	327	630	4	**6**	6	2	28	54
20	**233**	233	210	128	82	281	**20**	20	18	11	7	24

88/89	**a**	**b**	**c**	**d**	**e**	**x**	**a**	**b**	**c**	**d**	**e**	**x**
1	81	80	**421**	65	10	161	10	10	**51**	8	1	20
2	15	79	**350**	147	218	9	2	10	**43**	18	26	1
3	42	114	**214**	25	191	232	5	14	**26**	3	23	29
4	250	43	140	18	**172**	194	31	5	16	2	**21**	24
5	116	17	**195**	17	178	294	14	2	**24**	2	22	36
6	230	24	**90**	40	194	240	28	3	**11**	5	24	29
7	176	91	64	39	**138**	310	21	11	8	5	**17**	38
8	**25**	17	32	41	303	400	**3**	2	4	5	37	49
9	57	**171**	66	73	126	326	7	**21**	8	9	15	40
10	51	**351**	98	49	122	147	6	**43**	12	6	15	18
11	188	**121**	138	48	58	265	23	**15**	17	6	7	32
12	25	23	**184**	48	112	426	3	3	**22**	6	14	52
13	169	97	40	24	**198**	290	21	12	5	3	**24**	35
14	266	73	8	98	**67**	306	33	9	1	12	**8**	37
15	18	**182**	146	41	82	349	2	**22**	18	5	10	43
16	32	50	**56**	49	97	534	4	6	**7**	6	12	65

89/90	**a**	**b**	**c**	**d**	**e**	**x**	**a**	**b**	**c**	**d**	**e**	**x**
1	44	90	44	**158**	44	375	6	11	6	**21**	6	50
2	**243**	168	31	84	153	76	**32**	22	4	11	21	10
3	61	9	9	**516**	130	30	8	1	1	**69**	17	4
4	74	22	**408**	90	74	147	10	3	**54**	12	10	11
5	38	60	**266**	38	23	330	5	8	**35**	5	3	44
6	14	14	7	23	**333**	364	2	2	1	3	**44**	48
7	45	**243**	107	23	129	208	6	**32**	14	3	17	28
8	60	83	**319**	60	23	210	8	11	**42**	8	3	28
9	159	16	**538**	31	1	10	21	2	**72**	4	0	1
10	97	59	136	**340**	67	56	13	8	18	**45**	9	7
11	37	45	29	181	**129**	334	5	6	4	24	**17**	44
12	45	23	30	60	**144**	453	6	3	4	8	**19**	60
13	**288**	45	61	46	84	231	**38**	6	8	6	11	31
14	7	**29**	37	44	52	586	1	**4**	5	6	7	77
15	100	24	**159**	39	9	424	14	3	**21**	5	1	56
16	**37**	36	36	36	36	574	**5**	5	5	5	5	75

90/91	a	b	c	d	e	x	a	b	c	d	e	x
1	44	70	**535**	53	9	163	5	8	**61**	6	1	19
2	**426**	130	44	44	69	161	**41**	15	5	5	8	18
3	70	62	18	177	**370**	177	8	7	2	20	**43**	20
4	35	**106**	18	229	203	283	4	**12**	2	27	23	32
5	**504**	86	26	34	130	94	**57**	10	3	4	15	11
6	334	70	**79**	17	158	216	38	8	**9**	2	18	25
7	35	61	44	78	**252**	404	4	7	5	9	**29**	46
8	139	35	**52**	35	96	517	16	4	**6**	4	11	59
9	173	52	78	**130**	155	286	20	6	9	**15**	18	32
10	44	52	149	**158**	281	190	5	6	17	**18**	32	22
11	26	**420**	26	114	184	104	3	**48**	3	13	21	12
12	26	**44**	44	131	26	603	3	**5**	5	15	3	69
13	26	17	**202**	342	193	94	3	2	**23**	39	22	11
14	26	109	46	**196**	26	471	3	13	5	**22**	3	54
15	**88**	71	36	88	168	423	**10**	8	4	10	19	49
16	**44**	44	87	44	219	436	**5**	5	10	5	25	50

91/92	a	b	c	d	e	x	a	b	c	d	e	x
1	15	**89**	30	118	239	257	2	**12**	4	16	32	34
2	15	22	67	37	**575**	32	2	3	9	5	**77**	4
3	22	15	52	217	**398**	44	3	2	7	29	**53**	6
4	142	30	209	58	**102**	207	19	4	28	8	**14**	27
5	**151**	84	145	121	53	194	**20**	11	20	16	7	26
6	52	22	**98**	120	217	239	7	3	**13**	16	29	32
7	15	**277**	15	15	144	282	2	**37**	2	2	19	38
8	**453**	25	55	25	78	112	**61**	3	7	3	11	15
9	13	20	43	139	**73**	460	2	3	6	19	**9**	61
10	127	67	30	22	**149**	353	17	9	4	3	**20**	47
11	81	**125**	58	22	118	344	11	**16**	8	3	16	46
12	149	275	30	**104**	22	168	20	37	4	**14**	3	22
13	74	59	112	**142**	75	287	10	8	15	**19**	10	38
14	52	**58**	125	37	15	461	7	**8**	17	5	2	61
15	15	52	22	187	**269**	203	2	7	3	25	**36**	27
16	**82**	30	58	127	117	334	**11**	4	8	17	16	44

92/93	**a**	**b**	**c**	**d**	**e**	**x**	**a**	**b**	**c**	**d**	**e**	**x**
1	73	20	84	46	**532**	38	9	3	10	6	**67**	5
2	11	32	26	**475**	129	120	1	4	3	**60**	17	15
3	72	76	**532**	47	41	25	9	10	**67**	6	5	3
4	**174**	69	82	269	84	115	**22**	9	10	34	11	14
5	**221**	69	160	62	53	228	**28**	9	20	8	6	29
6	27	73	**272**	227	17	177	4	9	**34**	29	2	22
7	28	24	33	**63**	296	349	4	3	4	**8**	37	44
8	139	483	**36**	19	34	82	18	61	**5**	2	4	10
9	103	108	**233**	31	29	289	13	14	**29**	4	4	36
10	**117**	28	40	32	135	441	**15**	4	5	4	17	55
11	398	22	48	19	**136**	170	50	3	6	2	**17**	22
12	34	36	50	**92**	44	537	4	5	6	**12**	6	68
13	179	160	**102**	148	115	89	23	20	**13**	19	14	11
14	**44**	158	93	128	56	314	**6**	20	12	16	7	39
15	102	46	**184**	30	83	348	13	6	**23**	4	10	44
16	58	37	54	**50**	29	565	7	5	7	**6**	4	71

93/94	**a**	**b**	**c**	**d**	**e**	**x**	**a**	**b**	**c**	**d**	**e**	**x**
1	42	95	**426**	52	19	242	5	11	**49**	6	2	27
2	128	72	148	100	**262**	166	15	8	17	11	**30**	19
3	34	42	**473**	44	251	32	4	5	**54**	5	29	3
4	**407**	134	35	12	96	192	**46**	15	4	2	11	22
5	129	31	36	28	**633**	19	15	4	4	3	**72**	2
6	45	290	**110**	32	166	233	5	33	**12**	4	19	27
7	13	28	234	33	**418**	150	1	3	27	4	**48**	17
8	437	104	23	10	**272**	30	50	12	3	1	**31**	3
9	7	**344**	398	7	68	52	1	**39**	45	1	8	6
10	20	**186**	34	21	126	489	2	**21**	4	3	14	56
11	20	100	**176**	30	58	492	2	11	**20**	4	7	56
12	147	175	**392**	29	44	89	17	20	**45**	3	5	10
13	15	28	31	**117**	111	574	2	3	4	**13**	13	65
14	24	12	**236**	60	220	324	3	1	**27**	7	25	37
15	12	42	49	**87**	111	575	1	5	5	**10**	13	66
16	48	9	**43**	40	164	572	5	1	**5**	5	19	65

94/95	**a**	**b**	**c**	**d**	**e**	**x**	**a**	**b**	**c**	**d**	**e**	**x**
1	**135**	63	55	147	377	210	**14**	6	6	15	38	21
2	19	45	**745**	13	89	76	2	5	**75**	1	9	8
3	38	17	93	**417**	174	248	4	2	9	**42**	18	25
4	16	**605**	146	33	142	45	2	**61**	15	3	14	5
5	98	142	**641**	76	13	17	10	14	**65**	8	1	2
6	160	75	163	42	**237**	310	16	8	17	4	**24**	31
7	100	333	**323**	61	36	436	10	3	**33**	6	5	44
8	178	**81**	53	133	107	435	18	**8**	5	14	11	44
9	312	123	94	**290**	55	113	32	12	10	**29**	6	11
10	**127**	77	268	130	56	329	**13**	8	27	13	6	11
11	52	136	99	**451**	168	81	5	14	10	**46**	17	8
12	48	127	**46**	99	84	583	5	13	**5**	10	8	59
13	179	178	**80**	74	134	342	18	18	**8**	7	14	35
14	124	23	**117**	29	107	587	13	2	**12**	3	11	59
15	28	92	**107**	59	56	645	3	9	**11**	6	6	65
16	276	188	49	**63**	131	280	28	19	5	**7**	13	28

95/96	**a**	**b**	**c**	**d**	**e**	**x**	**a**	**b**	**c**	**d**	**e**	**x**
1	62	182	139	**303**	117	211	6	18	14	**30**	11	21
2	98	180	185	**186**	48	317	10	18	18	**18**	5	31
3	96	18	**695**	30	116	59	9	2	**69**	3	11	6
4	72	35	**80**	42	363	422	7	3	**8**	4	36	42
5	**123**	182	96	384	9	220	**12**	18	9	38	1	22
6	56	35	149	71	**562**	141	6	3	15	7	**55**	14
7	271	103	78	**158**	32	372	27	10	8	**15**	3	27
8	204	76	43	31	**122**	538	20	8	4	3	**12**	53
9	98	78	115	**40**	140	543	10	8	11	**4**	14	53
10	22	29	**93**	489	117	264	2	3	**9**	48	12	26
11	223	283	**201**	36	90	181	22	28	**20**	3	9	18
12	121	**63**	62	48	64	656	12	**6**	6	5	6	65
13	89	69	**60**	112	557	127	9	7	**6**	11	55	12
14	**102**	330	266	112	94	110	**10**	33	26	11	9	11
15	83	**138**	69	44	81	599	8	**14**	7	4	8	59
16	109	100	**105**	154	158	388	11	10	**10**	15	16	38

96/97	a	b	c	d	e	x	a	b	c	d	e	x
1	328	**578**	23	100	22	25	31	**54**	2	9	2	2
2	22	107	26	**329**	459	123	2	10	2	**31**	43	12
3	19	**662**	176	102	16	91	2	**62**	16	10	1	9
4	173	207	76	55	**536**	19	16	20	7	5	**50**	2
5	40	75	**509**	166	137	139	4	7	**48**	15	13	13
6	**503**	16	34	55	269	189	**47**	2	3	5	25	18
7	54	79	72	320	**61**	480	5	7	7	**30**	6	45
8	111	**194**	78	73	172	438	11	**18**	7	7	16	41
9	42	**860**	46	20	68	30	4	**81**	4	2	6	3
10	129	442	**205**	29	64	197	12	41	**19**	3	6	19
11	243	**218**	25	6	355	219	23	**20**	2	1	33	21
12	21	103	16	**554**	217	155	2	10	1	**52**	20	15
13	**107**	58	57	40	107	697	**10**	5	5	4	10	66
14	103	54	**95**	214	225	375	10	5	**9**	20	21	35
15	**109**	22	48	34	110	743	**10**	2	5	3	10	70
16	76	**89**	238	54	41	568	7	**9**	22	5	4	53

97/98	a	b	c	d	e	x	a	b	c	d	e	x
1	**324**	95	222	13	33	97	**41**	12	28	2	4	13
2	68	17	**465**	141	24	69	9	2	**59**	18	3	9
3	46	21	**626**	39	14	38	6	2	**80**	5	2	5
4	35	**464**	11	12	196	66	5	**59**	1	2	25	8
5	**358**	39	18	12	96	261	**46**	5	2	2	12	33
6	34	**541**	16	14	106	73	4	**69**	2	2	14	9
7	78	36	33	**543**	15	79	10	5	4	**69**	2	10
8	116	**287**	136	10	115	120	15	**37**	17	1	15	15
9	349	**129**	20	95	109	82	45	**16**	3	12	14	10
10	58	17	255	17	**355**	82	7	2	33	2	**45**	11
11	**627**	41	18	34	56	8	**80**	5	2	5	7	1
12	8	7	**624**	5	126	14	1	1	**79**	1	16	2
13	203	340	30	13	**89**	109	26	43	4	2	**11**	14
14	29	69	**134**	32	205	315	4	9	**17**	4	26	40
15	93	119	94	**131**	178	169	12	15	12	**17**	23	21
16	21	**83**	11	14	303	352	3	**10**	1	2	39	45

98/99	**a**	**b**	**c**	**d**	**e**	**x**	**a**	**b**	**c**	**d**	**e**	**x**
1	41	203	**351**	452	30	24	4	18	**32**	41	3	2
2	365	**345**	132	16	127	116	33	**31**	12	1	12	11
3	**769**	106	77	85	39	25	**70**	10	7	8	4	2
4	**127**	240	50	115	73	496	**12**	22	5	10	7	45
5	551	49	18	32	**208**	243	50	4	2	3	**19**	22
6	166	126	**229**	91	107	382	15	11	**21**	8	10	35
7	141	181	**102**	93	168	416	13	16	**9**	8	15	38
8	24	50	88	115	**387**	437	2	5	8	10	**35**	40
9	176	133	216	**158**	305	113	16	12	20	**14**	28	10
10	**420**	54	207	36	226	158	**38**	5	19	3	21	14
11	**344**	67	30	15	266	379	**31**	6	3	1	24	34
12	43	82	161	**124**	102	589	4	7	15	**11**	9	53
13	41	**91**	90	64	73	742	4	**8**	8	6	7	67
14	37	**80**	61	82	141	700	3	**7**	6	7	13	64
15	72	**188**	66	100	82	593	7	**17**	6	9	7	54
16	37	91	**348**	48	268	309	3	8	**32**	4	24	28

99/00	**a**	**b**	**c**	**d**	**e**	**x**	**a**	**b**	**c**	**d**	**e**	**x**
1	49	15	**783**	42	102	65	5	1	**74**	4	10	6
2	46	95	**636**	79	83	117	4	9	**60**	7	8	11
3	73	22	54	348	**402**	157	7	2	5	33	**38**	15
4	129	**434**	44	52	133	265	12	**41**	4	5	13	25
5	142	**268**	103	120	145	278	13	**25**	10	11	14	26
6	**86**	217	24	364	205	160	**8**	21	2	34	19	15
7	114	**180**	81	25	118	538	11	**17**	8	2	11	51
8	418	64	**184**	188	46	156	40	6	**17**	18	4	15
9	91	176	49	**499**	111	130	9	17	5	**47**	11	12
10	166	131	65	126	**124**	444	16	12	6	12	**12**	42
11	111	50	**372**	70	83	371	11	5	**35**	7	8	35
12	95	**238**	94	88	106	436	9	**23**	9	8	10	41
13	76	**293**	109	74	86	419	7	**28**	10	7	8	40
14	55	132	88	**114**	9	659	5	13	8	**11**	1	62
15	83	86	74	**207**	121	486	8	8	7	**20**	11	46
16	100	**80**	160	329	75	313	9	**8**	15	31	7	30

00/01	a	b	c	d	e	x	a	b	c	d	e	x
1	33	45	25	22	**908**	177	3	4	2	2	**75**	14
2	**664**	188	41	218	10	89	**55**	16	3	18	1	7
3	65	55	68	**159**	90	773	5	5	6	**13**	7	64
4	70	153	**780**	135	29	43	6	13	**64**	11	2	4
5	71	99	107	116	**417**	400	6	8	9	10	**34**	33
6	31	130	**383**	43	197	426	3	11	**32**	4	16	35
7	39	109	16	**229**	247	570	3	9	1	**19**	20	47
8	358	129	**223**	30	127	343	30	11	**18**	2	10	28
9	**505**	41	199	10	350	105	**42**	3	16	1	29	9
10	258	433	211	**105**	15	188	21	36	17	**9**	1	16
11	173	211	**258**	66	52	450	14	17	**21**	5	4	37
12	67	212	21	328	**98**	484	6	18	2	27	**8**	40
13	58	**491**	39	28	69	525	5	**41**	3	2	6	43
14	64	24	57	133	**465**	467	5	2	5	11	**38**	38
15	41	95	115	**87**	70	802	3	8	9	7	6	66
16	161	98	**89**	44	30	788	13	8	**7**	4	2	65

01/02	a	b	c	d	e	x	a	b	c	d	e	x
1	**412**	101	27	166	159	228	**38**	9	2	15	15	21
2	25	39	88	493	**194**	254	2	4	8	45	**18**	23
3	312	257	**196**	30	167	131	29	23	**18**	3	15	12
4	**159**	27	39	36	399	433	**15**	2	4	3	36	40
5	54	147	80	**371**	75	366	5	13	7	**34**	7	34
6	104	148	196	133	**193**	319	9	14	18	12	**18**	29
7	210	97	**74**	51	71	590	19	9	**7**	5	6	54
8	**76**	346	267	15	87	302	**7**	32	24	1	8	28
9	94	78	**351**	73	160	337	8	7	**32**	7	15	31
10	395	146	71	**194**	200	87	36	13	7	**18**	18	8
11	151	**98**	91	19	260	474	14	**9**	8	2	24	43
12	26	**231**	60	57	136	583	2	**21**	6	5	13	53
13	87	21	**179**	211	190	405	8	2	**16**	19	18	37
14	507	**36**	23	21	71	435	46	**3**	2	2	7	40
15	**48**	38	35	138	148	686	**4**	3	3	13	14	63
16	274	158	24	59	**295**	283	25	15	2	5	**27**	26

02/03	a	b	c	d	e	x	a	b	c	d	e	x
1	66	**464**	34	22	212	151	7	**49**	4	2	22	16
2	**465**	63	130	110	80	101	**49**	7	14	12	8	11
3	8	25	34	47	**739**	96	1	3	4	5	**78**	10
4	68	**244**	23	8	296	310	7	**26**	2	1	31	33
5	67	137	76	65	**241**	363	7	14	8	7	25	38
6	167	**421**	43	26	89	203	18	**44**	5	3	9	21
7	121	191	346	33	**155**	103	13	20	36	3	**16**	11
8	24	64	60	**132**	92	577	3	7	6	**14**	10	61
9	**313**	57	60	29	112	378	**33**	6	6	3	12	40
10	**79**	56	91	31	143	549	**8**	6	10	3	15	58
11	17	121	**118**	110	163	420	2	13	**12**	12	17	44
12	**126**	26	89	38	347	323	**13**	3	9	4	37	34
13	42	30	**256**	33	129	459	4	3	**27**	3	14	48
14	16	16	**62**	26	759	70	2	2	**7**	3	80	7
15	96	**393**	231	48	43	138	10	**41**	24	5	5	15
16	43	51	60	**77**	46	672	5	5	6	**8**	5	71

03/04	a	b	c	d	e	x	a	b	c	d	e	x
1	175	10	5	6	**668**	11	20	1	1	1	**76**	1
2	49	**291**	83	47	109	296	6	**33**	10	5	12	34
3	45	46	59	**318**	192	215	5	5	7	**36**	22	25
4	83	26	**435**	40	110	181	9	3	**50**	4	13	21
5	76	**247**	22	177	42	311	9	**28**	2	20	5	36
6	**186**	228	46	21	147	247	**21**	26	5	3	17	28
7	21	53	**322**	44	46	389	2	6	**37**	5	5	45
8	24	18	**411**	30	137	255	3	2	**47**	3	16	29
9	39	119	40	55	**338**	284	4	14	5	6	**39**	32
10	56	**156**	79	24	110	450	6	**18**	9	3	13	51
11	54	71	**116**	192	311	131	6	18	**13**	22	36	15
12	20	34	**105**	30	131	555	2	4	**12**	3	15	64
13	23	74	39	**102**	53	584	3	8	4	**12**	6	67
14	9	25	**260**	50	17	514	1	3	**30**	5	2	59
15	15	67	74	**81**	65	573	2	8	8	**9**	7	66
16	24	13	138	**87**	83	530	3	1	16	**10**	9	61

04/05	a	b	c	d	e	x	a	b	c	d	e	x
1	58	21	**350**	124	62	195	7	3	**43**	15	8	24
2	201	154	73	55	**132**	195	25	19	9	7	**16**	24
3	50	**151**	34	26	39	510	6	**19**	4	3	5	63
4	32	52	159	**479**	41	47	4	6	20	**59**	5	6
5	51	28	28	12	**653**	38	6	3	3	2	**81**	5
6	292	52	38	**51**	155	222	36	6	5	**6**	19	28
7	344	54	**43**	9	92	268	43	7	**5**	1	11	33
8	19	18	44	**435**	20	274	2	2	5	**54**	3	34
9	65	249	113	24	**81**	278	8	31	14	3	**10**	34
10	50	**267**	40	18	18	417	6	**33**	5	2	2	52
11	83	68	**70**	20	212	357	10	8	**9**	3	26	44
12	34	33	**308**	35	57	342	4	4	**38**	5	7	42
13	57	27	**33**	22	86	584	7	3	**4**	3	11	72
14	**24**	48	40	14	49	635	**3**	6	5	2	6	78
15	70	22	**38**	37	59	583	9	3	**5**	4	7	72
16	252	**34**	41	37	104	342	31	**4**	5	5	13	42

05/06	a	b	c	d	e	x	a	b	c	d	e	x
1	**615**	21	4	5	215	31	**69**	2	1	1	24	3
2	185	90	**240**	204	135	37	21	10	**27**	23	15	4
3	**348**	76	180	44	27	216	**39**	9	20	5	3	24
4	23	40	61	41	**125**	601	3	4	7	5	**14**	67
5	79	190	**524**	41	18	39	9	21	**59**	5	2	4
6	59	36	**378**	26	195	197	7	4	**42**	3	22	22
7	76	48	64	**146**	33	524	9	5	7	**16**	4	59
8	203	108	73	**77**	33	397	23	12	8	**9**	4	44
9	**146**	60	82	62	102	439	**16**	8	9	7	11	49
10	135	**315**	38	11	11	381	15	**36**	4	1	1	43
11	29	30	**189**	67	299	277	3	3	**21**	8	34	31
12	205	77	98	53	**259**	199	23	9	11	6	**29**	22
13	61	54	**93**	97	139	447	7	6	**10**	11	16	50
14	85	78	**48**	20	149	513	10	8	**5**	2	17	58
15	88	50	102	**45**	58	548	10	7	11	**5**	6	61
16	39	55	**39**	54	97	607	4	7	**4**	6	11	68

Second Round Answers

1983/1984

1. There are 6 such seven-digit numbers.

2. The distance from the water level to the table increases.

3. (b) Equality holds if and only if $x = \frac{1}{4}$.

4. The shortest distance between the lines AB and CD is $\frac{1}{\sqrt{2}}$.

5. (b) The second player can still ensure that the polynomial will have[1.]n distinct real roots.

1984/1985

1. There are 85 soldiers.

1985/1986

1. Of the students writing the contest this year, $\frac{5}{13}$ are girls.

3. (b) The minimum possible value of the inradius of the triangle is $\sqrt{2} - 1$.

4. (a) We can take $a_1 = a_2 = \cdots = a_{k-2} = 1$, $a + k - 1 = 3$ and $a_k = 2t + 1$ where $k = 4t + 1$.

1986/1987

2. The radii of the circles are $\frac{2}{3}$ and $\frac{5}{3}$.

3. There are six solutions, namely,

$$(x, y) = (-1, 1),\ (1, 1),\ (2, 4),\ (-3, 9),\ \left(\frac{-3 + \sqrt{5}}{2}, \frac{-3 - \sqrt{5}}{2}\right)$$

and

$$\left(\frac{-3 - \sqrt{5}}{2}, \frac{-3 + \sqrt{5}}{2}\right).$$

1987/1988

2. The minimum number of grooves Mark must re-color is 6.

3. The only solution is $(a, b, c) = (0, 0, 0)$.

4. There are three solutions, namely $(x, y) = (0, 0)$, (6,6) and (–6, –6).

1988/1989

1. The number of students in the library who do not wear glasses is 24.
2. (b) The result is true.
4. The square root of $X - Y$ has n digits, all of which are 6's.

1989/1990

1. (a) These consist of all even numbers not divisible by 4.
   (b) These consist of all odd numbers.
   (c) These consist of all multiples of 4.
2. (a) Any cut of length 25 centimeters such that one piece has 5 centimeters of the original perimeter will work.
   (b) The result is false.
3. We have $\angle ABD = 20°$.
4. (b) The number of Knights can be any odd number greater than 1.
5. The radius of the smaller ball is $4 - 2\sqrt{3}$.

1990/1991

1. The only special numbers are 19, 29, 39, 49, 59, 69, 79, 89 and 99.
3. The only such progressions are $\{1, 2, 4\}$, $\{1, 3, 9\}$, $\{1, 4, 16\}$, $\{2, 4, 8\}$, $\{3, 6, 12\}$, $\{4, 6, 9\}$, $\{4, 8, 16\}$ and $\{9, 12, 16\}$.
4. The rating of the new creature is 2.
5. The desired value is $a + b + c - ab - bc - ca + abc$.

1991/1992

1. CHEAP made either 800 or 2400 photocopies.
2. We have a $40\sqrt{5} \times 25\sqrt{5} \times 20\sqrt{5}$ block.
3. (b) The smallest positive integer with the 32-property is 8.
4. (b) Equality holds if and only if $x = y = z = 0$.
5. The minimum area is $\frac{1}{4}$ and the maximum area is $\frac{\sqrt{3}}{6}$.

1992/1993

2. (a) The probability is $\frac{3}{4}$. (b) The probability is $\frac{2}{3}$.
3. The radii are $\frac{bc}{2a}$, $\frac{ca}{2b}$ and $\frac{ab}{2c}$.
4. Steven can win if and only if n is congruent modulo 8 to 0, 2, 5, 6 or 7.

1993/1994

1. The only such polynomial is $P(x) = 2x + 3$.

2. There are four different amoebas.

3. (b) The smallest such integer N is $1993^2 + 2$.

5. (a) The only tadpoles of the form (a, a, a) are $(0, 0, 0)$ and $(2, 2, 2)$.

1994/1995

1. They are 520 meters from school.

2. The unique solution is $(a, b, c) = (\frac{\sqrt{13}-1}{2}, 1, \frac{\sqrt{13}+1}{2})$.

3. There are six solutions, namely, $(a, b, c) = (3, 2, 1)$, $(5, 1, 3)$, $(9, 2, 2)$, $(11, 4, 1)$, $(23, 1, 5)$ and $(27, 5, 1)$.

1995/1996

1. The desired sum must contain the digit 4 at least once.

2. We have ten mutually non-congruent pentagons, with $(BC, CD, DE, EA, AB) = (2, 3, 2, 3, 3)$, $(2, 3, 2, 2, 4)$, $(2, 3, 2, 3, 4)$, $(2, 3, 2, 4, 4)$, $(2, 3, 3, 3, 4)$, $(2, 3, 3, 4, 3)$, $(2, 3, 3, 4, 4)$, $(3, 3, 3, 4, 4)$, $(2, 3, 4, 4, 4)$ and $(3, 3, 4, 4, 4)$, respectively.

3. (b) Not every integer is expressible in the desired form.
   (c) The smallest such k is 6.

1996/1997

1. The inequality is satisfied if and only if $-7 < x < \frac{7}{3}$.

2. The desired sum is equal to $\frac{1}{15}(16^{1997} - 1)$.

4. (b) An example consists of $a = 50$, $b = 25$ and $p = 5$.

5. The total number of ways is 2^n.

1997/1998

1. One possible pair is $(\frac{1}{1}, \frac{199}{201})$.

2. This is impossible.

4. (c) This tournament cannot have seven teams.

1998/1999

1. There are 126 such numbers.

2. The maximum value is $135°$ and the minimum value is $90°$.

3. The only possible values are $(m, k) = (1, 2)$, $(1, 3)$, $(1, 5)$ and $(1, 9)$.

5. (a) The only possible values are $20°$ and $100°$.

1999/2000

1. (a) The weaker elf should fight first.
   (b) It does not matter which elf fights first.

3. The angles are $90°$, $75°$ and $15°$.

5. The maximum value is 4.

2000/2001

1. The only possible values are $(a, b) = (12, 4)$.

5. The minimum number of questions is eight.

2001/2002

2. The area of $ABCD$ is 29.

4. The largest possible value of M is 7531.

5. (b) We have $a = 2$, $b = 0$ and $c = -1$.

2002/2003

2. The total distance covered by the bee is 75 meters.

5. There are eight such sets, all containing $\{-\frac{1}{2}\}$ and any subset of $\{\frac{1}{2}, \frac{1}{4}, -\frac{1}{4}\}$.

2003/2004

1. The width of each brick is $\frac{5}{23}$ and the length of each brick is $\frac{5}{4}$.

2. We have $\lambda = 5$.

3. There are 2004 such numbers.

4. We have $\angle ABC = 105°$.

2004/2005

1. The two classes have 225 and 171 students respectively.

4. The radius is equal to $\frac{\sqrt{21}}{3}$.

5. We have $\frac{BE}{DE} = 2$.

2005/2006

1. We have $\ell = 4$.

2. The two price reductions are 9% and 13%.

3. The only numbers with the desired properties are 117, 156 and 195.

4. (b) We have $\frac{AB}{AD} = b^3$.

5. The maximum number of elements is 7.

First Round Solutions

November 15, 1983

1. **(d)** The first two primes greater than 50 are 53 and 59.

2. **(b)** Halving four times in succession reduces a quantity to one-sixteenth of its original value.

3. **(a)** The Pythagorean form of the equation is $(x-3)^2 + (y-4)^2 = 5^2$.

4. **(c)** We must have $a = 0$ as otherwise we have a quadratic polynomial. Similarly $b = 0$, but c can be any real number.

5. **(b)** Let the arithmetic progression be $a,\ a+d,\ a+2d,\ a+3d$ and $a+4d$. It follows from $a(a+4d) = (a+d)^2$ that $d = 2a$. Hence $\frac{a+d}{a} = 3$.

6. **(c)** For each face on the first die, there is exactly one face on the second die to yield a total of 7.

7. **(b)** The real part is $(-3)(3) + (4i)(-4i) = 7$. The imaginary part is $(-3)(-4i) + (3)(4i) = 24i$.

8. **(c)** Let Rose have m rows with n roses in each row. We are given that $mn - (m+1)(n-1) = 1$, which reduces to $m = n$.

9. **(a)** The third side cannot be less than 7 as $3^2 + 6^2 < 7^2$. It cannot be more than 7 as $3^2 + 7^2 < 8^2$.

10. **(e)** There are 6 ways of choosing a pair from the 4 horizontal lines and 10 ways to choose a pair from the 5 vertical lines, yielding a total of 60 rectangles.

11. **(b)** Let A and B be the respective centers of the circles. It follows from Pythagoras' Theorem that $PQ^2 = AB^2 - (BQ - AP)^2$.

12. **(e)** Adding up the two equations yields $10000x + 10000y = 100000$ or $x + y = 10$. Now subtract $3561x + 3561y = 35610$ from the first given equation.

13. **(e)** If $x > 2$, then $(x-2)^2 > 1$ and $x > 3$. If $x < 2$, then $(x-2)^2 < 1$ and $x > 1$. Hence the solution set is $\{x : 1 < x < 2 \text{ or } x > 3\}$.

14. **(d)** Each corner must be visited at least once while each of the other five intersections must be visited at least twice. However, in order for each of the four side-intersections to be visited twice, there must be an additional visit to the center or one of the corners.

15. **(b)** The limit L is the positive root of $L^2 = 1 + L$.

16. **(b)** The equation for $P(x, y)$ is a quadratic equation in x and y without the xy term and with the same coefficient for x^2 and y^2.

17. **(d)** We must have $\cos\theta > 0$ and $\sin\theta < 0$ as $\cos\theta < 1$.

18. **(d)** We must have $(-2p)^2 - 4(p-3)(6p) = 72p - 20p^2 > 0$ or $0 < p < 3.6$ in order for the equation to have distinct real roots. In order for the roots to be positive, we require in addition $(2p)^2 > 72p - 20p^2$.

19. **(a)** If y denotes the required value, then $\log_x y = \log_x(\log_x a)$.

20. **(b)** The top left corner must be occupied by an m while the bottom right corner must be occupied by a K. The word KSKS must be in the second row or column.

November 20, 1984

1. **(c)** The expression is approximately $0.05 \div (0.001 \div 0.005) = 0.05 \div 0.2 = 0.25$.

2. **(c)** The least common multiple of 15 and 25 is 75.

3. **(b)** We have 21=1+20=6+15=9+12.

4. **(c)** Note that $8^m = 2^{3m}$ and $16^n = 2^{4n}$.

5. **(c)** The exterior angle at vertex A is $180° - 35° = 145°$ and the sum of the three exterior anqles is $360°$.

6. **(b)** We have $ar^2 = \sqrt{15 \cdot 1215} = 15\sqrt{81} = 135$.

7. **(c)** The son is 12 years old in 1984.

8. **(a)** The space diagonal, in meters, of the room is $\sqrt{36 + 36 + 9} = 9$.

9. **(b)** Adding the four equations, we have $3(a + b + c + d) = 15$.

10. **(d)** There are three ways to fill in the first place and 1 can occupy any of three places. The remaining two places can only be filled in a single way. Hence the total is $3 \cdot 3 = 9$.

11. **(b)** The required probability is $\frac{3}{3} \cdot \frac{2}{3} \cdot \frac{1}{3} = \frac{2}{9}$.

12. **(a)** Use $\frac{a^2-b^2}{a+b} = a - b$.

13. **(a)** The polynomial is $4(x - \frac{1}{2})(x - 1)(x + 2) = 4x^3 + 2x^2 - 10x + 4$.

14. **(e)** The graph consists of the four segments $x + y = 1$ for $0 \le x \le 1$, $x - y = 1$ for $0 \le x \le 1$, $x + y = -1$ for $-1 \le x \le 0$ and $x - y = -1$ for $-1 \le x \le 0$.

15. **(b)** The midpoint of XY is at a constant distance $\frac{XY}{2}$ from the origin since the circle with XY as diameter passes through the origin.

16. **(b)** Note that $x = 1 - \frac{2b}{a+b} = \frac{2a}{a+b} - 1$ so that $\frac{1-x}{1+x} = \frac{b}{a}$. Similarly, $\frac{1-y}{1+y} = \frac{c}{b}$ and $\frac{1-z}{1+z} = \frac{a}{c}$.

17. **(b)** The numbers of spheres on the respective layers are 1, 3, 6, 10, 15, 21, 28 and 36.

18. **(e)** Let $ABCD$ be the parallelogram and M be the point of intersection of the diagonals. Let $BC = 5$, $CM = \frac{AC}{2} = 5$ and $BM = \frac{BD}{2} = 4$. We now have $CD^2 = 2CM^2 + 2BM^2 - BC^2 = 57$ by Apollonius's Theorem.

19. **(c)** If a column of the board is filled with 3 checkers, there can be at most 1 checker in each remaining column yielding a total of 10. However, we can put 2 checkers in each of three columns without forming any rectangle, and put 1 checker in each remaining column.

20. **(e)** There are infinitely many such polygons. For any positive integer n, start with a regular convex polygon with $4n + 2$ sides of length 1. Now push two opposite sides towards each other so that they remain parallel and the polygon remains convex with $4n + 2$ sides of length 1. Since the area of the variable polygon exceeds 1 initially and approaches 0 eventually, it has an area of 1 at some point.

November 19, 1985

1. **(c)** The ant follows the polygonal path from (0,0) through (0,1), (2,1), (2,–2), $(-2, -2)$, $(-2, -1)$, (0,–1) to (0,–2) where it crosses its own path for the first time. The total distance is 14 meters.

2. **(c)** Since the sum is 0, the integers are situated symmetrically about 0. Of the 15 numbers, 7 are positive. Hence the largest one is 7.

3. **(a)** From $2^m = 2^{2n}$, we have $k^4 = 2k^2$. Since $k > 0$, $k^2 = 2$ and $k = \sqrt{2}$.

4. **(d)** The rectangle determined by O, B and C has BC as one diagonal and a radius as the other diagonal. Hence BC is equal in length to the radius, which is 5 meters.

5. **(b)** Since all 9 teams are of equal strength, the fraction is $\frac{1}{2}$.

6. **(b)** We have $4y = 4t^2 + 4t = (2t + 1)^2 - 1 = x^2 - 1$.

7. **(d)** Someone is telling the truth. If it is a boy, then there are exactly two girls, and they are lying. If it is a girl, then there are exactly two boys, and they are lying. Hence the number of lying children is 2.

8. **(e)** The area in question is equal to that of two quadrants of a circle minus that of the square, which is $2\pi - 4$ square centimeters.

9. **(e)** We have $x + y + z = 1985(a - b) + 1985(b - c) + 1985(c - a) = 0$.

10. **(d)** The buses are $40 \cdot \frac{36}{60} = 24$ kilometers apart. Suppose the pedestrian had walked x kilometers before being overtaken. Then $\frac{x}{8} = \frac{x+24}{40}$ or $x = 6$. Hence the time elapsed is $\frac{6}{8}$ hours or 45 minutes.

11. **(a)** A number leaves the same remainder as its digit-sum when divided by 9. Hence $222\ldots2$ leaves the same remainder as 3970, 19, 10 and eventually 1 when divided by 9.

12. **(d)** Let $2x^2 + 18x + k = 2(x - a)(x - b)$ where a and b are integers. We have $a + b = -9$. The maximum of $k = 2ab$ occurs when a and b are as close to each other as possible, namely, $2(-4)(-5) = 40$.

13. **(b)** There are 5! ways of lining up the girls and one boy. The other boy can occupy any of 4 of the 6 positions along the line. Hence the number of ways is $120 \cdot 4 = 480$.

14. **(c)** The "land" part of the garden form four right triangles whose altitudes, and hence areas, are maximum when they are isosceles. All four maxima occur simultaneously when the garden is a square, which must be of side $\frac{4+6}{\sqrt{2}}$ meters. Thus the maximum area is 50 square meters.

15. **(c)** Let the radius be r meters. Then $(r+7)(r-7) = 12 \cdot 6$. Hence $r = 11$.

16. **(d)** We have $\alpha + \beta = -p$ and $\alpha\beta = q$. Hence $(\alpha^2+\alpha\beta)+(\alpha\beta+\beta^2) = (\alpha+\beta)^2 = p^2$ and $(\alpha^2+\alpha\beta)(\alpha\beta+\beta^2) = \alpha\beta(\alpha+\beta)^2 = p^2q$. Thus $x^2 - p^2x + p^2q = 0$ has the appropriate roots.

17. **(c)** Adding the equations in pairs, we have $a+d = 3$, $b+e = 5$, $c+a = 7$, $d+b = 9$ and $e+c = 6$. Adding them all yields $a+b+c+d+e = 15$. Hence $a = 15-5-9 = 1$. Similarly, $b = 6$, $c = 7$, $d = 3$ and $e = -1$, so that c is the largest.

18. **(c)** Note that $(\sqrt{n}+\sqrt{n+2})(\sqrt{n+2}-\sqrt{n}) = 2$. The sum in question is equal to

$$\frac{\sqrt{2}-\sqrt{0}}{2} + \frac{\sqrt{4}-\sqrt{2}}{2} + \cdots + \frac{\sqrt{400}-\sqrt{398}}{2} = \frac{\sqrt{400}}{2} = 10.$$

19. **(b)** Since the product is odd, all the integers are odd. Since the sum is odd, k is also odd. If $a+b+c = abc$ with $a \le b \le c$, we must have $a = 1$ as otherwise $a+b+c < 3c < ac < abc$. If $b = 1$ too, then $a+b+c = c+2 > c = abc$. If $b \ge 3$, then $a+b+c < 3c < abc$. Hence the smallest possible value of k is 5 since $1+1+1+3+3 = 1\cdot1\cdot1\cdot3\cdot3$.

20. **(d)** Let $ABCD$ and $A'B'C'D'$ be parallel faces with A on top of A', and so on. Let the chosen space diagonal be $A'C$. Then the plane in question passes through the midpoints of AB, AD, $C'B'$, $C'D'$, BB' and DD'. Thus the plane cross-section is a regular hexagon of side $\sqrt{2}$ centimeters and its area is $3\sqrt{3}$ square centimeters.

November 18, 1986

1. **(e)** Because the lawn has side length 1000 centimeters, the number of blades of grass is $1000 \cdot 1000 \cdot 10$.

2. **(e)** We have $1 < 1+2$, $-2 < -2-(-1)$, $1 < 1 \times 2$ and $-2 < -2 \div -1$.

3. **(a)** Equating coefficients, we have $A+B = -A$ and $AB = B$. Hence either $B = 0$ or $A = 1$. If $B = 0$, then $A = 0$ also, which is forbidden. Hence $A = 1$ and $B = -2$.

4. **(e)** Because $1+2+\cdots+2^{n-2} < 2^{n-1}$, the polygon does not exist.

5. **(c)** If r is a root of each polynomial, it is a root of their difference. From $(A-B)r + B - A = 0$, we have $r = 1$. Substituting $x = 1$ into $x^2+Ax+B = 0$, we have $A+B = -1$.

6. **(d)** We may have $T(P) = 1$ for all P, invalidating (a), (b) and (c). Suppose $T(A') - T(B') \ne 0$ for some diameter $A'B'$. Let PQ be a variable diameter which rotate from $A'B'$ 180° to $B'A'$. Now $T(P)-T(Q)$ varies continuously from $T(A')-T(B')$ to $T(B')-T(A')$. Hence there must be a position AB of PQ during the rotation in which $T(A)-T(B) = 0$.

7. **(c)** The term -1 can be omitted. Since $2^{10} \simeq 10^3$, $2^{1986} \simeq 10^{596}$. The number of digits is about 597.

8. **(c)** There are four planes of symmetry determined by the four lines of symmetry of a square end and the perpendicular to the square end. One more plane of symmetry is parallel to and halfway between the square ends.

9. **(c)** A sum of 6 can be obtained as 1+1+4 in 3 ways, as 1+2+3 in 6 ways and as 2+2+2 in 1 way.

10. **(d)** There are 27 distinct unit cubes and 8 distinct 2 by 2 by 2 cubes.

11. **(c)** We have $z^x = (xy)^x = (yz)xy = y^{xyz}$.

12. **(d)** Let $x = \sqrt[3]{1.0009}$ and $y = \sqrt[3]{1.00009}$. Then $x^3 - y^3 = 0.00081$ while $x^2 + xy + y^2 \simeq 3$. Hence $x - y = \frac{x^3-y^3}{x^2+xy+y^2} \simeq 0.00027$.

13. **(c)** From $x^2 + 2xy + 3y^2 + 4y + 3 = (x + y)^2 + 2(y + 1)^2 + 1$, the maximum value is seen to occur if and only if $x + y = 0$ and $y + 1 = 0$.

14. **(d)** Let $x + y + z = s$. Then $x = \frac{s}{8}$, $y = \frac{s}{3}$ and $z = \frac{s}{k}$. Hence $\frac{1}{8} + \frac{1}{3} + \frac{1}{k} = l$.

15. **(d)** The maximum is at most 8 as $3^2 - 1 = 8$. Since k is odd, both $k - 1$ and $k + 1$ are even. Moreover, one of them is divisible by 4 while the other is divisible by 2. Thus $k^2 - 1$ is always divisible by 8 for odd k.

16. **(e)** If $\frac{x+y+z}{x}$ is rational. Then so is $\frac{y+z}{x} = \frac{x+y+z}{x} - 1$ and conversely. Similarly, $\frac{x+y+z}{y}$ and $\frac{z+x}{y}$ are simultaneously rational or simultaneously irrational.

17. **(c)** We have $2 = \frac{1}{2}(\frac{a+b}{2} + \frac{c+d}{2}) \leq \frac{\sqrt{ab}+\sqrt{cd}}{2} \leq \sqrt[4]{abcd}$, with equality if and only if $a = b$, $c = d$ and $a + b = c + d$.

18. **(b)** Suppose $\frac{n!x^k}{k!(n-k)!} = \frac{n!x^{k+1}}{(k+1)!(n-k-1)!}$. Then $x = \frac{k+1}{n-k} > 0$. Thus we can have two consecutive terms equal. If the k-th, $(k + 1)$-st and $(k + 2)$-nd terms are equal, then $x = \frac{k+1}{n-k} = \frac{k+2}{n-k-1}$. This simplifies to $n = -1$, a contradiction.

19. **(d)** We have 95=6(14)+11, 87=5(15)+12, 98=5(17)+13 and 79=4(16) +15. In order to leave a remainder of 14, the sum of the two digits must be at least 15. None of 69, 78, 79, 87, 88, 89, 96, 97, 98 and 99 leave 14 as a remainder when divided by the sum of its digits.

20. **(a)** If $\sqrt{3} + \sqrt{2}$ is a root, the others are $\sqrt{3} - \sqrt{2}$, $-\sqrt{3} + \sqrt{2}$ and $-\sqrt{3} - \sqrt{2}$. B is then the sum of the products of the roots taken two at a time.

November 17, 1987

1. **(b)** We have $f(2) = 2$, $f(3) = 4$, $f(4) = 5$ and $f(5) = 10$.

2. **(b)** Since $1988 = 2 \cdot 2 \cdot 7 \cdot 71$, $2 + 2 + 7 + 71 = 82$.

3. **(d)** Pressing the square root button n times is equivalent to taking the $2n$-th root.

4. **(c)** Since $6 < \sqrt{x} < 7$, $27 < 36 < x < 49 < 64$.

5. **(b)** The neighbors of 1 must be 2 and 3. The other neighbor of 2 must be 4, and the other neighbor of 3 must be 5. It will turn out that the sum of any two diametrically opposite numbers is 11.

6. **(b)** If we set $x = -3$, the equation reduces to $\frac{1}{4} = \frac{1}{4}$. Similarly, if $x = -4$, we have $\frac{1}{3} = \frac{1}{3}$.

7. **(b)** The function $\frac{x}{(x-1)(x+1)}$ is positive if and only if 0 or 2 of the 3 factors are negative.

8. **(a)** Let $AH = x$. Then $AE = \sqrt{3}x$. Since $AE = 1 - x$ also, $x = \frac{1}{1+\sqrt{3}}$.

9. **(e)** The positive divisors of $87 - 19 = 68$ are 1, 2, 4, 17, 34 and 68.

10. **(e)** Let ABC be the triangle and $PQRS$ be the rectangle. We may place A inside $PQRS$ close to P, and make AB and AC intersect PQ and PS respectively. If AB and AC are long enough, R will be inside ABC. On the other hand, it is easy to have all of A, B and C outside $PQRS$.

11. **(c)** The number of handshakes between b boys is $\frac{b(b-1)}{2} = 66$ so that $b = 12$. Similarly, the number of girls is found to be 11.

12. **(c)** The sum of the digits of the number must be a multiple of both 2 and 9. To minimize the number, we may as well take this sum to be 18. Since each digit is at most 8, our number has at least three digits.

13. **(e)** There are 6 planes of symmetry, each containing one edge and passing through the midpoint of the opposite edge.

14. **(c)** From $\frac{1}{a+b} + \frac{1}{c+a} = \frac{2}{b+c}$, we have $b^2 + c^2 = 2a^2$.

15. **(a)** The equation may be rewritten as $|x-1||x+3| = x - 1$. Clearly $x = 1$ is a root. If $1 < x$, we have x+3=1 which is impossible. If $x < 1$, we have $|x+3| = -1$ which is also impossible.

16. **(e)** The coins may be loaded so that they always come up heads, favoring 3 heads over 2 heads and 1 tail. The situation is reversed if each coin is loaded so that its probability of coming up heads is just slight above $\frac{1}{2}$.

17. **(a)** Since AC and BD are diameters of the circumcircle, both PAC and PBD are right triangles. The value of $PA^2 + PC^2 + PB^2 + PD^2$ is constant.

18. **(e)** We have $\sqrt{x \pm 2\sqrt{x-1}} = \sqrt{1 \pm 2\sqrt{x-1} + (x-1)} = 1 \pm \sqrt{x-1}$.

19. **(b)** We have $PB = \frac{PC \cdot PD}{PA} = 12$. Let $CDEF$ be a rectangle inscribed in the circle. Then $DE = 12 - 4 = 8$ and $CE^2 = 14^2 + 8^2 = 260$.

20. **(a)** The digits 1, 9, 8 and 7 become 1, 2, 1 and 0 respectively when reduced modulo 7. Let the rearranged number be $1000a + 100b + 10c + d$. If it is divisible by 7, we must have $-a + 2b + 3c + d \equiv 0 \pmod{7}$. The only solutions (a, b, c, d) are (1,1,2,0) and (0,1,1,2).

November 15, 1988

1. **(c)** Since $9^2 < 99 < 10^2$ and $4^3 < 99 < 5^3$, $4 < \sqrt[3]{99} < 5 < 9 < \sqrt{99} < 10$.

2. **(c)** We may have 3 boxes of apples, 1 box of orange and 4 boxes of lemons. If we have 4 or more boxes of apples, the average cost of the other boxes will be under \$2, which is impossible.

3. **(c)** Let the acute angles be α and β with $\alpha \leq \beta$. Then α can assume any integer value of degrees from 1 to 45 inclusive.

4. **(e)** From the first equation, we have $(\frac{x}{y})^6 = 8$. From the second, we have $(\frac{x}{y})^6 = 9$.

5. **(c)** By the symmetry of x and $2 - x$, another solution is given by $2 - x = \frac{1}{2}$.

6. **(c)** We have $17^{105} > 16^{105} = 2^{420}$, $3^{210} < 4^{210} = 2^{420}$, $7^{140} < 8^{140} = 2^{420}$, $31^{84} < 32^{84} = 2^{420}$ and $127^{60} < 128^{60} = 2^{420}$.

7. **(e)** By symmetry, the area of the common part is equal to $\frac{1}{6}$ that of the hexagon, or that of an equilateral triangle of side 1.

8. **(a)** The constant term is $\binom{12}{3}(2x^3)^3\left(-\frac{1}{x}\right)^9$.

9. **(b)** Of the 48 equally likely outcomes, 6 result in a tie. By symmetry, half of the remaining 42 go to Arthur and the other half go to Betty.

10. **(b)** Reflect the midpoint of BC across CD and the resulting image across AD. The shortest distance the bug must crawl is equal to that between this final image and the midpoint of AB, or $\sqrt{12^2 + 9^2}$.

11. **(b)** Let $y = x + \frac{1}{x}$. Then $x^2 + \frac{1}{x^2} = y^2 - 2$. The equation now becomes $(2y-5)(y-1) = 0$. No real solutions exist for $x + \frac{1}{x} = 1$. From $x + \frac{1}{x} = \frac{5}{2}$, we have $x = 2$ or $\frac{1}{2}$.

12. **(c)** We have $1989 = 9 \cdot 13 \cdot 17$.

13. **(e)** Let the larger sphere touch the floor at D and the smaller sphere touch the floor at E. Then ACD and BCE are similar triangles. If $AC = x$, then $\frac{x-7}{3} = \frac{x}{4}$.

14. **(e)** From the identity $\sqrt[3]{x} - \sqrt[3]{y} = \dfrac{x - y}{\sqrt[3]{x^2} + \sqrt[3]{xy} + \sqrt[3]{y^2}}$, we have $\sqrt[3]{1.001001001} - \sqrt[3]{1.001001}$ approximately equal to $\frac{0.000000001}{3} = 3.3 \cdot 10^{-10}$.

15. **(b)** We have $224 = a_1 + a_8 + a_{12} + a_{19} = 4a_1 + 36d$ where d is the common difference of the arithmetic progression. Then $a_1 + 9d = 56$. Now $a_1 + a_2 + \cdots + a_{19} = 19a_1 + 171d = 19(a_1 + 9d)$.

16. **(c)** The volume of a tetrahedron is given by $\frac{1}{3}$ times its base area and its altitude. Instead of calculating the volume of the tetrahedron in question directly, we see that it may be obtained from the unit cube by removing four tetrahedra, each of which is congruent to the one with vertices at $(0, 0, 0)$, $(1, 0, 0)$, $(0, 1, 0)$ and $(0, 0, 1)$. This tetrahedron has base area $\frac{1}{2}$ and altitude 1, so that its volumes is $\frac{1}{6}$. Hence the volume of the central tetrahedron is $1 - 4(\frac{1}{6}) = \frac{1}{3}$.

November 21, 1989

1. **(d)** The smallest a is $2^{12} = 4096$.

2. **(a)** The distance from the center to a vertex of the square is $\frac{1}{\sqrt{2}}$.

3. **(d)** Sincc 48:36=4:3, the number of days the two stayed are also in the ratio 4:3. One of them must be at least 7 while the other must be at most 6.

4. **(c)** The desired average is given by $7 + 8.4 - 2.3$.

5. **(c)** The respective probabilities are $\frac{1}{4}$, $\frac{2}{6}$, $\frac{3}{8}$, $\frac{4}{12}$ and $\frac{5}{20}$.

6. **(e)** We have $4x^2+100xy+625y^2 = (2x+25y)^2 = 9$ when $x = 14$ and $y = -1$, for instance. Note that for $a = 28$, 60, 80 and 96, we will have $(2x + my)^2 + (ny)^2 = 9$ for $(m, n) = (7, 24)$, (15,20), (20,15) and (24,7) respectively. However, 9 cannot be expressed as a sum of two non-zero integral squares.

7. **(b)** We have $f(2) = 2^2 = 4$, $f(f(2)) = 2^4 = 16$ and $f(f(f(2))) = 2^{16} = 4^8$.

8. **(c)** The statement is not true for $n = 1$ or 2. It is easy to construct a circle tangent to exactly 1 or 2 sides of a regular polygon. The statement is true for $n = 3$. The circle tangent to three sides is the incircle of the triangle determined by these three sides, and is therefore unique. Hence it must also be the incircle of the regular polygon.

9. **(c)** We cannot have 7 dimes as otherwise there will be 6 worthless coins. We cannot have 6 dimes since no 7 coins add up to 10 cents. We can have 5 dimes, 3 nickels and 5 pennies.

10. **(d)** The statement is not true for $n = 3$ as 1+2+3=6. It is not true for n=4 as the sum of 4 consecutive integers is always even. It is not true for $n = 5$ as 2+3+4+5+6=20. Since the statement is true for $n = 2$, it must also be true for n=6.

11. **(e)** There are 2^6 sequences of 6 T's and F's. Each of 64 students can answer the 6 questions according to a different sequence. On the other hand, if there are more than 64 students, the Pigeonhole Principle will force two of them not to give opposite answers to any question. That students may leave questions unanswered is irrelevant.

12. **(e)** Suppose p students pass and f students fail. Then $pX + fY = (p + f)Z$. It follows that $\frac{Z-Y}{p} = \frac{X-Z}{f} = \frac{X-Y}{p+f}$ and $\frac{f}{p+f} = \frac{X-Z}{X-Y}$.

13. **(a)** We have $b+c < d+e < a+b < c+d < e+a$. It follows that $b < e < c < a$ and $b < d < a$.

14. **(b)** Let F be the center of $BCDE$. Triangles ABE and FBE have the same altitude BE and their areas are in the ratio 4:1. If $BE = BF = x$, then $AE = 4x$ and $AB^2 = 26x^2 = 1989$. The area of BCED is $2x^2 = 153$.

15. **(c)** Let the first term be a and the common ratio be r. Then $\dfrac{a(r^{n-1})}{r-1} = 2n$ and $\dfrac{a(r^{n-1})}{r-1} = n$. If we divide the second equation by the first, we have $r^n + 1 = \frac{1}{2}$

or $r^n = -\frac{1}{2}$. Thus we have

$$\begin{aligned}\frac{a(r^{3n}-1)}{r-1} &= \frac{a(r^n-1)(r^{2n}+r^n+1)}{r-l} \\ &= 2n(\frac{1}{4}-\frac{1}{2}+1) \\ &= \frac{3n}{2}.\end{aligned}$$

16. **(a)** The altitude of the triangle to the 48-centimeter base is 32 centimeters. Hence the area of the triangle is 768 square centimeters. The semi-perimeter of the triangle is 64 centimeters. Hence the inradius of the triangle is $\frac{768}{64}$ or 12 centimeters. The ratio of the radii of the two circles is equal to $32 : (32 - 2 \cdot 12)$ or 4:1.

November 20, 1990

1. **(c)** The old number of the new #1 is $1 + 33 - 7 = 27$. The new number of the old #26 is $26 + 18 - 8 = 36$.

2. **(a)** Rotate the circle so that the vertices of the smaller square are at the midpoints of the sides of the larger square.

3. **(e)** The minimum height of each stack is the least common multiple of 2, 3 and 5. The numbers of the bricks in the stacks are $\frac{30}{2}$, $\frac{30}{3}$ and $\frac{30}{5}$ respectively.

4. **(b)** Introducing a number equal to the average of a set of numbers yields a new set with the same average.

5. **(a)** Five markers are necessary as each can take care of at most 4 squares. Five markers are sufficient if they are placed in knight moves apart, starting from a corner square.

6. **(c)** We have $(x - y)(x + y) = 1 \cdot 99 = 3 \cdot 33 = 9 \cdot 11$. Hence $(x, y) = (50, 48)$, (18,15) or (10,1).

7. **(e)** The ratio of the areas of ADE and $ABCE$ is $DE : (CE + 5)$. Hence $DE = CE + 5$. Also, $DE + CE = 8$. Hence $DE = \frac{13}{2}$ and $CE = \frac{3}{2}$.

8. **(c)** We have $2^{2x} = \dfrac{2^{x+y}2^{x+z}}{2^{y+z}} = 15$.

9. **(d)** There is one cube in which every pair of opposite faces is of one color. There are three cubes in which exactly one pair of opposite faces is of the same color. There are two cubes in which no pair of opposite faces is of the same color.

10. **(d)** There is one way of forming a triangle with three short straws, and one way with three long ones. There are nine ways of forming a triangle with two long straws and one short one.

11. **(b)** Note that $x + y - \sqrt{3xy} > x + y - 2\sqrt{xy} = (\sqrt{x} - \sqrt{y})^2 \geq 0$, $(x + y)^2 - (x^2 + y^2) = 2xy > 0$ and

$$\begin{aligned}(x + y)(x^3 + y^3) - (x^4 + x^2y^2 + y^4) &= xy(x^2 + y^2 - xy) \\ &> xy(^2{+}y^2 - 2xy) \\ &= xy(x - y)^2 \\ &\geq 0.\end{aligned}$$

12. **(b)** Note that $\angle B_1C_1A_2 = \angle A_2B_2C_2$. It follows that $C_2D_2 = A_2B_2 = C_1A_2 = \frac{C_1D_1}{2}$. Hence the total length is $1 + \frac{1}{2} + \frac{1}{4} + \cdots = 2$.

13. **(c)** Alice, Brian and Caroline wake up every 6, 8 and 10 minutes respectively. The first time they all wake up together is after 120 minutes.

14. **(d)** A square cannot end in a 3. Also, when divided by 4, a square always leaves a remainder of 0 or 1. To find this remainder, only the number formed of the last two digits needs to be considered.

15. **(a)** We have

$$\sqrt{200} - \sqrt{199} = \frac{1}{\sqrt{200} + \sqrt{199}} > \frac{1}{\sqrt{201} + \sqrt{200}} = \sqrt{201} - \sqrt{200}$$

from the identity $A^2 - B^2 = (A - B)(A + B)$, which is equivalent to $A - B = \frac{A^2 - B^2}{A + B}$. Similarly, we also have

$$\sqrt{200} - \sqrt{199} > \frac{1}{2\sqrt{200}} = \frac{5\sqrt{2}}{200} > \frac{6}{200} = 0.03.$$

Since $A^3 - B^3 = (A - B)(A^2 + AB + B^2)$, we have

$$\sqrt[3]{304} - \sqrt[3]{300} < \frac{1}{3\sqrt[3]{300^2}} = \frac{2}{15\sqrt[3]{90}} < \frac{3}{105} < 0.03$$

since $(\frac{14}{3})^3 = \frac{2744}{27} > 90$. Finally, from the identity $A^4 - B^4 = (A - B)(A^3 + A^2B + AB^2 + B^3)$,

$$\sqrt[4]{409} - \sqrt[4]{400} < \frac{9}{4\sqrt[4]{400^3}} = \frac{9}{40\sqrt[4]{6400}} < \frac{9}{320} < 0.03.$$

16. **(a)** If the two lines intersect, the locus is a rectangle. Suppose they are parallel and at a distance D apart. If $D > d$, the locus is the empty set. If $D = d$, the locus is an infinite strip. If $D < d$, the locus is a pair of parallel lines.

November 19, 1991

1. **(b)** The prime numbers which divide 1991 are 11 and 181. Hence $N = 192$. The prime numbers which divide 192 are 2 and 3.

2. **(e)** The equation simplifies to $x - 1 = 1$.

3. **(e)** Each term is of the form $1 - \frac{1}{2^k}$ for some k, $1 \le k \le 10$. Hence the sum is

$$10 - \left(\frac{1}{2} + \frac{1}{4} + \frac{1}{8} + \cdots + \frac{1}{1024}\right) = 9 + \frac{1}{1024}.$$

4. **(e)** The border of the cardboard contributes 2(40+60) centimeters towards the total perimeter of the pieces. The cut contributes 2(2000) centimeters. Hence the total perimeter is 4200 centimeters.

5. **(a)** If $x = 10$, then $y = 10^{10}$. Hence $x^y = 10^{10000000000}$. On the other hand, we only have $100000^{100000} = 10^{500000}$.

6. **(c)** First suppose ℓ_2 is between ℓ_1 and ℓ_3. Drop perpendiculars from A and B to ℓ_2 and ℓ_3 at E and F respectively. Then triangles ABE and CBF are congruent. Hence $BF = 7$ and $CF = AE = 5$. The cases where ℓ_2 is not between ℓ_1 and ℓ_3 can be handled in a similar manner and lead to the same answer.

7. **(b)** Let $p = a + b + c$, $q = a - b - c$, $r = a - b + c$ and $s = a + b - c$. Then

$$\begin{aligned} p + q + r + s &= 4a, \\ pq &= a^2 - (b + c)^2 \\ rs &= a^2 - (b - c)^2. \end{aligned}$$

Now $(p+x)(q+x) = (r-x)(s-x)$ can be simplified to $(p+q+r+s)x = rs - pq$.

8. **(a)** Let x, y and z denote the respective numbers of years in which there were 25, 16 and 20 problems. Then we have $x + y + z = 12$ and $25x + 16y + 20z = 213$, which lead to $9x + 4z = 21$. Now z is divisible by 3.

9. **(e)** We denote the five factors initially by 00000. We insert four 1's to divide them into five parts which are not necessarily non-empty. These parts will correspond to the five variables. For instance, 100110100 stands for b^2de^2. The total number of terms is equal to the number of sequences consisting of five 0's and four 1's, or $\binom{9}{4}$.

10. **(e)** Suppose the King collected x dollars and bought y firecrackers that day. Then we have $x - 10y = 1000$ and $y - \frac{4x}{100} = 200$, yielding $x = 5000$ and $y = 400$.

11. **(b)** The given expression can be factored into $(1 + x)(1 + x^2 + x^4 + \cdots + x^{1990})$. Since the second factor is always positive, the only real root is -1.

12. **(d)** We have $\frac{w}{x} = \frac{AE}{EC} = \frac{z}{y}$ so that $wy = xz$ is always true. If $ABCD$ is say a square, $w + y = x + z$. On the other hand, if $AB = BC = CD = 1$ and $DA = 0.00001$, then w, y and z are all very small but x is not, so that $w + y \neq x + z$.

13. **(d)** Brenda's best strategy is to aim at the region labeled 4. If she hits it, she does not throw the second dart. If not, she aims at the region which would bring her score to 4. Her winning probability is $40\% + 60\%(40\%) = 64\%$.

14. **(b)** The expression is equal to $\dfrac{x^2-x+1}{\sqrt[3]{9}}$, where $x=\sqrt[3]{2}$. Now

$$x^2-x+1=\frac{x^3+1}{x+1}=\frac{3}{\sqrt[3]{2}+1}.$$

Note that $\frac{3}{\sqrt[3]{9}}=\sqrt[3]{3}$.

15. **(e)** Since $\angle APB=\angle BPC=\angle CPA=120°$, the rays PA, PB and PC are coplanar. So are PB, PC and PD so that PA and PD coincide.

16. **(a)** If (a) and (d) are both true, then $m+7n=(m+n)+6m$ is greater than 3 and divisible by 3, hence not a prime. If both (a) and (e) are true, then $m+n=3n+5$ cannot be divisible by 3. It follows that (a) is false. In fact, we must have $m=17$ and $n=6$.

November 17, 1992

1. **(e)** Since $2^{x+1}>8$, $4^{x+2}=4(2^{x+1})^2>256$.

2. **(d)** The areas of the four triangles are $\frac{1}{2}$, 1, 2 and 4.

3. **(c)** Apart from Andy, Candy and Randy, the other four ate 18 peanuts. The average value is 4.5. However, if Sandy only ate 5, there would be a tie for first place.

4. **(a)** The compound fraction simplifies to $\frac{ad}{bc}$, and is uniquely determined when a and d are chosen. The number of ways is $\binom{4}{2}$.

5. **(a)** The triangles ABC and ACD are similar. Hence $\frac{AD}{AC}=\frac{AC}{AB}$.

6. **(c)** Take the four vertices of a square. Then P is true and Q is false. Conversely, if three points are on a line, there is a plane which contains this line and any given fourth point.

7. **(d)** If the sides of the square are parallel to the grid of the lattice, its area must be a square number. If not, each side is the hypotenuse of a right triangle with integer sides, so that the area is a sum of two square numbers.

8. **(c)** If there is no change in the tens digit, exactly 1 in every 5 consecutive positive integers is a fiver. In a block of 7, the tens digit can change at most once. Thus we have at most 3 fivers. The block from 49999 to 50005 contains 3 fivers.

9. **(c)** The figure consisting of the smallest two circles and the two lines is similar to the figure consisting of the largest two circles and the two lines. Hence the three radii are in geometric progression.

10. **(a)** If $x=-2$, the inequality in (b) holds. If $x=-\frac{1}{2}$, those in (c) and (d) both hold. We now solve the inequality in (a). If $x<0$, then $1-x>0$ and the inequality fails. Hence $x>0$, so that $1-x\geq 0$.

11. **(e)** The expression $a + 4b + 11c$, where each of a, b and c is one of -1, 0 and 1, can assume the values 1, 3, 4, 5, 6, 7, 8, 10, 11, 12, 14, 15 and 16. Clearly, the weight n of the object is not equal to any of these. Now $n \neq 2$ as otherwise it can be determined from $1 < n < 3$. Similarly, $n \neq 9$ or 13.

12. **(d)** Clearly, one of the primes must be 47. Let the other two be x and y with $x < y$. Then we have $47xy = 47(x + y + 47)$ or $(x - 1)(y - 1) = xy - x - y + 1 = 48$. Now $(x - 1, y - 1)$ must be one of (1,48), (2,24), (3,16), (4,12) or (6,8).

13. **(c)** The points of contact of two spheres with the carpet must be at least 100 apart, distance measured in centimeters. Divide the carpet into three rectangles of respective dimensions 14 by 99, 85 by 50 and 85 by 49. The diagonals of each rectangle is less than 100. Hence no more than 3 spheres can touch the carpet. By placing a sphere on top of a corner of the carpet, it is easy to arrange for two more to touch the carpet.

14. **(a)** Let $x = \sqrt[3]{1 + \left(\frac{2}{3}\right)\sqrt{\frac{7}{3}}}$, $y = \sqrt[3]{1 - \left(\frac{2}{3}\right)\sqrt{\frac{7}{3}}}$ and $s = x + y$. Then $xy = -\frac{1}{3}$. It follows that $2 = x^3 + y^3 = (x + y)(x^2 - xy + y^2) = s(s^2 + 1)$. Hence $(s - 1)(s^2 + s + 2) = 0$. Since s is real, $s = 1$.

15. **(c)** The bottom three teams play three games among themselves, worth a total of 6 points. Hence the minimum point totals for the top three teams are 9, 8 and 7 respectively. Since there are 30 points available overall, these values are exact. Hence the bottom three lose all games against the top three teams. We may have the first place team beating the third place team and the fourth place team beating the sixth place team, with all other games ties.

16. **(d)** Note that $1 + x^2 + x^4 = 1 + 2x^2 + x^4 - x^2 = (1 + x^2)^2 - x^2 = (1 - x + x^2)(1 + x + x^2)$. Then

$$\frac{x}{1 + x^2 + x^4} = \frac{1}{2}\left(\frac{1}{1 - x + x^2} - \frac{1}{1 + x + x^2}\right).$$

Hence the given sum becomes

$$\frac{1}{2}\left(\left(1 - \frac{1}{3}\right) + \left(\frac{1}{3} - \frac{1}{7}\right) + \left(\frac{1}{7} - \frac{1}{13}\right) + \cdots + \left(\frac{1}{9703} - \frac{1}{9901}\right)\right) = \frac{1}{2} - \frac{1}{19802}.$$

November 16, 1993

1. **(c)** On the average, you will get one question out of three, scoring 5 points. Thus your average score per question is $\frac{5}{3}$. *Note that you score 2 points for each question if you do not guess.*

2. **(e)** The total volume of rain in cubic millimeters is $10^7 \cdot 10^7 \cdot 10 = 10^{15}$. Since the volume of each drop is 10 cubic millimeters, the number of rain drops is 10^{14}.

3. **(c)** Since $0.\overline{3} = \frac{1}{3}$ and $0.\overline{6} = \frac{2}{3}$, their product is $\frac{2}{9} = 0.\overline{2}$.

4. **(a)** The tablecloth is 4 meters from corner to opposite corner. Hence the length of a side in meters is $\frac{4}{\sqrt{2}} = 2\sqrt{2}$.

5. **(e)** Suppose Danny has x brothers and x sisters. Then his sister has $x+1$ brothers and $x-1$ sisters. From $x+1=2(x-1)$, $x=3$.

6. **(c)** Note that if one of the four statements is true, then so are all following ones. Therefore, A and B are liars. Hence C's statement is true, so that C and D tell the truth.

7. **(e)** There are three ways to arrange the seven rectangles into a large one. The two obvious ways yield rectangles of perimeters 50 and 62. By arranging four rectangles into a 4×12 one and three rectangles into a 3×12 one, we have a 7×12 rectangle with perimeter 38.

8. **(e)** Since the number of non-natives does not change but the percentage of non-natives has doubled, the total population must have halved. Since the original population was 1000, 500 natives must have emigrated.

9. **(b)** Samira covers 6 kilometers in $\frac{1}{3}+\frac{1}{2}+\frac{2}{3}=\frac{3}{2}$ hours. Hence her average speed is $6\div\frac{3}{2}=4$ kilometers per hour.

10. **(b)** Let the backbencher's age and IQ be x, and the number of cabinet members be n. Then $x-51=n(51-50)$ and $111-x=n(114-111)$. Addition yields $60=4n$ or $n=15$.

11. **(c)** The units-digit of one number is 5 and its hundreds-digit is odd. If this digit is 9, the product will exceed 450000. If it is 5 or less, then the product will be under 360000. Hence it is 7. Let the tens-digit be x. Then $(5+7)x+3$ leads to the tens-digit 5 of the product. Hence x is 1 or 6, and $567\cdot 765=433755$.

12. **(c)** We must not have addition and subtraction back to back, nor multiplication and division. Otherwise the answer can only be a. To maximize the value of the expression, addition should come after division but before multiplication. Hence the optimal expression is $(\frac{a-2}{2}+2)2$ or $(\frac{a}{2}+2)2-2$, either of which yields $a+2$.

13. **(d)** By symmetry, C lies on FG and the area of $DEHC$ is twice that of DEH. By Pythagoras' Theorem, $EA=12$, so that $BE=3$. Since triangles ADE and BEH are similar, $HE=5$. Hence the area of DEH is $\frac{75}{2}$ and the area of $DEHC$ is 75.

14. **(c)** You get $3\cdot 2^{30}-2^{29}-2^{28}-\cdots-2-1=3\cdot 2^{31}-(2^{30}-1)=2^{31}+1$.

15. **(d)** Let $ABCD$ be the building and the peg P be off the wall AB. Extend PA to E and PB to F such that $AE=10=BF$. The goat can eat the grass in triangle PAB, three quadrants of the circle with center P and radius 20 between PE and PF, an octant of the circle with center A and radius 10 between AD and AE, and an octant of the circle with center B and radius 10 between BC and BF. The total area is

$$50+300\pi+\frac{25\pi}{2}+\frac{25\pi}{2}=325\pi+50.$$

16. **(c)** Let $a(x)=x+2\sqrt{x-1}$ and $b(x)=x-2\sqrt{x-1}$ for $1\le x\le 2$. Then $a(x)+b(x)=2$ and $a(x)b(x)=x^2-4(x-1)=(2-x)^2$. Now

$$(\sqrt{a(x)}+\sqrt{b(x)})^2=a(x)+b(x)+2\sqrt{a(x)b(x)}=2x+2(2-x)=4.$$

Hence

$$F(x) = \frac{1}{\sqrt{a(x)}} + \frac{1}{\sqrt{b(x)}} = \frac{\sqrt{a(x)} + \sqrt{b(x)}}{\sqrt{a(x)b(x)}} = \frac{2}{2-x}.$$

It follows that $F(\frac{3}{2}) = 4$.

November 15, 1994

1. **(a)** Of the 10 possible combinations, only (20,30,40), (20,40,50) and (30,40,50) satisfy the Triangle Inequality.

2. **(c)** Denote the height by h. Then $7 \cdot 12 \cdot 15 = 7 \cdot 18 \cdot h$ or $h = \frac{12 \cdot 15}{18} = 10$.

3. **(d)** Put Lil's cottage at point 0 and Ted's at point 3. One of Jack's and Jill's cottages is at point 2 and the other at point 6. Hence the distance between them is 4 kilometers.

4. **(b)** There are at least one King and two Queens. We may have the King of Spades to the left of the Queen of Spades, which is to the left of the Queen of Hearts. It follows that the minimum number of cards is three.

5. **(c)** Ace, Bea, Cec and Dee get respectively \$20, \$26, \$37 and \$37 under the father's plan and \$24, \$24, \$32 and \$40 under the mother's. Thus Bea and Ced prefer their father's plan while Ace and Dee prefer their mother's.

6. **(e)** The volume of a sphere is proportional to the cube of its radius. Hence $n \geq \frac{1000}{27}$. The minimum value of n is $\lceil \frac{1000}{27} \rceil = 38$ and it exceeds 20.

7. **(c)** The sum of all positive integers up to 600 is 300(600+1) =180300, and the sum of all positive multiples of 3 up to 600 is 100(600+3) =60300. Hence the sum of all positive integers up to 600 which are not multiples of 3 is $180300 - 60300 = 12000$.

8. **(b)** The five lines can be divided into classes of mutually parallel lines in 7 ways, namely, a single class of 5, a class of 4 and a class of 1, a class of 3 and a class of 2, one class of 3 and two classes of 1, two classes of 2 and one class of 1, one class of 2 and three classes of 1, and five classes of 1. They yield respectively 0, 4, 6, 7, 8, 9 and 10 points of intersection, and these 7 values are distinct.

9. **(d)** We have $h = 12(1 + \frac{2}{3} + (\frac{2}{3})^2 + \cdots + (\frac{2}{3})^5) = 36(1 - (\frac{2}{3})^6)$. It follows that $h < 36$. Since $1 - (\frac{2}{3})^6 > \frac{8}{9}$, $h > 32$.

10. **(a)** We have $(r+1)(r+2)(r-4) = r(r^2 - r - 10) - 8 = -8$, an integral value.

11. **(d)** Tim can get at most \$2 more than Sarah, and Ursula at most \$5 more than Tim. Hence the largest amount of money, in dollars, that Ursula can get is $N + 7$. If Sarah gets \$8, then Tim gets \$10 and Ursula \$15. Thus Ursula can get \$7 more than what Sarah gets.

12. **(c)** The equation may be rewritten as $(2x-1)^3 + (y+2)^3 = 28$. Note that 28 is expressible as an ordered sum of two positive cubes in only two ways, namely $1^3 + 3^3$ and $3^3 + 1^3$. We may have $2x - 1 = 1$ and $y + 2 = 3$, yielding $9x, y) = (1, 1)$. Alternatively, $2x - 1 = 3$ and $y + 2 = 1$, yielding $(x, y) = (2, -1)$. It remains

to be shown that 28 is not expressible as a sum of a positive cube and a negative cube. Suppose $28 = a^3 - b^3 = (a-b)(a^2+ab+b^2)$ for integers $a > b > 0$. Then $a - b = 1$, 2 or 4, leading respectively to $b^2 + b = 9, 3b^2 + 6b = 10$ or $b^2 + 4b + 3 = 0$, none of which has positive integral solutions.

13. **(c)** The set $\{1, 2, \ldots, 24\}$ may be partitioned into 16 equivalence classes, four of which contains at least two numbers: $\{2, 4, 8, 16\}, \{3, 9\}, \{6, 12, 18, 24\}$ and $\{10, 20\}$. By taking one number from each class, we have the set $\{1, 2, 3, 5, 6, 7, 10, 11, 13, 14, 15, 17, 19, 21, 22, 23\}$ which consists of 16 numbers non-equivalent to one another.

14. **(c)** The region consists of the square with vertices $(2, 2)$, $(2, -2)$, $(-2, 2)$ and $(-2, -2)$, minus four symmetrically placed right isosceles triangles, one of which has vertices $(1, 0)$, $(2, 1)$ and $(2, -1)$. The area of the square is 16, and the area of each right isosceles triangle is 1. Hence the area of the region is 12.

15. **(c)** Either disc can be divided into 6 unit equilateral triangles each of area $\frac{\sqrt{3}}{4}$, plus 6 circular segments each of area $\frac{\pi}{6} - \frac{\sqrt{3}}{4}$. The intersection consists of 2 unit equilateral triangles and 4 circular segments. The total area is

$$\frac{\sqrt{3}}{2} + \frac{2\pi}{3} - \sqrt{3} = \frac{2\pi}{3} - \frac{\sqrt{3}}{2}.$$

16. **(d)** There are 16 possible outcomes when Roberta tosses her 4 coins. In 1 case, she will have 0 heads and loses half of the time when Wilma has 1 head. In 4 other cases, Roberta will have 1 head, and wins half of the time when Wilma has 0 heads. In the remaining 11 cases, Roberta will have at least two heads and wins automatically. Hence the odds of Roberta winning are 11+2 to 1/2 or 26 to 1. If Wilma has the nickel, she risks 5 cents for the chance of winning 136 cents, and the ratio 136:5 is closer to 26:1 than if Wilma has a different coin.

November 21, 1995

1. **(d)** The parabola divides the plane into two regions. The points of intersection of the circle and the parabola divides the circle into arcs, each of which carves an existing region into two. Since the maximum number of points of intersection of a circle and a parabola is four, the maximum number of regions is six.

2. **(d)** We have $71^2 - 37^2 - 51 = (71+37)(71-37) - 51 = 3 \cdot 17(36 \cdot 2 - 1) = 3 \cdot 17 \cdot 73$, which has three ditsint odd prime factors.

3. **(c)** Suppose the height was 100 two years ago. Then it was 120 a year ago and 132 now. Thus the percentage increase is 32.

4. **(c)** All prime factors of 1995 are distinct and odd, with the largest one being 19. Hence the last even number used is 38.

5. **(a)** Let O be the center of the large circle, P that of one of the small circles, and Q the point of tangency of the small circles. Then $\angle PQO = 90°$, $PQ = 1$ and $OP = 1 + 2 = 3$. By Pythagoras' Theorem, $OQ = 2\sqrt{2}$. Hence the width of the rectangle is $3 + 2\sqrt{2}$.

6. **(e)** Suppose Mary Lou usually works x hours per day but y on that day. All we know is $x + \frac{3}{2}(y - x) = 12$ or $3y - x = 24$. Even if x and y are required to be integers satisfying $0 < x < y < 24$, we can still have $(x, y) = (6, 10)$ or $(9,11)$. Thus the number of hours Mary Lou usually works per day cannot be uniquely determined.

7. **(d)** Clearly, the desired probability is less than 1. To see that it is greater than $\frac{3}{4}$, partition the tosses into consecutive groups of three, discarding the last toss. If we never get 3 heads in a row, none of the 3333 groups can consist of 3 heads. The probability of this is $(\frac{7}{8})^{3333}$, which is clearly less than $\frac{1}{4}$. In fact, $(\frac{7}{8})^{18} < \frac{1}{4}$ since $(\frac{7}{8})^3 = \frac{343}{512} < \frac{3}{4}$, $(\frac{3}{4})^3 = \frac{27}{64} < \frac{1}{2}$ and $(\frac{1}{2})^2 = \frac{1}{4}$.

8. **(e)** The sum of the angles of this polygon is exactly $(n - 2)180°$. Hence

$$180 < \frac{n^2}{n-2} = n + 2 + \frac{4}{n-2}.$$

It is easy to verify that we must have $n > 6$, so that $\frac{4}{n-2} < 1$. From $180 < n + 3$, we have $n > 177$.

9. **(d)** Let $Q(x) = P(x) - x = ax^3 + bx^2 + cx + d$. Then we have $0 = Q(1) = a + b + c + d, 0 = Q(2) = 8a + 4b + 2c + d, 0 = Q(3) = 27a + 9b + 3c + d$ and $1 = Q(4) = 64a + 16b + 4c + d$. Solving this system of equations, we have $a = \frac{1}{6}, b = -1, c = \frac{11}{6}$ and $d = -1$. Hence $P(6) = Q(6) + 6 = 16$.

10. **(c)** We have $\frac{1}{x^4} + \frac{1}{4y^4} = (\frac{1}{x^2} - \frac{1}{2y^2})^2 + \frac{1}{x^2y^2} \geq 1$, with equality if and only if $x^2 = 2y^2$.

11. **(c)** Since (0,0), (1,2) and (2,4) are collinear, a circle passes through at most two of them. Since (2,0), (3,1), (3,3) and (4,3) are not concyclic, a circle passes through at most three of them. It follows that we have to leave out at least two points. The circle with center (1,2) and passing through (0,0) also passes through (2,0), (3,1), (3,3) and (2,4). Hence the maximum is five.

12. **(b)** Subtraction yields $24 = x + y^2 - x^2 - y = (y - x)(y + x - 1)$. Note that one factor is odd and the other even, and that the first is smaller than the second. Hence either $y - x = 1$ and $y + x - 1 = 24$, or $y - x = 3$ and $y + x - 1 = 8$. They lead to $(x, y, z) = (12, 13, 57)$ and $(3, 6, -85)$ respectively. However, we must have $z > 0$. Hence there is only one solution.

13. **(c)** Let BCC' be a triangle with $\angle BCC' = \angle BC'C > 60°$. Let D be a point on the circumcircle of triangle BCC' such that it is on the opposite side of BC' to C, and $\angle BC'D > \angle CBC'$. Complete the parallelogram $ABC'D$. Then $ABCD$ is a convex quadrilateral, and we have $AD = BC' = BC$. Also, $\angle BAD = \angle BC'D = \angle BCD$ by Thales' Theorem. However, $ABCD$ is not a parallelogram. The given conditions in (b) guarantee that the opposite sides of $ABCD$ are parallel. If the given conditions in (a) or (d) hold, triangles BAD and BCD are congruent and again the opposite sides of $ABCD$ are parallel.

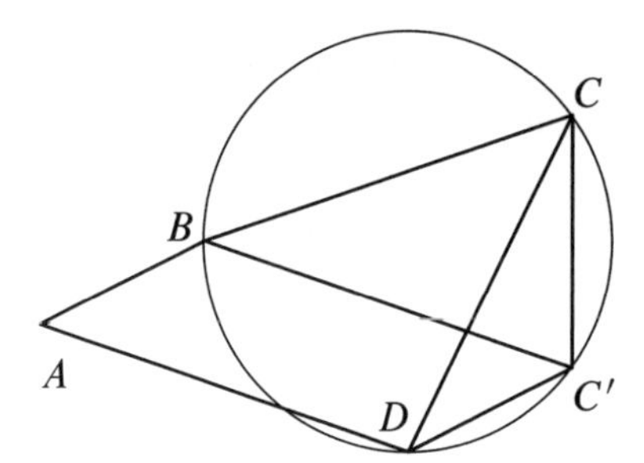

14. **(a)** If $x = -\frac{1}{4}$ and $y = -\frac{1}{3}$, then $x^3 + y^4 = -\frac{1}{64} + \frac{1}{81} < 0$. If $x = 0$ and $y = -1$, each of the other three expressions is equal to -1. Hence none of these expressions is always positive for $x > y$.

15. **(b)** Note that $\angle CAD = 90° - \angle BCA = \angle CBH$. Hence triangles CAD and HBD are similar, so that $\frac{HD}{BD} = \frac{CD}{AD}$. It follows that $HD = \frac{BD \cdot CD}{AD} = \frac{3}{2}$.

16. **(c)** The given expression factors into

$$\frac{(2-1)(2^2+2+1)(3-1)(3^2+3+1)(4-1)(4^2+4+1)\cdots(100-1)(100^2+100+1)}{(2+1)(2^2-2+1)(3+1)(3^2-3+1)(4+1)(4^2-4+1)\cdots(100+1)(100^2-100+1)}.$$

Since $(n+2)-1 = n+1$ and $(n+1)^2-(n+1)+1 = n^2+n+1$, cancellations yield

$$\frac{(2-1)(3-1)(100^2+100+1)}{(2^2-2+1)(99+1)(100+1)} = \frac{10101}{15150},$$

which is very close to $\frac{2}{3}$.

November 19, 1996

1. **(b)** The area of a circle is proportional to the square of its radius. It follows that we are comparing $\frac{16}{3} = 5.\overline{3}$, $\frac{25}{4} = 6.25$, $\frac{36}{6} = 6$ and $\frac{49}{8} = 6.125$. Hence we should take a slice from the ten-inch pizza.

2. **(d)** The total expenditure is $10 = \frac{m}{4} + \frac{n}{3} + \frac{m}{4} + \frac{n}{6} = \frac{m+n}{2}$. Hence $m + n = 20$ and we bought $2(m+n) = 40$ plums.

3. **(b)** Clearly, F, E, C, B and A are in that order from top to bottom. If D is pointing up, it is under A. If it is pointing down, it is under B. In either case, it is under C. Hence C is the third piece to be placed.

4. **(e)** Both the customer's way and the clerk's way are the same since $1.07 \cdot 0.79 = 0.79 \cdot 1.07$. This product is 0.8103, so that the discount is approximately 19%. This is far better than the 14% of the compromise.

5. **(c)** We have $m - 2n = m - 2n + 2m - n - 3 = 3(m-n-1)$. Hence it must be a multiple of 3, and can be any multiple of 3.

6. **(a)** Since y is $y\%$ of z, $z = 100$. The situation is possible if and only if $y = 100$ also, in which case $x > 0$ is arbitrary.

7. **(e)** The axes of rotational symmetry are the line through the center and perpendicular to the hexagon, the three lines joining opposite vertices and the three lines joining the midpoints of opposite sides.

8. **(b)** By symmetry, ACD is an equilateral triangle. Hence its centroid is the center O of the circle. Since $AO = 1$, $AE = AO + OE = \frac{3}{2}$.

9. **(b)** Kerry is clearly lying, and is the one who tells the truth on Saturday. Hence the conversation takes place Monday, Tuesday or Wednesday, and Kelly's statement is only true on the last day.

10. **(c)** Either both m and $m + 1$ are powers of 2, or both are negatives of powers of 2. The two solutions are $(m, n) = (1, 1)$ and $(m, n) = (-2, 1)$.

11. **(b)** Note that $5^2 + 12^2 = 13^2$. By the converse of Pythagoras' Theorem, only the second triangle has a right angle between the sides of lengths 5 and 12. For each triangle, take the side of length 12 as the base. In the second triangle, the altitude on the base has length 5, while in each of the others, the altitude on the base has length less than 5. Thus the second triangle has the greatest area among the five.

12. **(d)** If $y \geq 0$, $x - y = x - |y|$ is the minimum. If $y < 0$, $x + y = -|x + y| = x - |y|$ is the minimum. It follows that $x - |y|$ is never greater than any of the others.

13. **(a)** Let z be the length of the shorter side of a blue rectangle. Then the longer side has length $z + \frac{1}{2}(y - x)$. From $8z(z + \frac{1}{2}(y - x)) = xy$, we have $z = \frac{1}{4}(x - y \pm \sqrt{x^2 + y^2})$. Clearly, the negative square root is to be rejected, so that $z = \frac{1}{4}(x - y + \sqrt{x^2 + y^2}$.

14. **(c)** If card number 1 is initially in the 1-st or 2-nd position from the top, you will win. You can also win if it is in the 3-rd, and card number 2 is not in the 1-st. Hence the winning probability is $\frac{1}{10} + \frac{1}{10} + \frac{1}{10} \cdot \frac{8}{9} = \frac{13}{45}$.

15. **(a)** The sum of the exterior angles of the five given angles is $5(180° - 108°) = 360°$. Hence these five angles are the only angles of the convex polygon. It follows that the maximum angle is also $108°$, and it lies in the interval $(105°, 120°)$.

16. **(b)** Suppose $k = m^2 - n^2 = (m - n)(m + n)$. If k is odd, we can set $m - n = 1$ and $m + n = k$. If $k = 4\ell$, we can set $m - n = 2$ and $m + n = 2\ell$. However, if $k = 4\ell + 2$, then $m - n$ and $m + n$ cannot have the same parity. It follows that a number is not expressible as a difference of two squares if and only if it is even but not divisible by 4. Only the second and the fourth numbers are even, and a number is divisible by 4 if and only if so is the number formed of its last two digits. Since 68 is divisible by 4 but 66 is not, 314159266 cannot be expressed as a difference of two squares.

November 18, 1997

1. **(a)** The smallest number of different digits is 1. An example is 43210+56789=99999.

2. **(c)** If Ace ate the tarts, then all three told the truth. If Bea ate them, then all three lied. If Cec ate them, then Ace and Bea told the truth and Cec lied. Hence Cec ate the tarts.

3. **(c)** Since $\frac{1}{2}+\frac{1}{4}+\frac{1}{7}=\frac{25}{28}$, the number of birds in the flock must be a multiple of 28. The smallest positive multiple of 28 is 28 itself. If there were 28 birds, then 3 flew north. This is the minimum number of birds which flew north.

4. **(b)** We have $4^a=(2^a)^2=(b^3)^2=b^6$.

5. **(a)** If n is odd, the last digit of 14^n is 4, and so is that of $11\cdot 14^n$. Hence $11\cdot 14^n+1$ is divisible by 5. Since this number is greater than 5 for any n, it is never a prime number.

6. **(b)** Clearly, $b\le 2$ as otherwise $a^2+4b^2\ge 37$. When $b=2$, the only solution is $(a,b)=(3,2)$. When $b=1$, there are no solutions. Hence the value of $a+b$ must be $2+3=5$.

7. **(d)** The total length of the circumferences is π times the total length of the diameters, which is given to be 1 meter. Hence the total length of the circumferences is π meters, regardless of whether the circles have equal size and how they are arranged.

8. **(b)** The sum may be rewritten as $(100-99)+(98-97)+\cdots+(2-1)=1+1+\cdots+1=50$.

9. **(b)** We have $(10^{50}-1)^2=10^{100}-2\cdot 10^{50}+1$. This number has 100 digits when written out in base 10. It starts with 49 9's followed by an 8. This is in turn followed by 49 0's and a 1. Hence it has 49 zeros when written out in base 10.

10. **(e)** Since the corner lots are not empty, any pigs there are counted towards the total on two sides of the farm. Since each side has 9 pigs, the total must be less than 36. Each of the other four numbers is a possible total, as shown in the diagram below.

4	1	4
1		1
4	1	4

3	3	3
3		3
3	3	3

4	4	1
4		5
1	5	3

1	7	1
7		7
1	7	1

11. **(a)** It is not possible to visit all four corner squares plus the central square in any maze, but it is possible to visit eight squares, skipping one of these five. The highest total from the mazes are 40, 39, 39 and 39 respectively. Thus we should choose maze A.

12. **(c)** Each piece can cover at most one of the four protruding squares of Figure C, so four pieces will be needed. Each of the other three figures can easily be covered with three pieces.

13. **(e)** Suppose $1\le x\le 2$. Then $|x-1|+|x-2|=1=|x-4|-|x-3|$. Hence there are infinitely many solutions.

14. **(c)** We have $x^2+\dfrac{1}{x^2}=\left(x+\dfrac{1}{x}\right)^2-2=7$ and $x^3+\dfrac{1}{x^3}=\left(x+\dfrac{1}{x}\right)\left(x^2+\dfrac{1}{x^2}\right)$ $-\left(x+\dfrac{1}{x}\right)=18$.

15. **(d)** Since CF bisects $\angle C$ and $BC>AC$, we have $BF>AF$. On the other hand, we have $\angle ABC=\angle ACB=30°$, $\angle DAC=\angle DCA=30°$ and $\angle AFG=\angle AGF=45°$. It follows that the other three equalities are correct.

16. **(b)** Suppose the first term is a and the common ratio is r. Then $q = ar^{p-1}$ and $p = ar^{q-1}$. Hence $ar^{2p-q-1} = \dfrac{(ar^{p-1})^2}{ar^{q-1}} = \dfrac{q^2}{p}$.

November 17, 1998

1. **(c)** Let the old buying price be $2x$. Then the new buying price is x and the selling price is $4x$. The latter exceeds the former by $3x$.

2. **(b)** The only solution is $n = 17$ since the inequality can be rewritten as $0 \geq (n-17)^2$.

3. **(a)** The new ratings are 75 for MacLuck, 72 for MacLock, 80 for MacLick, 81 for MacLeck and 76 for MacLack. *This problem illustrates that quality and ranking are two different things.*

4. **(a)** Suppose the length to width ratio for the original rectangle is $a : b$ with $a > b$. Then the ratio for the strips is $b : \frac{a}{n} = a : b$. Hence $\frac{a}{b} = \sqrt{n}$.

5. **(e)** The largest possible value of a is 50 and the smallest is 46.

6. **(c)** We have $S = \frac{1}{2}10^n(10^n + 1) = 2^{n-1}5^n(10^n + 1)$.

7. **(c)** We have $(1 + x + x^2) + (x^3 + x^4 + x^5) = (1 + x + x^2)(1 + x^3) = (1 + x + x^2)(1 + x)(1 - x + x^2)$. The two quadratic factors are irreducible over polynomials with integral coefficients since the only possible linear factors are $1 + x$ and $1 - x$, but neither divides them.

8. **(e)** The line joining the midpoints of AB and BC is parallel to AC. Hence $\angle CAB = 90°$ and the area of triangle ABC is $\frac{1}{2}AB \cdot AC = 100$.

9. **(d)** We have $f(f(x)) = (x^x)^{(x^x)} = x^{xx^x}$.

10. **(a)** When the sequence is performed, the A-face first goes to the left, pointing up, then stays at the left but pointing to the front, and finally goes to the back, pointing to the left. When the sequence is performed again, the A-face first goes to the right, pointing to the back, then stays at right, pointing up, and finally returns to the front, pointing up.

11. **(a)** We have $1 + z^2 = x(2 - x)$ which is equivalent to $(x - 1)^2 + z^2 = 0$. Hence $x = y = 1$ and $z = 0$.

12. **(d)** Triangles PRA and SRC are similar. Since $PR = 2RS$ and $AC = 3$, we have $CR = 1$. Let the foot of perpendicular from R to BC be T. Since $\angle ACB = 60°$, we have $RT = \frac{1}{2}\sqrt{3}$ and $CT = \frac{1}{2}$, so that $BT = \frac{5}{2}$. By Pythagoras' Theorem, $BR^2 = RT^2 + BT^2$.

13. **(b)** We have $a+b+c+d = 8$ while $abcd = -160$. The desired expression is equal to $\frac{a+b+c+d}{abcd}$.

14. **(b)** Since 2^b is even, so is b. When b is even, 2^b ends alternately in 4 and 6. For $2 \leq b \leq 20$, the only matches are $b = 14$ and $b = 16$. Since everything repeats in a cycle of 20, Wei's list is 14, 16, 34, 36, ... The $(2n-1)$-st number is $10(2n-1)+4$ and the $2n$-th number is $10(2n-1)+6$.

15. **(b)** Note that $x+1 < \sqrt{x^2+2x+4} < x+2$ while $2x < \sqrt{4x^2+2x+1} < 2x+1$. Hence the number of integers between the two radicals is $x-1$.

16. **(c)** Let the three side lengths be $x \le y \le z$. Then $x^2+y^2=z^2$ since we have a right triangle. Hence $\frac{1}{2}xy=120$ and $xy=2^4 3\cdot 5$. Of $3^2+80^2, 5^2+48^2, 15^2+16^2, 6^2+40^2, 10^2+24^2$ and 12^2+20^2, we only have an integral value for z when $x=10$ and $y=24$, namely, $z^2=2^2(5^2+12^2)=(2\cdot 13)^2$.

November 16, 1999

1. **(c)** Since multiplication is commutative, $d=0$.

2. **(c)** We have $7\cdot 11\cdot 13=1001$ and $11\cdot 13\cdot 17=2431$.

3. **(e)** Since there are no ends to the primes, there is no largest possible size of the smallest of the three primes.

4. **(b)** Let the digits be x and y respectively. From $10x+y-xy=12$, we have $(x-1)(10-y)=2$. This leads to 28 or 39.

5. **(b)** Let the y-intercept be b. Suppose the slope m is equal to b. Then the equation of the line is $y=bx+b$. Hence $0=ba+b$ and $a=-1$. Conversely, suppose $a=-1$. Then the equation of the line is $\frac{y}{x+1}=m$. Hence $\frac{b}{0+1}=m$ and we do have $b=m$.

6. **(a)** Take $A=9900000000001$ and $B=99999999999$.

7. **(b)** Let $OD=x$. Then $(x^2+9)(9x^2+1)=CD^2AF^2=BE^4=(4x^2+4)^2$. This simplifies to $(x^2-7)(7x^2-1)=0$.

8. **(c)** We have $12-\sqrt{143}=\frac{1}{12+\sqrt{143}}$. Note that $11<\sqrt{143}<12$. It follows that we have $4<\frac{100}{24}<\frac{100}{12+\sqrt{143}}<\frac{100}{23}<4.5$.

9. **(d)** The six possible products are 2, 2, -1, -2, -2 and -4. Only two of them are positive and even.

10. **(e)** The divisors of $9^9=3^{18}$ which are cubes are 3^0, 3^3, 3^6, 3^9, 3^{12}, 3^{15} and 3^{18}.

11. **(c)** We have $p+q=14$, so that one of them is 3 and the other is 11.

12. **(b)** We have $\frac{AB}{XY}=\frac{XY}{CD}$ so that $XY=6$. We also have $\frac{AB}{XY}=\frac{BX}{CX}=\frac{BX}{BC-BX}$ so that $4(9-BX)=6BX$.

13. **(b)** Let the worth of the three kinds of coins be a, b and c respectively. From Matthew's collection, we have $2a+b+c=28$. From Daniel's, we have either $a+2b+2c=21$ or $a+b+3c=21$. In the former case, addition yields $3a+3b+3c=49$, but 49 is not a multiple of 3. Subtracting $a+b+3c=21$ from $2a+b+c=28$, we have $a-2c=7$. If $c=1$, then $a=9$ and $b=9$, which must be rejected. If $c\ge 3$, then $a\ge 13$ and $b<0$. Hence $c=2$, $a=11$ and $b=4$.

14. **(d)** We have $3=f(4)=f(f(1))=f(3)-3$ so that $f(3)=6$. It follows that we have $6=f(3)=f(f(4))=f(6)-3$ so that $f(6)=9$. Hence $9=f(6)=f(f(3))=f(5)-3$ so that $f(5)=12$.

15. **(d)** The sum is $S = \frac{b(b+1)}{2} - \frac{a(a-1)}{2} = \frac{(b+a)(b-a+1)}{2}$. Note that $12 \leq a + b \leq 30$. If $S = 91$, then $2S = 2 \times 7 \times 13$. We may have $a + b = 14$, $a = 1$ and $b = 13$. If $S = 92$, then $2S = 2 \times 2 \times 2 \times 23$. We must have $a + b = 23$, $a = 8$ and $b = 15$. If $S = 95$, then $2S = 2 \times 5 \times 19$. We must have $a + b = 19$, $a = 5$ and $b = 14$. For $S = 99$, we can take $a = 4$ and $b = 14$. However, if $S = 98$, then $2S = 2 \times 2 \times 7 \times 7$. One of $a + b$ and $b - a + 1$ is odd and the other is even. Hence $2a - 1 = (b + a) - (b - a + 1) \geq 28 - 7 = 21$, and $a \geq 11$. This contradicts $a \leq 10$.

16. **(b)** First, note that we may have one point joined to six other points by segments of lengths 1, 2, 4, 8, 16 and 32 respectively. Suppose the task can be accomplished with six points. We draw the segments one at a time. At any point during this process, two points are said to be in the same group if we can travel from one to the other along one or more segments already drawn. At the start, each point is in a group by itself. When a segment is drawn, it joins two points either in different groups or already in the same group. The latter is impossible since this means that some subset of the segments of lengths 1, 2, 4, 8, 16 and 32 forms a polygon. However, the longest of them will be longer than the total length of the others. It follows that whenever a segment is drawn, it always joins two points in different groups. However, after five segments have been drawn, the six points would have merged into a single group, and it is no longer possible to draw the last segment.

November 21, 2000

1. **(e)** The number of pennies Amy has must be a multiple of 5. If she has 55 pennies, then the maximum value of the other 3 coins is 30 cents, and the total is at most 85 cents. If she has 45 pennies, then the minimum value of the other 13 coins is 65 cents, and the total is at least 120 cents. If she has fewer pennies, the total will be higher. Hence she has 50 pennies. If the other 8 coins are all nickels, the total will be 10 cents short. Hence 2 of the nickels must be changed into dimes.

2. **(a)** Nima covers 15 kilometers in 2.5 hours.

3. **(d)** The number is equal to 10^{2000} times $5^4 = 625$.

4. **(c)** Since $1 + 2 + \cdots + 10 = 55$ and $1 + 2 + \cdots + 11 = 66$, the bug is on its eleventh segment 1 minute after the start.

5. **(e)** The minimum sum of the smallest thirteen of the numbers is $1+2+\cdots+13 = 91$. Hence the maximum sum of the largest two is $13 \cdot 15 - 91 = 104$. It follows that the maximum value of the second largest number is 51.

6. **(c)** When expanded, $(p(x))^2$ becomes

$$a^2x^{2k} + b^2x^{2\ell} + c^2x^{2m} + d^2x^{2n} + 2abx^{k+\ell} + 2acx^{k+m}$$
$$+ 2adx^{k+n} + 2bcx^{\ell+m} + 2bdx^{\ell+n} + 2cdx^{m+n}.$$

By choosing $k = 1$, $\ell = 10$, $m = 100$ and $n = 1000$, all the terms in the above expansion are distinct.

7. **(d)** Long division yields $x^3-2x+1=(x^2-x-1)(x+1)+2$. Since $x^2-x-1=0$, $x^3-2x+1=2$.

8. **(c)** In the vertical plane containing the rod and the center of the sphere, the cross-section of the sphere is a circle. From the bottom of the rod, two equal tangents can be drawn to this circle, one running along the rod and the other connecting the bottom of the sphere and the bottom of the rod.

9. **(a)** After each game in which he lost, David had half as much money as he had before. After each game in which he won, David had one and a half times as much money as he had before. No matter which three games he won, he would have $\$64(\frac{1}{2})^3(\frac{3}{2})^3=\$27$ at the end.

10. **(d)** There are three cases. First, $n+2=0$ while $n^2-n-1\neq 0$. This yields $n=-2$. Second, $n^2-n-1=-1$ while $n+2$ is a non-zero even integer. This yields $n=0$. Finally, $n^2-n-1=1$ while $n+2$ is any non-zero integer. This yields $n=-1$ and $n=2$.

11. **(c)** Let $\angle ACF=y$ and $\angle BCE=z$, as illustrated in the diagram. Then $x+y+z=90°$. Since $AC=AE$, $\angle AEC=\angle FEC=(x+y)$. Similarly, $\angle EFC=x+z$. It follows that $x+(x+y)+(x+z)=180°$ so that $2x=90°$.

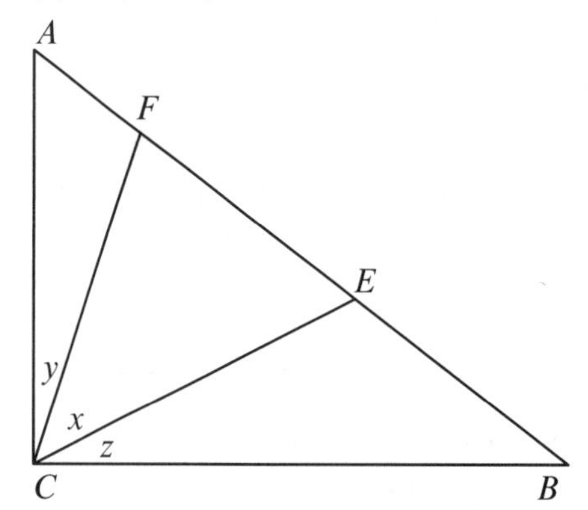

12. **(e)** The integer parts of these two rational numbers must be 5 and 1, but there are many possible choices for their fractional parts, such as $\frac{1}{3}$ and $\frac{3}{4}$. Now $(5+\frac{1}{3})(1+\frac{3}{4})=\frac{28}{3}$ while $(5+\frac{3}{4})(1+\frac{1}{3})=\frac{23}{3}$.

13. **(b)** When the reciprocal of a divisor of 120 is multiplied by 120, the product is also a divisor of 120. When the sum of the reciprocals of all the divisors of 120 is multiplied by 120, the product is the sum of all the divisors of 120, which is given to be 360.

14. **(e)** We have $\frac{a-b}{a+b}+\frac{b-c}{b+c}=\frac{2b(a-c)}{(a+b)(b+c)}$ while

$$\frac{c-a}{c+a}\left(1+\frac{(a-b)(b-c)}{(a+b)(b+c)}\right)=\frac{c-a}{c+a}\cdot\frac{2b(c+a)}{(a+b)(b+c)}=\frac{2b(c-a)}{(a+b)(b+c)}.$$

Hence the original expression simplifies to 0.

15. **(d)** The chart shows the numbers of digits of a positive integer n in bases 2 and 3. Since the number of digits in base 2 increases every second integer while that in base 3 every third, the largest number with one digit more in base 2 than in base 3 is 9.

n	1	2	3	4	5	6	7	8	9	10	11
Base 2	1	2	2	3	3	4	4	5	5	6	6
Base 3	1	1	2	2	2	3	3	3	4	4	4

16. **(c)** In the diagram, triangles BCG and FAG are similar, as are triangles ABF and DEF. Hence $\frac{EF}{BF} = \frac{DF}{AF} = \frac{BC-AF}{AF} = \frac{BG}{FG} - 1 = 2$.

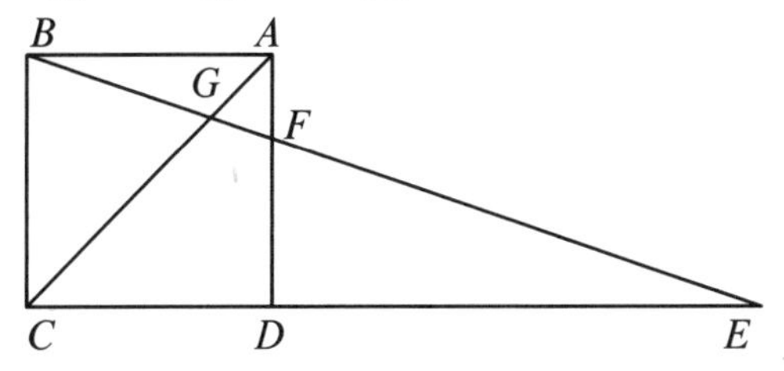

November 20, 2001

1. **(a)** If $p = 2$, then $2001 + p^2 = 2005$ is not prime. If $p > 2$, then $2001 + p^2 > 2$ is even and not prime.

2. **(e)** If 96.5% were found guilty, then 3.5% were not. It is not possible for the percent of male tourists and the percent of female tourists found not guilty to be both less than 3.5%.

3. **(c)** We have $\frac{2n+1}{n+1} = 2 - \frac{1}{n+1}$. Hence $n+1$ must divide 1, so that $n+1 = \pm 1$. Thus either $n = 0$ or $n = -2$.

4. **(a)** Note that $|x| - 1$ must be 0 or ± 1. If $|x| - 1 = 0$, then $x = \pm 1$. If $|x| - 1 = \pm 1$, then $x = 0$ or ± 2. The same goes for $|y| - 1$. However, at least one of $|x| - 1$ and $|y| - 1$ must be 0. Hence we have $-2 \le x \le 2$ and $-2 \le y \le 2$, x and y not both even.

5. **(d)** Since we are only interested in the last digit, we may treat 13 as though it is 3. The last digits of successive powers of 3 end in 3, 9, 7 and 1 cyclically. Hence $3^{101+103} = 3^{204}$ ends in 1. The last digits of successive powers of 7 end in 7, 9, 3 and 1 cyclically. Hence 7^{102} ends in 9.

6. **(e)** For dogs, either Sunera buys one, Dima buys one, or both. There are three possibilities. This is the same for cats and for hamsters.

7. **(c)** Since $y(x+z) = 29$, we must have $y = 1$. Then $x + z = 29$ and $z(x+1) = 81$. From the second equation, z can only be 1, 3, 9 or 27. However, the values of x from the two equations match only when $z = 3$ or 27.

8. **(a)** Let $AB = b$, $BC = c$, $CD = d$, $DE = e$, $GH = h$, $HI = i$ and $IJ = j$. We also have $FG = 1$ and $JK = 25$. Since AF, BG, CH, DI and EJ are parallel

to one another, $b = \frac{c}{h} = \frac{d}{i} = \frac{e}{j}$. Since AG, BH, CI, DJ and EK are parallel to one another, $\frac{b}{h} = \frac{c}{i} = \frac{d}{j} = \frac{e}{25}$. Hence $\frac{1}{h} = \frac{h}{i} = \frac{i}{j} = \frac{j}{25}$. Multiplying together $i = h^2$, $h^2j^2 = i^4$ and $25i = j^2$, we have $i^2 = 25$.

9. **(c)** For $1 \le k \le 1000$, at least one of $x < 2k$ and $x > 2k + 1$ is false. Hence the total number of true claims is at most 1001. This can be attained by taking any $x > 2001$.

10. **(d)** In the first half of 2002, Andy pays \$90 less rent than Monika. In the second half, Jane pays \$30 less. In the first half of 2003, he pays \$30 more. In the second half, he pays \$90 more and catches up with Monika in total accumulated rent. In January 2004, their rents are the same, but Andy's will rise above Monika's the next month.

11. **(b)** Since $x^2 - 3x + 1 = (x - r)(x - s) = x^2 - (r + s)x + rs$, we have $r + s = 3$ and $rs = 1$. Hence $(\sqrt{r} + \sqrt{s})^2 = r + s + 2\sqrt{rs} = 5$.

12. **(b)** Let $a = BC$, $b = CA, c = AB$ and $s = \frac{a+b+c}{2}$. By Heron's Formula, the area of ABC is given by $\sqrt{s(s-a)(s-b)(s-c)}$. Now $c = 5$ and $a + b = 7$ so that $s = 6$, $s - c = 1$ and $(s - a) + (s - b) = 5$. Since $0 \le (\sqrt{s-a} + \sqrt{s-b})^2 = 5 - 2\sqrt{(s-a)(s-b)}$, the area is $\sqrt{6(s-a)(s-b)} \le \frac{5}{2}\sqrt{6}$, with equality if $a = b = \frac{7}{2}$. Alternatively, since $AC + BC = 7$, C lies on the ellipse with foci A and B, and the maximum distance from C to AB occurs when $AC = BC = \frac{7}{2}$. By Pythagoras' Theorem, this distance is $\sqrt{(\frac{7}{2})^2 - (\frac{5}{2})^2} = \sqrt{6}$ and the maximum area is $\frac{5}{2}\sqrt{6}$.

13. **(c)** Let the numbers given Xenia, Yvonne and Zeke be x, y and z respectively. Then x and y must be odd, and $x \ge 9$. Hence $(x, y, z) = (9, 1, 6)$, (9,3,4), (9,5,2), (11,1,4), (11,3,2) or (13,1,2). Since Zeke can determine all three numbers, we must have $z = 6$.

14. **(b)** Solving $x^2 - x + y^2 - y = 0$ as a quadratic equation in x, we have $x = \frac{1 \pm \sqrt{2-(2y-1)^2}}{2}$. To make x as large as possible, we take the plus sign and set $y = \frac{1}{2}$.

15. **(a)** We have $a + 2 = f(5) = 6 - b$ and $2a + 2 = f(8) = 3 - b$. Hence $a = -3$ and $b = 7$ so that $f(13) = 3a + 2 = -7$.

16. **(e)** Let the lines be ℓ_0, ℓ_1 and ℓ_2. We may choose a coordinate system so that ℓ_0 lies on the plane $z = 0$ and ℓ_1 lies on the plane $z = 1$. Now ℓ_2 has at most two points with z-coordinates equal to 0 or 1. Let P be any other point on ℓ_2 and let Π be the plane determined by P and ℓ_0. The planes Π and $z = 1$ intersect at a line ℓ, which is not parallel to ℓ_1 as otherwise ℓ_0 and ℓ_1 would be parallel to each other. Hence ℓ_1 intersects Π at a point Q. Since P and Q have different z-coordinates, PQ is not parallel to ℓ_0, and these two coplanar lines intersect at some point R. Now PQR is a line with P on ℓ_2, Q on ℓ_1 and R on ℓ_0.

November 19, 2002

1. **(b)** Since addition and subtraction have equal priority, the brackets may be rearranged in any way. Thus $(1 - 2) + (3 - 4) + \cdots + (2001 - 2002) = -1001$.

2. **(a)** If the first selection does not produce a match, retain either the key or the lock, but not both, in the selection. This will produce a match.

3. **(e)** There is only 1 such number with 1 digit, namely 5. There are 10 such numbers with 2 digits, namely, those from 50 to 59. Similarly, there are 100, 1000 and 10000 such numbers with 3, 4 and 5 digits respectively, yielding a total of 11111.

4. **(b)** The units digit of $3a + 3b$ is 2+5=7, the units digit of $a + b$ is $27 \div 3 = 9$, and the units digit of a is $12 - 9 = 3$.

5. **(e)** All the lateral faces of the pyramid must be equilateral triangles, and at most 5 of them can fit around the top vertex.

6. **(b)** Since the objective is to downsize 2002 to 1, we should be halving as much as we can, and increasing only when necessary. For instance, the sequence (2002,2003, 2004,1002) may be shortened to (2002,1001,1002). It follows that the most efficient sequence is

$$(2002, 1001, 1002, 501, 502, 251, 252, 126, 63, 64, 32, 16, 8, 4, 2, 1).$$

7. **(e)** Of the four numbers in a line with 6, two are odd and two are even. Hence the number at the bottom of the line with 7 is odd. it must be one less than the number at the center of the letter H. Hence these two numbers must either be 1 and 2 or 3 and 4, respectively. In the first case, the number directly below 7 must be 5. In the second case, the number directly below 7 must be 2. The two numbers in a vertical line with 6 can be placed either way.

8. **(d)** Note that $(4 + 2\sqrt{2})^2 = 24 + 16\sqrt{2}$ and $(24 - 16\sqrt{2})(24 + 16\sqrt{2}) = 2^6$.

9. **(a)** The slopes of the lines are $-2a$ and $-2(a + 1)$. If the lines are perpendicular, the product of the slopes is -1. Note that $(-2a)(-2(a + 1)) - (-1) = (2a + 1)^2$.

10. **(a)** The diagonal crosses 21 horizontal grid lines and 49 vertical grid lines. However, since the greatest common divisor of 22 and 50 is 2, it crosses exactly one corner of a square, namely the one at the middle of the table. Thus it has $21 + 49 - 1 = 69$ points of intersections on it, and they divide it into 70 segments of positive length.

11. **(c)**. If there were 10 other participants, they would have played 45 games among themselves. If there were 8 other participants, they would have played only 28 games among themselves. Hence there are 8 other participants who have played 36 games among themselves. Adding the 6 games played by the retiring players, the total is 42. Thus 2 of these 6 games must be between two retiring players.

12. **(a)** For $x = 1$, we have $f(1) + 2f(0) = 3$. For $x = 0$, we have $f(0) + 2f(1) = 0$. Hence $f(0) = 2$ and $f(1) = -1$.

13. **(c)** Let the radii be r and R. Then the parallel sides of the quadrilateral have lengths $2r$ and $2R$. The distance between them is at most the distance between the centers, which is $r + R$. Hence the maximum area of the quadrilateral is $144 = \frac{1}{2}(2r+2R)(r+R) = (r + R)^2$.

14. **(c)** If the first marble is red, the probability for the second to be red is $\frac{4}{7} \times \frac{4+k}{7+k} = \frac{16+4k}{7(7+k)}$. If the first marble is white, the probability for the second to be red is $\frac{3}{7} \times \frac{4}{7+k} = \frac{12}{7(7+k)}$. The overall probability is $\frac{16+4k+12}{7(7+k)} = \frac{4}{7}$.

15. **(b)** Denote the expression by E. If $x < -1$, $E = (-x-1) + (-x) - (-x+1) = -x-2 > -1$. If $x = -1$, $E = 0+1-2 = -1$. If $-1 < x \leq 0$, $E = (x+1) + (-x) - (-x+1) = x > -1$. If $0 < x \leq 1$, $E = (x+1) + (x) - (-x+1) = 3x > 0$. Finally, if $x > 1$, then we have $E = (x+1) + (x) + (1-x) = x+2 > 3$.

16. **(d)** We have $121 + 19 \times 3000 = 239^2$, and $239 = 12 \times 19 + 11$. Let $121 + 19n = m^2$. Then $19n = (m-11)(m+11)$. If 19 divides $m-11$, then the first acceptable value of m is 11 and the last is 239, for a total of 13 values. If 19 divides $m+11$, then the first acceptable value of m is 27 (not 8 as that would make n negative) and the last is 236, for a total of 12 values.

November 18, 2003

1. **(e)** We may take the number of students in the class to be 100. Of the 40 students who did not like mathematics before, 10 changed their minds. Thus the percentage is $\frac{10}{40} \times 100\% = 25\%$.

2. **(c)** On the day when Daniel's watch is slow, he gets to school by walking 2 minutes and running 5 minutes since the whole trip takes $10 + 2 - 5 = 7$ minutes. Thus the trade off is $5 - 2 = 3$ minutes running for $10 - 2 = 8$ minutes of walking so that the ratio of his running and walking speeds is 8:3.

3. **(d)** The greatest common divisor of 80 and 72 is 8. Hence there are 8 moments when both Lily and Lala start saying "Yakkity-yak". Hence the total number of moments is $80 + 72 - 8 = 144$.

4. **(c)** The slope is $\frac{b-a}{\frac{1}{a}-\frac{1}{b}} = ab$.

5. **(b)** If all three of a, b and c are odd or all three are even, then each of $\frac{b+c}{2}$, $\frac{c+a}{2}$ and $\frac{a+b}{2}$ is an integer. Otherwise, two of a, b and c, say a and b, are of one parity while the third, in this case c, is of the opposite parity. Then $\frac{a+b}{2}$ is the only integer among the three fractions.

6. **(a)** Squaring the expression, we have $2+\sqrt{3}+2\sqrt{(2+\sqrt{3})(2-\sqrt{3})}+2-\sqrt{3} = 6$.

7. **(c)** Multiplying the four equations together, we have $(abcd)^3 = 64$ so that $abcd = 4$. Now $a+b+c+d = abcd(\frac{1}{bcd} + \frac{1}{cda} + \frac{1}{dab} + \frac{1}{abc}) = \frac{15}{2}$.

8. **(e)** Coloring the 4×6 rectangle in checkerboard fashion, we have 12 white and 12 black squares. Each of the first four pieces covers 2 white and 2 black squares. The fifth covers either 1 white and 3 black squares or 1 black and 3 white squares. Hence the sixth piece must be a duplicate of the fifth. The following diagram shows that the construction is possible.

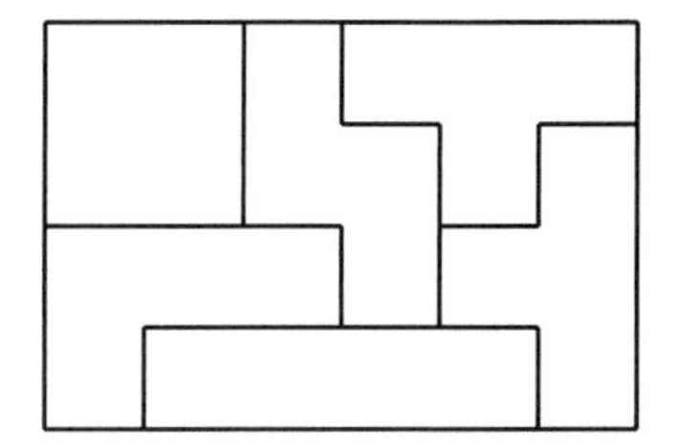

9. **(c)** If $a = 1$, we have $f(f(f(2))) = 2 < 222$. If $a = 2$, we have $f(f(f(2))) = 256 > 222$. Hence $1 < a < 2$.

10. **(b)** Let $y = x^2$. The remainder when $f(y) = y^{50} - 2y^{26} + 2$ is divided by $y + 1$ is $f(-1) = 1$.

11. **(c)** If the Championship is over in four games, the winning team would have won exactly two home games. If it is over in five games and the team which starts at home wins the Championship, it can win Games 2, 3, 4 and 5. Then only Games 2 and 5 are won by the home team. If it is over in seven games and the team which starts on the road wins the Championship, it can win Games 3, 4, 6 and 7. Then only Games 5 and 6 are won by the home team. In a six-game Championship, each team has three home games. If one team wins the two games won at home, then it must win all three road games, but the Championship would have been awarded before the series goes 5-1. If each team wins one of the two games won at home, then each must win two road games, but the Championship could not have been awarded when the series is tied at 3-3.

12. **(c)** Let the radii be $a > b > c$. Then the distance between the centers of the largest two balls is $a + b$, the horizontal distance is 4 and the vertical distance is $a - b$. By Pythagoras' Theorem, $(a + b)^2 = 4^2 + (a - b)^2$ so that $ab = 4$. Similarly, $bc = 1$ and $ca = \frac{9}{4}$. Hence $(abc)^2 = 9$ and $abc = 3$.

13. **(d)** Setting $x = 1$, we have $2f(1) - f(0) = 6$. Setting $x = 0$, we have $2f(0) - f(1) = 0$. Subtracting the second equation from twice the first, we have $3f(1) = 12$ so that $f(1) = 4$. Similarly, setting $x = 2$ and then $x = -1$, we have $f(2) = 6$. It follows that $f(1) + f(2) = 10$.

14. **(c)** We have $a + \frac{b}{c} = 11$ and $b + \frac{a}{c} = 14$. Adding the two equations yields $(a + b)\frac{c+1}{c} = 25$ or $(a + b)(c + 1) = 25c$. Since $c + 1$ and c are relatively prime, $c + 1$ must divide 25. Hence $c = 4$ or $c = 24$. If $c = 24$, then $a + b = 24$ but at least one of $\frac{a}{c}$ and $\frac{b}{c}$ is not an integer. Hence $c = 4$ and $a + b = 20$ so that the correct answer is $\frac{20}{4} = 5$.

15. **(d)** Let BE and CF intersect at G. Then $BG = \frac{2}{3}BE = 12$ while $CG = \frac{2}{3}CF = 16$. The area of GBC is one-third that of ABC. Hence the area of ABC is maximized if and only if so is that of GBC. This occurs when BG is perpendicular to CG, yielding an area of $\frac{1}{2} \times 12 \times 16 = 96$ for GBC. Hence the maximum area of ABC is $96 \times 3 = 288$.

16. **(d)** We have

$$\begin{aligned}\sin^6 x + \cos^6 x &= \sin^6 x + 3\sin^4 x\cos^2 x + 3\sin^2 x\cos^4 x + \cos^6 x\\ &\quad -3\sin^2 x\cos^2 x(\sin^2 x + \cos^2 x)\\ &= (\sin^2 x + \cos^2 x)^3 - 3\sin^2 x\cos^2 x\\ &= 1 - \frac{3}{4}\sin^2 x.\end{aligned}$$

The maximum value 1 occurs at $x = 0°$ and the minimum value $\frac{1}{4}$ occurs at $x = 45°$. Hence the desired sum is $\frac{5}{4}$.

November 16, 2004

1. **(c)** The volume of Mount Rundle, in cubic meters, is $\frac{1}{2} \times 5000 \times 1800 \times 10000 = 45 \times 10^9$.

2. **(e)** If there were five unanswered questions, the score would be the same if two of them were answered correctly and three incorrectly. Hence we may assume that the number of unanswered questions is 0, 1, 2, 3 or 4. Checking each case, the unattainable scores are 21, 96, 23, 93, 98 and 99.

3. **(b)** Since $x^{2004} - 1$ is divisible by $x^2 - 1$, the desired remainder is $-2x + 2 + 1 = -2x + 3$.

4. **(d)** If the number begins in 1, we have 3 choices for the last digit, 2 for the third digit and 1 for the second, yielding a total of 6. If the number begins with 2, the choices are 2, 2 and 1 respectively, yielding a total of 4. If the number begins with 3, the total is also 4. Hence the grand total is 6+4+4=14.

5. **(e)** The number of true statements is non-increasing as n increases. Since there are 4 true statements when $n = 3$ and only 3 true statements when $n = 4$, there are no possible values for n.

6. **(d)** In order for the prediction to be false, someone must be born after 1936 and die before 2025, yielding a maximum age of 87.

7. **(c)** Since $y \neq 0$, $x + y \neq x - y$, so that we must have $xy = \frac{x}{y}$. Since $x \neq 0$, we have $y^2 = 1$ or $y = \pm 1$. However, $y = 1$ leads to a contradiction whether $x + y = xy$ or $x - y = xy$. Hence $y = -1$. If $x - 1 = -x$, then $x = \frac{1}{2}$. If $x + 1 = -x$, then $x = -\frac{1}{2}$.

8. **(d)** Note that the sum of the first and the fifth numbers is also -4. Since their product is -12, they are -6 and 2 respectively. Hence the common difference is 2 and the numbers are $-6, -4, -2, 0, 2, 4, 6$ and 8.

9. **(e)** We may approximate the area of the smaller regular polygon by a circle with circumference 100 and that of the larger regular polygon by a circle with circumference 400.

10. **(b)** Suppose none of the digits of $999n$ is 9, where n is some positive integer. Note that $999n = 1000n - n$. Just to ensure that there are no 9s among the last three digits of $999n$, the last three digits of n must form a number at least 112. Hence $n \geq 112$. Taking $n = 112$, we have $999n = 111888$. This is the smallest positive multiple of 999 which does not have any digit equal to 9.

11. **(c)** If $x \geq 0$, the equation is $x^2 - x - 1 = 0$. The roots are $x = \dfrac{1 \pm \sqrt{5}}{2}$, but we must reject the negative root $\dfrac{1-\sqrt{5}}{2}$. If $x < 0$, the equation is $x^2 + x - 1 = 0$. The roots are $x = \dfrac{-1 \pm \sqrt{5}}{2}$, but we must reject the positive root $\dfrac{-1+\sqrt{5}}{2}$.

12. **(c)** Let $SP = SR = x$. Since the area of triangle SAR is 1, its height is $\frac{2}{x}$. Complete the parallelogram $CRST$. Then triangle PST is congruent to triangle QRC, so that triangle BST has area 4. Since it is similar to triangle SAR which has area 1, its height x is double the height $\frac{2}{x}$ of the latter. Hence $x = 2$.

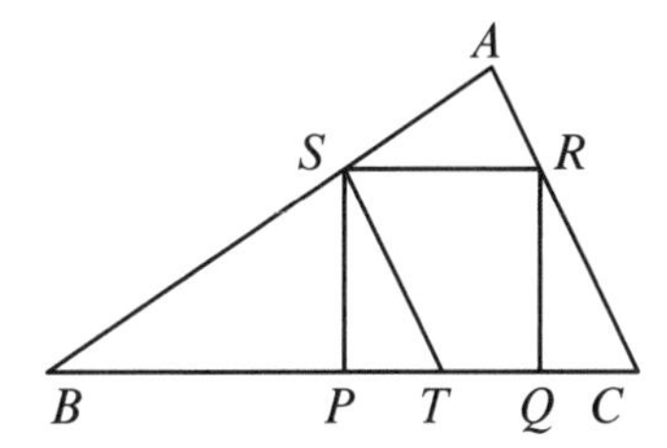

13. **(c)** If $x^2 - 13x + 1 = 0$, then $x + x^{-1} = 13$. Squaring yields $x^2 + 2 + x^{-2} = 169$ so that $x^2 + x^{-2} = 167$. Squaring again, $x^4 + 2 + x^{-4}$ will be a number ending in 9. Hence the units-digit of $x^4 + x^{-4}$ is 7.

14. **(a)** First, note that $6125 = 5^3 7^2$. Of the numbers from 1 to 2004, the number of multiples of 5 is $\lfloor \frac{2004}{5} \rfloor = 400$, the number of multiples of 25 is $\lfloor \frac{400}{5} \rfloor = 80$, the number of multiples of 125 is $\lfloor \frac{40}{5} \rfloor = 16$ and the number of multiples of 625 is $\lfloor \frac{16}{5} \rfloor = 3$. The total number of factors of 5 in 2004! is 400+80+16+3=499. If 5^{3n} is to divide 2004!, we must have $n \leq \frac{499}{3}$ or $n \leq 166$. Similarly, of the numbers from 1 to 2004, the numbers of multiples of 7, 49 and 343 are 286, 40 and 5 respectively. If 7^{2n} is to divide 2004!, we must have $n \leq \frac{331}{2}$ or $n \leq 165$. Thus the largest value of n is 165.

15. **(c)** Drop perpendiculars AH and DK to BC from A and D, respectively. Then triangle HAB is half a square with diagonal $\sqrt{6}$, so that $HA = HB = \sqrt{3}$. Also, triangle KCD is half an equilateral triangle with side 6, so that $CK = 3$ and $DK = 3\sqrt{3}$. Hence $HK = 8$ while the difference of AH and DK is $2\sqrt{3}$. The length of AD is $\sqrt{8^2 + (2\sqrt{3})^2} = 2\sqrt{19}$.

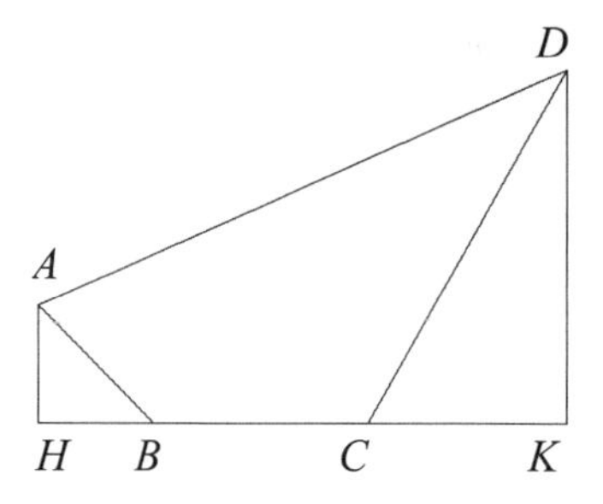

16. **(b)** Since $x + y = 1$, we have

$$\begin{aligned} x^4y + xy^4 &= xy(x+y)(x^2 - xy + y^2) \\ &= xy(x+y)((x+y)^2 - 3xy) \\ &= \frac{1}{3}(3xy)(1 - 3xy). \end{aligned}$$

This is less than or equal to $\frac{1}{3}\left(\frac{3xy + 1 - 3xy}{2}\right)^2 = \frac{1}{12}$ by the Arithmetic-Geometric Means Inequality, with equality if and only if $xy = \frac{1}{6}$.

November 15, 2005

1. **(a)** We have $2005\times20042004-2004\times20052005 = 2004\times2005(10001-10001) = 0$.

2. **(b)** The number of different single cones is 20, as is the number of different double cones with identical scoops. The number of different double cones with non-identical scoops is $\frac{20\times19}{2} = 190$. Hence the grand total is $20 + 20 + 190 = 230$.

3. **(a)** Since $BD = CD$, the area of triangle BAD is 75. Since $ED = 2AE$, the area of triangle ABE is 25.

4. **(e)** The first given inequality may be rewritten as $a - b + 3c \geq 2$. Adding three times this to two times $b - 2a + c \geq 3$, we have $x \geq 12$.

5. **(c)** If Penelope takes only 5 socks, she may have 3 red, 1 yellow and 1 blue, forming only one matching pair. If she leaves out 3 socks, she can leave out at least 2 socks of at most one color. Hence of the 6 socks she takes, she can guarantee to have two matching pairs.

6. **(c)** Half the diagonal of the square base has length $\frac{3\sqrt{2}}{2}$. Since the pyramid is half of a regular octahedron, its height has the same length. Hence the height of the top of the building from the ground is $2 + \frac{3\sqrt{2}}{2}$.

7. **(d)** We have $f(2) + 3f(\frac{1}{2}) = \frac{7}{2}$ and $f(\frac{1}{2}) + 3f(2) = -\frac{11}{2}$. Subtracting the second equation from three times the first, we have $8f(\frac{1}{2}) = 16$ so that $f(\frac{1}{2}) = 2$.

8. **(d)** Since $44^2 < 2005 < 45^2$, $12^3 < 2005 < 13^3$ and $3^6 < 2005 < 4^6$, the numbers of squares, cubes and sixth powers less than 2005 are 44, 12 and 3 respectively. Hence 2005 sits in position $2005 - 44 - 12 + 3 = 1952$.

9. **(a)** The six solutions for x must be 1, 2, 3, 4, 5 and 6. If $a = 42$, $3x \leq 17$ and $x = 6$ is not a solution. Larger values of a eliminate further solutions. If $a = 40$, $3x \leq 21$ and $x = 7$ is also a solution. Smaller values of a allow for further solutions. Hence we must have $a = 41$. Then $3x \leq 19$, and we have the six solutions above.

10. **(b)** The total area of triangles PAD and PCB is half that of the parallelogram. Hence $\frac{1}{2} - \frac{1}{3} = \frac{1}{6}$ of the area of the parallelogram is 36, and its area is 36.

11. **(c)** If we have a gold medalist and a bronze medalist, we can convert them to two silver medalists. This way, we will end up with either no gold medalists or no bronze medalists. Altogether, $3 \times 60 = 180$ correct solutions are received. If all students were silver medalists, they would have turned in $2 \times 100 = 200$ correct solutions. Since we are $200 - 180 = 20$ short, we have 20 bronze medalists and 0 gold medalists. This difference is not affected by our conversions.

12. **(e)** Divide the infinite piece of graph paper into 3×3 regions. If we paint the nine square in each region in 9 different colors in exactly the same way, the tile cannot cover two squares of the same color. If we use less than 9 colors, two squares in the same region will have the same region. Both can be covered by a suitably placed tile.

13. **(c)** We have $\frac{a}{b+c} + \frac{b}{c+a} + \frac{c}{a+b} = \frac{7-(b+c)}{b+c} + \frac{7-(c+a)}{c+a} + \frac{7-(a+b)}{a+b} = 7(\frac{1}{b+c} + \frac{1}{c+a} + \frac{1}{a+b}) - 3 = 7$.

14. **(c)** Adding $x + y + 2$ to both sides of the given equation, and letting $u = x + 1$ and $v = y + 1$, the equation becomes $u^2 + v^2 = uv + 1$. If $uv = 0$, the solutions are $(u, v) = (0, \pm 1)$ or $(\pm 1, 0)$. If $uv > 0$, then $(u - v)^2 = 1 - uv$ and we must have $uv = 1$. The solutions are $u = v = \pm 1$. If $uv < 0$, let one of them be a and the other be $-b$, where a and b are positive. Then the equation becomes $a^2 + ab + b^2 = 1$, which has no solutions. Hence there are six solutions for (u, v), and six corresponding solutions for (x, y).

15. **(d)** Clearly, $n = 2^{2005} - 1$ is odd. Since $2^2 \equiv 1 \pmod 3$, we have $n \equiv 1^{1002}2 - 1 = 1 \pmod 3$. Since $2^4 \equiv 1 \pmod 5$, we have $n \equiv 1^{501}2 - 1 = 1 \pmod 5$. However, n is divisible by $2^5 - 1 = 31$.

16. **(c)** Since the three equal spheres are pairwise tangent, their centers form an equilateral triangle of side 2. The distance of the center of this triangle from any vertex is $\frac{2\sqrt{3}}{3}$. The projection of this center onto the plane is the center of the large sphere. Its distance from the center of one of the spheres is $\sqrt{1 + (\frac{2}{\sqrt{3}})^2} = \sqrt{\frac{7}{3}}$. Since the large sphere is also tangent to the others, its radius is $\sqrt{\frac{7}{3}} + 1$.

Second Round Solutions

February 14, 1984

1. Let n be a seven-digit number with digit-sum 61. The possible digit combinations for n are (8,8,9,9,9,9,9) and (7,9,9,9,9,9,9). Let a be the sum of the digits of n in the odd positions and b be the sum of the digits in the even positions. Then $a + b = 61$ and $a - b$ is a multiple of 11, say $a - b = 11k$. It follows that $2a = 61 + 11k$. Since $34 \leq a \leq 36$, we must have $k = 1$ and $a = 36$, so that the digits in the odd positions are all 9's. Thus there are six numbers with the desired properties, namely 9998989, 9899989, 9898999, 9999979, 9997999 and 9799999.

2. We only have to consider the planar section through the axis of the cylindrical glass. In the diagram, let P be the fixed point of contact between the glass and the table, and H be a variable point which coincides with the midpoint of the water surface. Note that we may assume that the glass is of negligible thickness. Because of the conservation of volume and the shape of the glass, the distance PH is fixed, so that the locus of H is a circle. Clearly, as the glass is tilted slightly, the distance of H from the table increases.

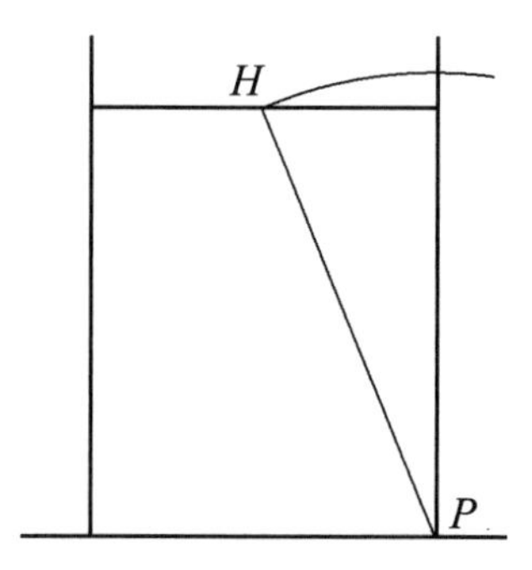

3. The inequality is equivalent to $3(16^x + 1) - 4(1-x)(16^x + 4x) \geq 0$ or $(4x-1)(16^x + 4x - 3) \geq 0$. For $x < \frac{1}{4}$, both factors are negative. When $x = \frac{1}{4}$, both factors are 0. For $x > \frac{1}{4}$, both factors are positive. Hence this inequality holds for all x with equality when $x = \frac{1}{4}$.

4. The diagram shows two parallel planes, one containing AB and the other CD. Since AB and CD are not parallel, the shortest distance between them is equal to the distance between the two planes. This is half the diagonal of a unit square or $\frac{1}{\sqrt{2}}$.

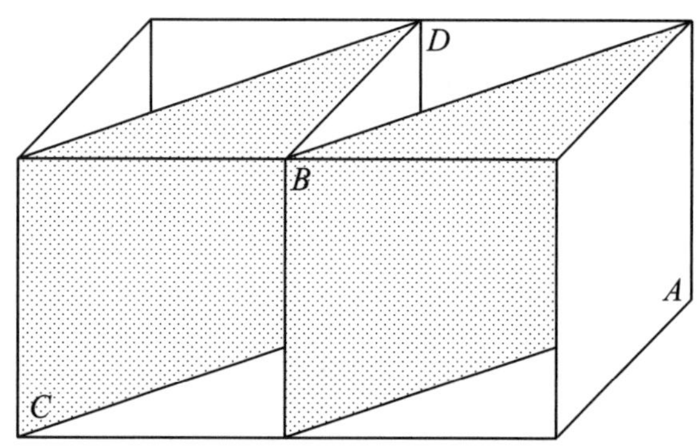

5. The answer to (b) is affirmative, and we give a proof by induction on the degree n of the polynomial. For $n = 1$, the result is trivial. For $n > 1$, let the first player choose a sequence S of n signs. By induction hypothesis, the second player can find a polynomial
$$f(x) = x^{n-1} + a_1x^{n-2} + \cdots + a_{n-2}x + a_{n-1}$$
such that

(1) it has $n - 1$ distinct real roots;

(2) the a's are nonzero;

(3) the sequence of signs of the a's is identical to S if the last sign of S is ignored.

Let $a_0 = 1$ and ϵ be a real number such that

(4) ϵ is distinct from any root of $f(x)$;

(5) $a_k - a_{k-1}\epsilon$ is nonzero for $1 \le k \le n - 1$;

(6) $a_k - a_{k-1}\epsilon$ has the same sign as a_k for $1 \le k \le n - 1$;

(7) the sign of $a_{n-1}\epsilon$ is the same as the last sign of S.

Condition (7) poses no problem while conditions (4), (5) and (6) are satisfied if the absolute value of ϵ is sufficiently small. The second player can now complete the polynomial
$$(x-\epsilon)f(x) = x^n+(a_1-a_0\epsilon)x^{n-1}+(a_2-a_1\epsilon)x^{n-2}+\cdots+(a_{n-1}-a_{n-2}\epsilon)x+a_{n-1}\epsilon.$$
By (1) and (4), this polynomial has n distinct roots. By (2) and (5), none of its coefficients is zero. By (3), (6) and (7), the sequence of signs of its coefficients is identical to S. This completes the proof, which includes (a) as a special case.

February 12, 1985

1. Counting the General, there are 21 officers with $5 \times 21 = 105$ subordinates. Since 20 of these subordinates are officers, there are $105 - 20 = 85$ soldiers.

2. **First Solution:** We use induction on n. Clearly, $r_0 = 0 < \frac{1}{2}(1+\sqrt{1+4t})$. Suppose $r_k < \frac{1}{2}(1 + \sqrt{1+4t})$. Then $r_{k+1} = \sqrt{t + r_k} < \sqrt{t + \frac{1}{2}(1 + \sqrt{1+4t})}$. Observe that
$$(\frac{1}{2}(1 + \sqrt{1+4t}))^2 = \frac{1}{4}(1 + 2\sqrt{1+4t} + 1 + 4t) = t + \frac{1}{2}(1 + \sqrt{1+4t}).$$
Hence $r_{k+1} < \frac{1}{2}(1 + \sqrt{1+4t})$, completing the induction.

Second Solution: Comparing $\frac{1}{2}(1+\sqrt{1+4t})$ with $\frac{1}{2a}(-b+\sqrt{b^2-4ac})$, we have $a = 1, b = -1$ and $c = -t$. Hence $x = \frac{1}{2}(1 + \sqrt{1+4t})$ is the positive root of the equation $x^2 - x - t = 0$, whence $x = \sqrt{t+x}$. We now use induction on n. Clearly, $r_0 = 0 < x$. If $r_k < x$, then $r_{k+1} <= \sqrt{t + r_k} < sqrtt + x < x$, completing the induction.

3. If two of a, b and c, say a and b, are of opposite parity, then $a-b = 2k+1$ for some integer k. Since $2k+1 = (k+1)^2-k^2$, we have $a+k^2 = b+(k+1)^2$. If not, then two of them, say a and b, differ by a multiple of 4. From $a-b = 4k = (k+1)^2-(k-1)^2$, we have $a+(k-1)^2 = b+(k+1)^2$. -

4. **First Solution:** Extend CE to G so that $CE = EG$. Since $CF = FD$ as well, we have $GD = 2EF$. Now $BC = AG$ since triangles BCE and AGE are congruent by the S.A.S. Postulate. By the triangle inequality, $2EF = GD \leq AD + AG = AD + BC$.

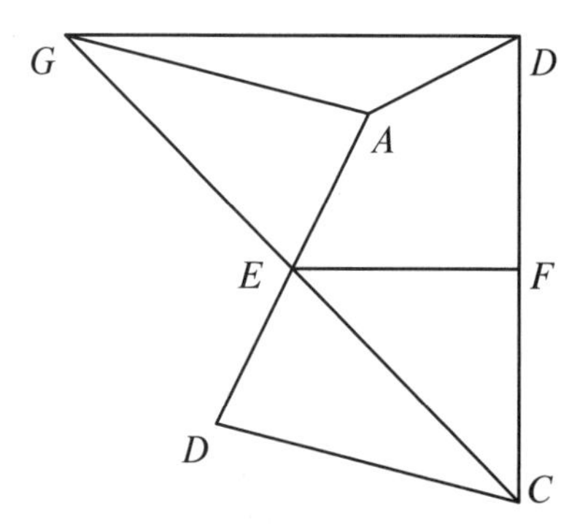

Second Solution: Let **a, b, c, d, e** and **f** be vectors from a fixed point to the points A, B, C, D, E and F respectively. Then $2\mathbf{e}-\mathbf{a}+\mathbf{b}$ and $2\mathbf{f} = \mathbf{c}+\mathbf{d}$. Now $2\overline{EF} = 2\mathbf{f}-2\mathbf{e}$ and

$$2EF = |\mathbf{c}+\mathbf{d}-\mathbf{a}-\mathbf{b}| \leq |\mathbf{d}-\mathbf{a}| + |\mathbf{c}-\mathbf{b}| = AD + BC.$$

Remark: Equality holds if and only if AD is parallel to BC.

5. We may consider the corresponding planar problem with circles instead of poles. Assume to the contrary that 101 circles are completely visible. In the diagram, O is the position of the observer and OX is perpendicular to the line joining the centers of the circles. Of the 101 completely visible circles, we may assume that at least 51 have centers on or to the right of X. Let B be the center of the last completely visible circle on the right side, and let A be the center of the circle immediately to its left. Let Y be the point on XA extended such that OY is tangent at Z to the circle with center A. Since the circle with center B is completely visible, it has at most one point in common with OY. It follows that $AB \geq 2AY$. Since at least 51 centers are at or between X and B and the centers are equally spaced, $XY \geq 49AB + AY \geq 99AY$. On the other hand, the triangles OXY and AZY are similar so that $\frac{OY}{OX} = \frac{AY}{AZ}$. Since $OX = 99$ and $AZ = 1$, Hence $OY = 99AY \leq XY$. This is a contradiction as OY is the hypotenuse of the right triangle OXY.

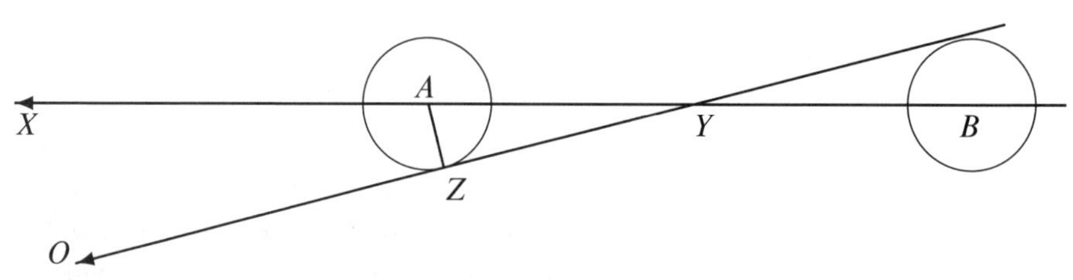

Remark: We may replace 99 by n and 100 by $n+1$.

February 11, 1986

1. Let x and y be the numbers of boys and girls, respectively, writing the contest in 1980. Then the corresponding numbers for this year are $\frac{6x}{5}$ and $\frac{3y}{2}$ with $\frac{6x}{5}+\frac{3y}{2}=\frac{13(x+y)}{10}$. Solving this equation, we have $x=2y$ so that the required fraction is $\frac{3y}{2}\div\frac{13(x+y)}{10}=\frac{5}{13}$.

2. Represent each lake by a bead and each river by a thread. If a river joins two lakes, tie the two ends of the corresponding thread to the two corresponding beads. If a river joins a lake and the sea, tie one end of the corresponding thread to the corresponding bead and tie a knot at the other end of the thread. Now hold up each string which consists of connected beads, threads and knots. If the string has a knot, hold it up by the knot. Otherwise, hold it up by any bead. We claim that in any string, there is a unique bead dangling at the lower end of each thread. We cannot have a bead dangling at the lower ends of two threads, as a closed loop would be formed. The lakes and rivers corresponding to this loop would enclose an area separated from the remaining part of the island. Similarly, we cannot have a knot dangling at the lower end of a thread as there would be two knots in the string. The lakes and rivers corresponding to the part of this string joining the two knots would separate the island into two mutually inaccessible parts. It follows easily from the claim that there are at least as many beads as threads in each string. Hence there are at least as many lakes as rivers on the island.

3. (a) More generally, let ABC be any triangle in which the altitude from C ha@ length 1. Now the diameter of the incircle perpendicular to AB is shorter than the altitude from C. Hence the inradius is less than $\frac{1}{4}$.

   (b) Draw a quadrant Q of a circle with center C and radius 1, with A and B lying respectively on the extensions of the two perpendicular radii. Since the altitude of triangle ABC from C is of length 1, AB is tangent to Q. The incircle of ABC must have points in common with Q, as otherwise it will not be tangent to AB. Thus it is at least as large as the circle K which is tangent to AB, AC and Q internally. Simple calculation yields $\sqrt{2}-1$ as the radius of K. Moreover, K is in fact the incircle of ABC if, and only if, $CA=C$B. Hence the minimum inradius is $\sqrt{2}-1$, achieved uniquely when the triangle is isosceles.

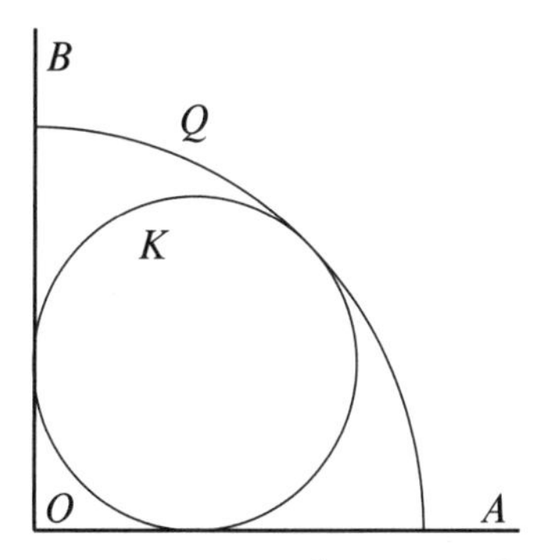

4. (a) We can take $a_1=a_2=\cdots=a_{k-2}=1, a_{k-1}=3$ and $a_k=2t+1$ where $k=4t+1$. Then both their sum and their product are equal to the odd number $6t+3$.

(b) Let $a_1, a_2, \ldots, a_k$ be the positive integers with sum s and product p, where $s = p$ is odd. Since p is odd, each a_i is odd. Since s is odd, k is also odd. It remains to rule out the case where k is of the form $4t - 1$. Suppose $k = 4t - 1$. Let x be the number of i's such that $a_i \equiv 1 \pmod 4$. Then $s \equiv x - (k - x) \equiv 2x + 1 \pmod 4$ while $p \equiv (-1)^{x-k} \equiv (-1)^{x+1} \pmod 4$. If x is odd, $s \equiv -1 \pmod 4$ while $p \equiv 1 \pmod 4$. If x is even, $s \equiv 1 \pmod 4$ while $p \equiv -1 \pmod 4$. Either case yields a contradiction.

5. Draw the normal to each circle through its center. These two normals lie on the plane passing through the midpoint of the common chord of the two circles and normal to this chord. Since the two circles are not coplanar, the normals are not parallel. Hence they intersect at some point O. Now O is equidistant to all points on each circle. Since the two circles have common points, O is the center of the sphere containing both circles.

February 10, 1987

1. Since 3 and 8 have no common factor greater than 1, a number divisible by both 3 and 8 is divisible by 24. To prove that $ab(a-b)(a+b)$ is divisible by 3, we consider three cases. If $a \equiv 0 \pmod 3$ or $b \equiv 0 \pmod 3$, then 3 divides ab. If $a \equiv b \equiv 1 \pmod 3$ or $a \equiv b \equiv 2 \pmod 3$, then 3 divides $a - b$. Finally, if $a \equiv 1 \pmod 3$ and $b \equiv 2 \pmod 3$, or $a \equiv 2 \pmod 3$ and $\equiv 1 \pmod 3$, then 3 divides $a + b$. To prove that $ab(a-b)(a+b)$ is divisible by 8, we note that since a and b have no common factor greater than 1 and $a + b$ is even, both a and b are odd. We consider two cases. If $a \equiv b \equiv 1 \pmod 4$ or $a \equiv b \equiv 3 \pmod 4$, then 2 divides $a + b$ and 4 divides $a - b$. If $a \equiv 1 \pmod 4$ and $b \equiv 3 \pmod 4$, or $a \equiv 3 \pmod 4$ and $b \equiv 1 \pmod 4$, then 2 divides $a - b$ and 4 divides $a + b$.

2. Consider a vertical section of the table through the center O of the sphere and the center C of the disc. Let A and B be the points of contact with the floor of the disc and the sphere respectively, and let D be the point of intersection of AB and CO. Note that we have $\angle ADC = \angle ODB$ and $\angle ACD = 90° = \angle OBD$. Hence triangles ADC and ODB are similar, so that $\frac{BD}{OB} = \frac{CD}{AC}$. Let $BD = x$. Note that $OB = \frac{1}{2}$ and $AB = AC = 1$. By the Pythagorean Theorem, we have $CD = \sqrt{(1+x)^2 - 1} = \sqrt{x(2+x)}$. Hence $2x = \sqrt{x(2+x)}$ or $4x^2 = x(2+x)$, yielding $x = \frac{2}{3}$. It follows that the two desired radii are $BD = \frac{2}{3}$ and $AD = \frac{5}{3}$.

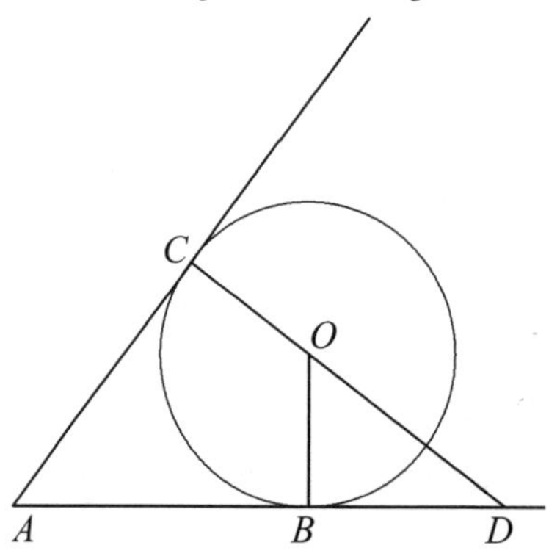

3. If either $x = 0$ or $y = 0$, the other is equal to 0 also, but then 0^0 is undefined. Multiplying $x^{x+y} = y^3$ and $y^{x+y} = x^6y^3$, we have $(xy)^{x+y} = (xy)^6$. We may have $x + y = 6$. Substituting back into $x^{x+y} = y^3$, we have $x^6 = y^3$ or $y = x^2$. From $x + x^2 = 6$, we have $0 = x^2 + x - 6 = (x - 2)(x + 3)$. If $x = 2$, $y = 4$, and if $x = -3$, $y = 9$. We may also have $xy = 1$. Then $y = \frac{1}{x}$ and $x^{x+\frac{1}{x}+1} = 1$. Either $x = 1$ or $x + \frac{1}{x} + 1 = 0$. The solutions to $x^2 + 3x + 1 = 0$ are $x = \frac{1}{2}(-3 \pm \sqrt{5})$. If $x = 1$, $y = 1$. If $x = \frac{1}{2}(-3 + \sqrt{5})$, $y = \frac{1}{2}(-3 - \sqrt{5})$. If $x = \frac{1}{2}(-3 - \sqrt{5})$, $y = \frac{1}{2}(-3 + \sqrt{5})$. Finally, we may have $xy = -1$ with $x + y$ an even integer. Then $y = -\frac{1}{x}$ and $x^{x-\frac{1}{x}+3} = -1$. We must have $x = -1$ whence $y = 1$. In summary, there are six solutions.

4. Let the first guess be $B = 6$. If $B = A$, Bernice has achieved her objective. Suppose Bernice is not close on her first guess. Then $A \in \{1, 2, 10, 11, 12\}$. The second guess is $B = 10$. Suppose $B \neq A$. If she is not close, then $A \in \{1, 2\}$. If she is close, then $A \in \{11, 12\}$. With two guesses remaining, Bernice will have no problem achieving her objective. Suppose Bernice is close on her first guess. Then $A \in \{3, 4, 5, 7, 8, 9\}$. The second guess is $B = 4$. Suppose $B \neq A$. If she is not close, then $A \in \{8, 9\}$ and she has two guesses remaining to settle the issue. If she is close, then $A \in \{3, 5, 7\}$. The third.guess is $B = 7$. Suppose $B \neq A$. If she is not close, then $A = 3$. If she is close, then $A = 5$. The remaining guess allows Bernice to achieve her objective.

5. Since $\angle GBF = \angle FBA$ and $\angle GFB = 90° = \angle FAB$, triangles GFB and FAB are similar. Hence $\frac{BG}{BF} = \frac{BF}{BA}$ so that $\frac{BG}{BC} = \frac{BC}{DC}$. It follows that triangles GBC and BCD are also similar, so that $\angle GCB = \angle BDC$. Hence $\angle GCB + \angle CBD = 90°$ and CG is perpendicular to BD.

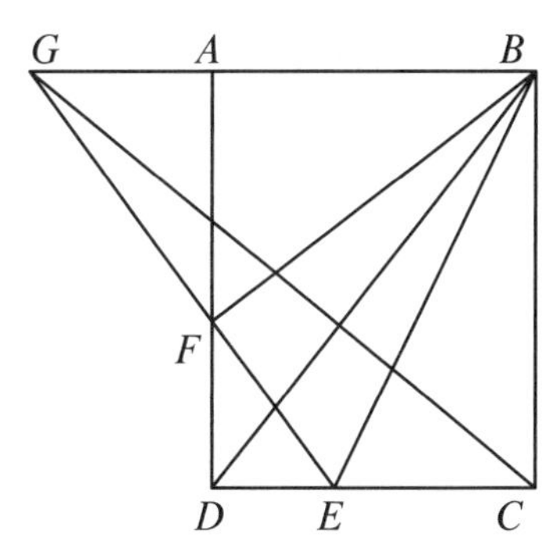

February 9, 1988

1. Let h, m and s denote the respective fractions of a videotape taken up by a hockey game, a movie and an episode of Star Trek. From the given information, we have $2h+s \leq 1$ and $2m+3s \leq 1$. Adding these and dividing by 2, we have $h+m+2s \leq 1$. It now follows that $7h + 11m + 20s = (h + m + 2s) + 3(2h + s) + 5(2m + 3s) \leq 9$, so that 9 videotapes are sufficient. On the other hand, if all inequalities are in fact equalities, then exactly 9 videotapes will be required, and 8 will not be enough.

2. If fewer than 5 grooves are recolored, we will have less than 10 pairs of colored grooves. Hence re-coloring 5 grooves is necessary. Suppose it is also sufficient. Denote by a, b, c and d the four distances between consecutive colored grooves. Then the other distances between two colored grooves are $a+b, b+c, c+d, a+b+c, b+c+d$ and $a+b+c+d$. The sum of these 10 distances is $4a+Gb+6c+4d$ on the one hand, and 1+2+3+4+5+6+7+8+9+10=45 on the other. Since one value is even and the other odd, this is not possible. Hence re-coloring 6 grooves is necessary. To see that it is also sufficient, label the 11 grooves from 0 to 10 in that order. It is easy to verify that re-coloring the grooves numbered 0, 1, 2, 5, 8 and 10 will accomplish the task.

3. We have

$$\begin{aligned} 0 &= (ab+bc+ca)^2-(abc)^2 \\ &= (ab)^2+(bc)^2+(ca)^2+2abc(a+b+c)-(abc)^2 \\ &= (ab)^2+(bc)^2+(ca)^2+(abc)^2. \end{aligned}$$

Since each term in the last expression must be 0, at least two of a, b and c are 0. From $a+b+c=abc=0$, so is the third one. Hence $(a,b,c)=(0,0,0)$ is the only solution.

4. Clearly, (0,0) is a solution. If (x,y) is a solution and either is 0, both are 0. If (x,y) is a non-zero solution, then $xy>0$ and $(-x,-y)$ is also a solution. Hence we need only search for positive solutions. Let $x=3^m a$ and $y=3^n b$ where a and b are not divisible by 3. We may assume that $m \geq n$. Then the equation becomes $8((3^{m-n}a)^2+b^2)=3^{3m+n-4}a^3b^3$. Since $(3^{m-n}a)^2 \equiv 0$ or 1 (mod 3) while $b^2 \equiv 1$ (mod 3), the left side is not divisible by 3. It follows that $3m+n=4$, and in view of $m \geq n \geq 0$, we must have $m=n=1$. The equation now simplifies to $8(a^2+b^2)=a^3b^3$. We may assume that $a \geq b$. Then $16a^2 \geq a^3b^3$ or $16 \geq ab^3$. Hence $b \leq 2$. If $b=2$, we must have $a=2$. If $b=1$, we have $a^3-8a^2-8=0$. The only possible positive integer roots are 1, 2, 4 and 8, but none of them checks out. It follows that we must have $(a,b)=(2,2)$ so that $(x,y)=(6,6)$. In summary, the original equation has tbree solutions, (0,0), (6,6) and $(-6,-6)$.

5. We may assume that the sphere is of unit radius. The diagram on the left shows a vertical section through the center O of the sphere. Suppose the base of the tetrahedron lies on the horizontal plane represented by EF, at a distance x below O. Now the volume of a tetrahedron is equal to one-third of the product of its height and base area. From a horizontal base below O, the maximum height occurs when the apex of the tetrahedron is at the highest point A of the sphere. Let H be the point of intersection of AO and EF. By Pythagoras' Theorem, we have $EH=\sqrt{1-x^2}$ and $AE=\sqrt{2+2x}$. The diagram on the right shows the horizontal section of the sphere on which the base BCD of the tetrahedron lies. The base area is maximum when BCD is equilateral, so that H is the centroid and $BH=\sqrt{1-x^2}$. Let G be the point of intersection of BH and CD. Then $HG=\frac{1}{2}\sqrt{1-x}^2$ and $CG=\frac{1}{2}\sqrt{3-3x^2}$. Hence $CD=\sqrt{3-3x^2}$ and the area of BCD is $\frac{3\sqrt{3}}{4}(1-x^2)$. It follows that the volume of the tetrahedron is

$\frac{\sqrt{3}}{4}(1+x)(1-x^2)$, which may be rewritten as $\frac{\sqrt{3}}{8}(1+x)(1+x)(2-2x)$. The product of the three numbers $1+x$, $1+x$ and $2-2x$, with constant sum 4, is maximum if and only if all three are equal. From $1+x=\frac{4}{3}$, we have $x=\frac{1}{3}$. Hence $AB = AC = AD = \sqrt{2+2x} = \sqrt{\frac{8}{3}}$ and $BC = BD = CD = \sqrt{3-3x^2} = \sqrt{\frac{8}{3}}$ also. Thus the maximum volume is attained if and only if the tetrahedron is equilateral.

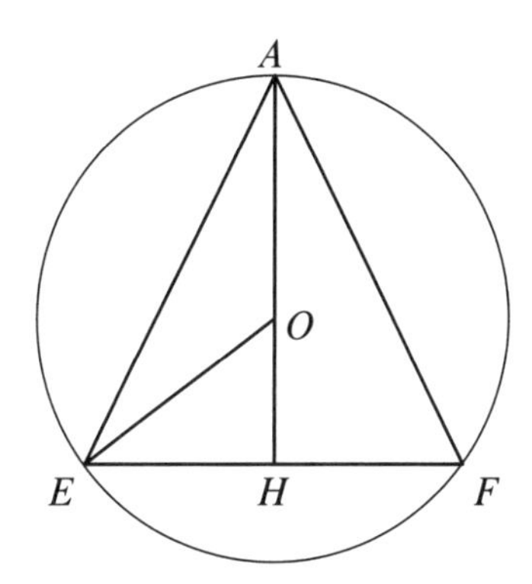

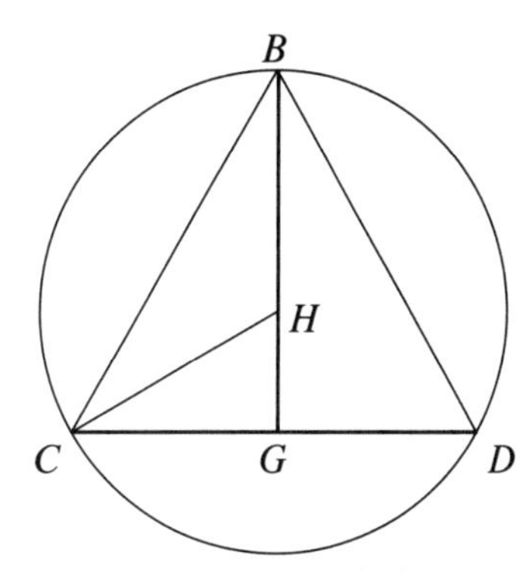

February 14, 1989

1. Let there be in the library x students who wear glasses and y who do not. Then $\frac{x-1}{x+y-1}=\frac{1}{5}$ and $\frac{x+1}{x+y+1}=\frac{1}{4}$. Simplification leads to $4x-y=4$ and $3x-y=-3$. Subtracting four times the second equation from three times the first, we obtain $y = 24$.

2. **First Solution:**

   (a) We have $b^5 = b^3b^2 = (b+1)b^2 = b^3+b^2 = (b+1)+b^2 = b^2+b+1$. On the other hand, $b^4+1 = b^3b+1 = (b+1)b+1 = b^2+b+1$. Hence $b^5 = b^4+1$.

   (b) Multiplying $b^5 = b^4+1$ by b, we have $b^6 = b^5+b = b^4+1+b$. Hence $b^6-1 = b^4+b$ or $(b^3-1)(b^3+1) = b(b^3+1)$. Now the only real cube root of -1 is -1, which does not satisfy $b^5 = b^4+1$. Hence $b^3+1 \neq 0$ so that $b^3-1=b$ or $b^3=b+1$.

   Second Solution: We have $b^5-b^4-1 = (b^3-b-1)(b^2-b+1)$.

   (a) Clearly, if $b^3 = b+1$ or $b^3-b-1=0$, then $b^5-b^4-1=0$ or $b^5=b^4+1$.

   (b) If $b^5=b^4+1$ or $b^5-b^4-1=0$, either $b^3-b-1=0$ or $b^2-b+1=0$. In the latter case, we must have $b=\frac{1\pm\sqrt{1-4}}{2}$ by the Quadratic Formula. However, this is impossible since b is a real number. Therefore, $b^3-b-1=0$ or $b^3=b+1$.

3. The point O may be inside or outside triangle PQR. In the first diagram, $AOCQ$, $AOBR$ and $BOCP$ are cyclic quadrilaterals. It follows that $\angle QAO = \angle OCP$, $\angle RAO = \angle OBP$ and $\angle OCP + \angle OBP = 180°$. Hence $\angle QAO + \angle RAO = 180°$ and Q, A and R are collinear. In the second diagram, $ACOQ$, $ABOR$ and $BOCP$ are cyclic quadrilaterals. It follows that $\angle QAO = \angle QCO$, $\angle RAO = \angle RBO$ and

$\angle OCP + \angle OBP = 180°$. Hence $\angle QAO + \angle RAO = 180°$ and Q, A and R are collinear.

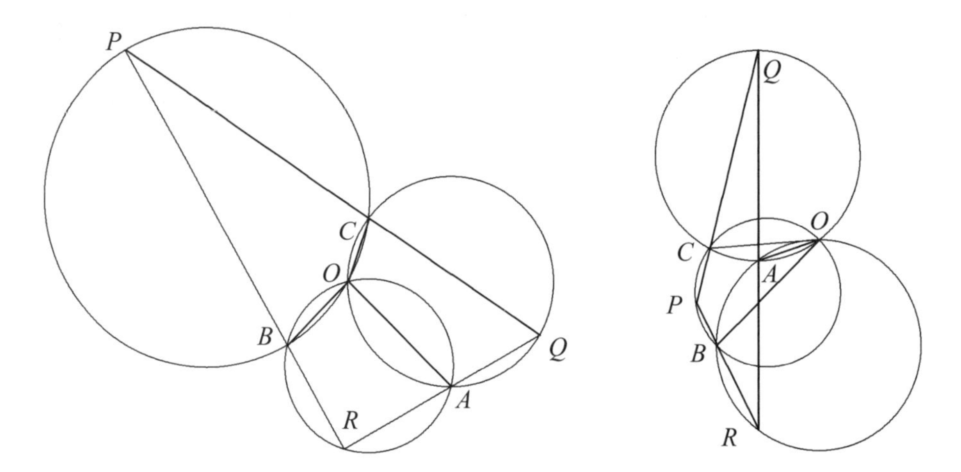

4. **First Solution:** We have

$$\begin{aligned} X - Y &= 4(1 + 10 + 10^2 + \cdots + 10^{2n-1}) - 8(1 + 10 + 10^2 + \cdots + 10^{n-1}) \\ &= 4((10^n + 10^{n+1} + \cdots + 10^{2n-1}) - (1 + 10 + 10^2 + \cdots + 10^{n-1})) \\ &= 4(10^n - 1)(1 + 10 + 10^2 + \cdots + 10^{n-1}) \\ &= 4(10 - 1)(1 + 10 + 10^2 + \cdots + 10^{n-1})^2 \\ &= (6(1 + 10 + 10^2 + \cdots + 10^{n-1}))^2. \end{aligned}$$

Hence $\sqrt{X - Y}$ is a positive integer with n digits all of which are 6's.

Second Solution: Calculations for small values of n suggest a definite pattern. Let x_n be the positive integer with $2n$ digits all of which are 4's, y_n be the positive integer with n digits all of which are 8's, and z_n be the positive integer with n digits, all of which are 6's. We use mathematical induction on n to prove that $2z_1z_n = 9y_n$ and $z_n2 = x_n - y_n$ for all positive integers n. For $n = 1, 2 \cdot 6 \cdot 6 = 9 \cdot 8$ and $6^2 = 44 - 8$ so that the results are true. Suppose they also hold for some positive integer n. Then

$$\begin{aligned} 2z_1z_{n+1} &= 2z_1(10z_n + z_1) \\ &= 10(2z_1z_n) + 2z_1z_1 = 10(9y_n) + 9y_1 \\ &= 9(10y_n + y_1) \\ &= 9y_{n+1}. \end{aligned}$$

Moreover,

$$\begin{aligned} z_{n+1}^2 &= (10z_n + z_1)^2 \\ &= 100z_n^2 + 20z_1z_n = z_1^2 \\ &= 100(x_n - y_n) + 90y_n + (x_1 - y_1) \\ &= (100x_n + x_1) - (10y_n + y_1) \\ &= x_{n+1} - y_{n+1}. \end{aligned}$$

This completes the induction argument. Now $\sqrt{X - Y} = \sqrt{x_n - y_n}$ is a positive integer with n digits, all of which are 6's.

5. Suppose $n \geq 5$. Take A_i to be the point such that PA_i have constant length for $1 \leq i \leq 5$. Then the triangles $A_i PA_j$ are congruent to one another for $1 \leq i < j \leq 5$. It follows that $A_i A_j$ have constant length for $1 \leq i < j \leq 5$. This means that $A_1A_2A_3A_4$ and $A_2A_3A_4A_5$ are regular tetrahedra. Since the rays PA_1 and $PA - 5$ are different, these two tetrahedra are back-to-back with $A_2A_3A_4$ as their common base. It is easy to see that A_1A_5 is longer than A_1AA_2, a contradiction.

February 13, 1990

1. (a) If $n \equiv 2 \pmod 4$, then the sum of n consecutive integers is always odd. This is because we have a sum,of an odd number of odd integers and some even integers.

   (b) If $n \equiv 1 \pmod 2$, then the sum of n consecutive integers is sometimes odd and sometimes even. This is because the parity of the sum changes when we drop the first integer of the block and pick up the first integer after the block.

   (c) If $n \equiv 0 \pmod 4$, then the sum of n consecutive integers is always even. This is because we have a sum of an even number of odd integers and some even integers.

2. (a) Any cut of length 25 centimeters such that one piece has 5 centimeters of the original perimeter will work. One example is shown in the diagram below.

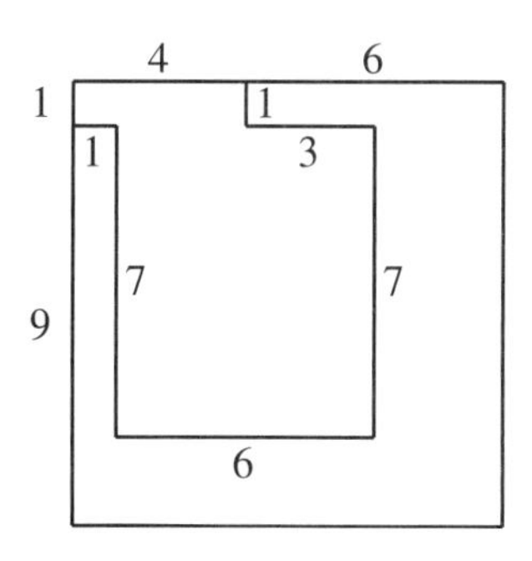

   (b) Since the combined perimeters of the two pieces is 150 centimeters and the original perimeter is 40 centimeters, the cut must be of length $\frac{1}{2}(150 - 40) = 55$ centimeters. Thus it is impossible for one of the pieces to have perimeter 50 centimeters.

3. Extend DA to E and DC to F. Then $\angle BAE = 180° - \angle BAC - \angle CAD = 50° = \angle BAC$ and $\angle BCF = 180° - \angle BCA - \angle ACD = 70° = \angle BCA$. Hence B is the point of intersection of the bisectors of two of the exterior angles of triangle DAC, so that it is one of the excenters of the triangle. It follows that $\angle BDA = \angle BDC$. Now $\angle CDA = 180° - \angle CAD - \angle ACD = 60°$. Hence $\angle BDA = \frac{1}{2}\angle CDA = 30°$. Finally, $\angle ABD = 180° - \angle BDA - \angle CAD - \angle BAC = 20°$.

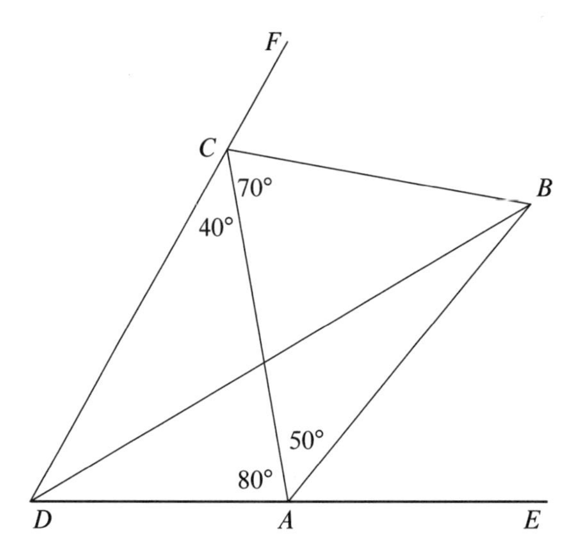

4. Construct a graph, representing each Knight by a vertex. Join two vertices by an edge if and only if the Knights they represent were opponents. Since each Knight had two opponents, the degree of each vertex is two. Hence the graph is a union of disjoint cycles. Moreover, since no two Knights had the same two opponents, there are no four-cycles. Suppose there is an even cycle $(1, 2, \ldots, 2m)$, $m \geq 3$. Then the Knights $2, 4, 6, \ldots, 2m$ must be seated consecutively in that order, and with $2m$ seated next to 2. There is no room for any other Knights, not even $1, 3, 5, \ldots, 2m-1$. Clearly, this is impossible. Hence all cycles are odd. Consider an odd cycle $(1, 2, \ldots, 2m+1)$, $m \geq 1$. Then the Knights $2, 4, 6, \ldots, 2m, 1, 3, 5, \ldots, 2m+1$ must be seated consecutively in that order, and with $2m+1$ seated next to 2. Again, there is no room for any other Knights. It follows that the whole graph is a single odd cycle. Thus we have proved that

   (a) the number of Knights cannot be even;

   (b) the number of Knights can be any odd number greater than one.

5. Let O and C be the centers of the large and the small spheres and P and D be their points of contact with the floor, respectively. Let A be the corner of the room nearest to the spheres. Then A, C and O lie on a straight line. Note that triangles ACD and AOP are similar. Let $CD = r$ be the radius of the smaller sphere. We have $OP = 2$, $OC = r + 2$ and $OA = 2\sqrt{3}$. Hence $\frac{r}{2} = \frac{CD}{OP} = \frac{AC}{AO} = \frac{2\sqrt{3}-r-2}{2\sqrt{3}}$, which yields $r = 4 - 2\sqrt{3}$.

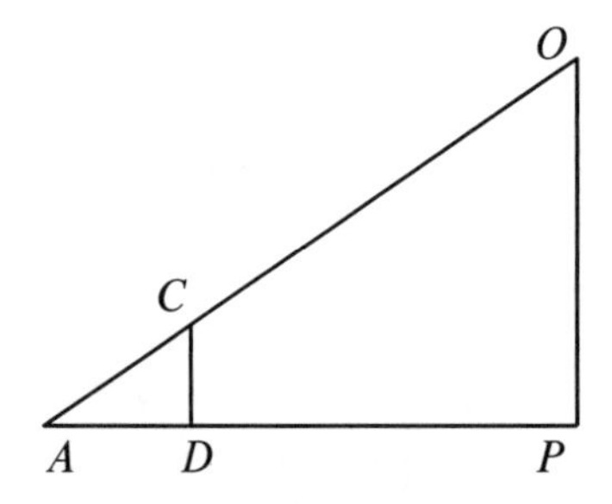

February 12, 1991

1. Let $x = 10^n x_n + 10^{n-1}x_{n-1} + \cdots + 10x_1 + x_0$ be a special number, where x_i is a single-digit number for $0 \le i \le n$. Clearly, we must have $n \ge 1$. The sum of the digits of x is $S(x) = x_n + x_{n-1} + \cdots + x_1 + x_0$ and the product of its digits is $P(x) = x_n x_{n-1} \cdots x_1 x_0$. The latter is less than or equal to $9^n x_n$, with equality if and only if $x_0 = x_1 = \cdots = x_{n-1} = 9$. Note that

$$\begin{aligned} x - S(x) - P(x) &= (10^n - 1)x_n + (10^{n-1} - 1)x_{n-1} + \cdots + (10 - 1)x_1 - P(x) \\ &\ge (10^n - 1)x_n - 9^n x_n \\ &= (10^n - 9^n - 1)x_n \\ &\ge 0. \end{aligned}$$

Equality holds if and only if $n = 1$ and $x_0 = 9$. Hence the only special numbers are 19, 29, 39, 49, 59, 69, 79, 89 and 99.

2. **First Solution:** Let $\angle A = \alpha$, $\angle B = \beta$ and $\angle C = \gamma$. Since $BD = BE$, we have $\angle BDE = \frac{\beta}{2}$. Since $CF = CG$, we have $\angle CFG = \frac{\gamma}{2}$. Since $AE = AF$, we have $\angle AFE = \frac{\pi-\alpha}{2}$ so that $\angle CFE = \frac{\pi+\alpha}{2}$. It follows that $\angle GDE + \angle EFG = \frac{\beta+\pi+\alpha+\gamma}{2} = \pi$, whereby $DEFG$ is a cyclic quadrilateral.

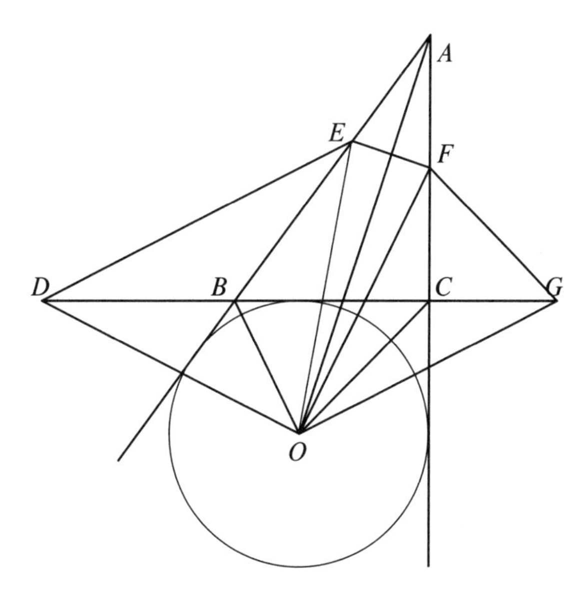

Second Solution: Let O be the center of the circle which is tangent to BC and the extensions of AB and AC. Then OA bisects $\angle BAC$, OB bisects $\angle ABD$ and OC bisects $\angle ACG$. Since $AE = AF$, triangles OAE and OAF are congruent, so that $OE = OF$. Now triangles OBD and OBE are also congruent, so that $OD = OE$. Similarly, we have $OF = OG$. Hence D, E, F and G all lie on a circle with O as center.

3. Let d be the common difference. Then $a_n = a_1 + (n-1)d$ for $1 \le n \le 16$. We have $a_1 + 15d = a_16 = a_7 + a_9 = a1_+6d + a_1 + 8d = 2a_1 + 14d$. Hence

$a_1 = d$ and $a_n = nd$. Since d is a common divisor of all terms, its value does not affect whether three terms form an geometric progression or not. Hence we may assume that $d = 1$. Thus the problem reduces to that of finding all three-term geometric progressions contained in $\{1, 2, \ldots, 16\}$. Let $\{b_1, b_2, b_3\}$ be such a progression, with common ratio r. We need only consider $r > 1$. Then $\frac{b_3}{b_1} = r^2$. If b_1 and b_3 are relatively prime, then both must be squares. We have $(b_1, b_3) = (1, 4)$, (1,9), (1,16), (4,9) or (9,16). If they are not relatively prime, then they must reduce to one of the above pairs when their greatest common divisor is factored out. Only three more pairs (2,8), (3,12) and (4,16), obtained from (1,4), fall within range. Hence there are 8 three-term geometric progressions contained in $\{1, 2, \ldots, 16\}$, namely, $\{1, 2, 4\}$, $\{1, 3, 9\}$, $\{1, 4, 16\}$, $\{2, 4, 8\}$, $\{3, 6, 12\}$, $\{4, 6, 9\}$, $\{4, 8, 16\}$ and $\{9, 12, 16\}$.

4. Let A be the new creature of rating r, B be the creature of rating 1 and C be the one of rating 6. Note that the battle lasts at most two rounds, because C will knock off B in the first round and, if it survives, knock off A in the second. It is a win for the two creatures if and only if they knock off C in the first round, and a tie if and only if A knocks off C in the second. Now in the first round, A can knock off C with probability $\frac{r}{6}$ and B can do so with probability $\frac{1}{6}$. From the sum, we must subtract the overlap to obtain the probability of their winning, which is $\frac{r}{6} + \frac{1}{6} - \frac{r}{6} \cdot \frac{1}{6} = \frac{5r+6}{36}$. If $r \geq 3$, this expression exceeds $\frac{1}{2}$ so that the advantage is clearly with the two creatures. Suppose $r = 2$. Then this winning probability is $\frac{4}{9}$. The battle proceeds to the second round with probability $\frac{5}{9}$, and the probability of A not knocking off C is $\frac{2}{3}$. Hence C's winning probability is $\frac{10}{27}$, less than $\frac{4}{9}$. For $r = 1$, the respective winning probabilities are $\frac{11}{36}$ and $(1 - \frac{11}{36})(1 - \frac{1}{6}) = \frac{125}{216}$, favorable to C. Note that $\frac{4}{9} - \frac{10}{27} = \frac{16}{216} < \frac{57}{216} = \frac{125}{216} - \frac{11}{36}$. Hence the fairest battle occurs when the new creature has rating 2.

5. **First Solution:** To obtain the desired quantity $a^3x + b^3y + c^3z$, we multiply $a^2x + b^2y + c^2z$ by $a + b + c$, which yields $a^3x + b^3y + c^3z + a^2(b+c)x + b^2(c+a)y + c^2(a+b)z$. To get rid of the extra terms, we subtract $(ab + bc + ca)(ax + by + cz)$, which is equal to $a^2(b+c)x + b^2(c+a)y + c^2(a+b)z + abc(x+y+z)$. It follows that $a^3x + b^3y + c^3z = (a+b+c)(a^2x + b^2y + c^2z) - (ab+bc+ca)(ax+by+cz) + abc(x + y + z)$. Since $a^2x + b^2y + c^2z = ax + by + cz = x + y + z = 1$, $a^3x + b^3y + c^3z = a + b + c - ab - bc - ca + abc$.

Second Solution: We have

$$x + y + z = 1, \tag{1}$$

$$ax + by + cz = 1, \tag{2}$$

$$a^2x + b^2y + c^2z = 1. \tag{3}$$

Multiplying (1) by c and subtracting it from (2), we have

$$(a - c)x + (b - c)y = 1 - c. \tag{4}$$

Multiplying (2) by c and subtracting it from (3), we have

$$a(a - c)x + b(b - c)y = 1 - c. \tag{5}$$

Multiplying (4) by b and subtracting it from (5), we have $(a-b)(a-c)x=(1-b)(1-c)$, from which we obtain $x=\frac{(1-b)(1-c)}{(a-b)(a-c)}$. By symmetry, $y=\frac{(1-c)(1-a)}{(b-c)(b-a)}$ and $z=\frac{(1-a)(1-b)}{(c-a)(c-b)}$. Substituting these values into $a^3x+b^3y+c^3z$, we have

$$\frac{a^3(1-b)(1-c)(c-b)+b^3(1-c)(1-a)(a-c)+c^3(1-a)(1-b)(b-a)}{(a-b)(b-c)(c-a)}.$$

This expression is valid as long as a, b and c are distinct. Suppose two of them are equal, say $b=c$. Let $w=y+z$. Then we have $x+w=1$, $ax+bw=1$ and $a^2x+b^2w=1$. As before, we can get $(a-b)x=1-b$, so that $x=\frac{1-b}{a-b}$ and $w=\frac{1-a}{b-a}$. Substituting these values into $a^2x+b^2w=1$, we have $\frac{a^2(1-b)-b^2(1-a)}{a-b}=a+b-ab=1$. Hence $0=ab-a-b+1=(a-1)(b-1)$ so that either $a=1$ or $b=1$. If $a=1$, then $x+w=1=x+bw$. Since w is not always 0, we must have $b=1$. Similarly, if $b=1$, then $a=1$ also. Hence $a^3x+b^3y+c^3z=x+y+z=1$.

Remark: When a, b and c are distinct, it is possible to reduce the long expression in the Second Solution to the simpler one in the First Solution.

February 11, 1992

1. Suppose CHEAP made x photocopies. If $13x \leq 2000$, then the expenditure of the new committee is exactly 13 times that of CHEAP, which is impossible unless $x=0$. This is clearly not the case from the context of the problem. It follows that $13x > 2000$, and the new committee spent $(13x-2000)0.05+140=0.65x+40$ dollars. Suppose $x \leq 2000$. Then CHEAP's expenditure is $0.07x$ dollars. From $0.65x+40=0.7x$, we have $x=800$. Suppose $x > 2000$. Then CHEAP's expenditure is $(x-2000)0.05+140=0.05x+40$ dollars. From $0.65x+40=0.5x+400$, we have $x=2400$. It follows that CHEAP either made 800 or 2400 photocopies.

2. Let the dimensions, in centimeters, of the block be x, y and z. Since the rise in water-level is different each time, we may assume that $x>y>z$. When the $y \times z$ face rests on the bottom, the volume submerged is $4yz=10000$. When the $z \times x$ face rests on the bottom, the volume submerged is $5zx=20000$. When the $x \times y$ face rests on the bottom, the volume submerged is $6xy=30000$ provided that the block reaches the surface of the water. Assuming this for now, we have $yz=2500$, $zx=4000$ and $xy=5000$. Multiplying the first two equations and dividing by the third, we have $z^2=2000$ or $z=20\sqrt{5}$. Similarly, $y=25\sqrt{5}$ and $x=40\sqrt{5}$. Suppose now that the block does not reach the surface of the water when the $x \times y$ face rests on the bottom. The volumes submerged is $xyz=30000$. Dividing this by $yz=2500$, we have $x=12$. However, $1728=x^3>xyz=30000$ is a contradiction.

3. (a) When divided by 23, 24, 25, 26 and 27, 26 leaves remainders of 3, 2, 1, 0 and 26 respectively. They add up to 32.

   (b) For $1 \leq k \leq 6$, the maximum sum of the five remainders when k is divided by five consecutive integers is $5k \leq 30 < 32$. Hence no number less than 7 has the 32-property. Consider 7. If all five dividers are greater than 7, the sum of the remainders is $5 \cdot 7 = 35 \neq 32$. If at least one divider is 7, the maximum sum of the remainders is $4 \cdot 7 = 28 < 32$. If the dividers are 2, 3, 4, 5 and 6, the sum of the remainders is $8 \neq 32$. If the dividers are 1, 2, 3, 4 and 5, the sum of the remainders is $7 \neq 32$. Hence 7 does not have the 32-property either. When divided by 8, 9, 10, 11 and 12, 8 leaves remainders of 0, 8, 8, 8 and 8 respectively. Their sum is 32. Hence 8 is the smallest positive integer with the 32-property.

(c) The least common multiple of 8, 9, 10, 11 and 12 is 3960. Consider the number $3960t + 8$ where t is an arbitrary positive integer. When divided by 8, 9, 10, 11 and 12, the term $3960t$ leaves no remainders so that $3960t + 8$ leaves remainders of 0, 8, 8, 8 and 8 respectively, with sum 32. Since t is arbitrary, there are infinitely many positive integers with the 32-property.

4. **First Solution:**

(a) Since $ax + by + cz = 0$ and a, b and c are all positive, x, y and z cannot all be of the same sign. Hence at least one of xy, yz and zx must be non-positive. By symmetry, we may assume that $xy \leq 0$. Then $0 \leq (x + y - z)^2 - 4xy = x^2 + y^2 + z^2 - 2xy - 2yz - 2zx$. The desired inequality now follows.

(b) For equality to hold, we must have $x + y - z = 0$ and $xy = 0$. By symmetry, we may assume that $x = 0$, so that $y = z$. If $y = z \neq 0$, then $ax + by + cz \neq 0$. Hence equality holds if and only if $x = y = z = 0$.

Second Solution:

(a) Since $ax + by + cz = 0$ and a, b and c are all positive, x, y and z cannot all be of the same sign. Hence at least two of xy, yz and zx must be non-positive. By symmetry, we may assume that $yz \leq 0$ and $zx \leq 0$. Now $0 \leq (x - y)^2 + z^2 \leq x^2 - 2xy + y^2 + z^2 - 2yz - 2zx$. The desired inequality now follows.

(b) For equality to hold, we must have $x = y$ and $z = 0$. If $x = y \neq 0$, then $ax + by + cz \neq 0$. Hence equality holds if and only if $x = y = z = 0$.

5. **First Solution:** By symmetry, we may assume that $AE \leq AF$. Let A falls on the point A' on arc BD. Let G be the midpoint of AA'. Then $AG = \frac{1}{2}$. Now EF is the perpendicular bisector of AA'. Hence it passes through G. Note that the area of triangle AEF is given by $\frac{EF \cdot AG}{2} = \frac{EF}{4}$. Note that since $\angle EA'F = 90°$, A, E, A' and F all lie on a circle ω. Since EF is a diameter and AA' is a chord of ω, $EF \geq AA' = 1$. Hence the minimum length of EF is 1, attained when $AEA'F$ is a square. It follows that the minimum value of the area of triangle AEF is $\frac{1}{4}$. To find the maximum length of EF, note that AA' is a chord of ω of fixed length. The diameter of ω increases as $\angle AFA'$ decreases, and this happens as AF increaes. It follows that EF is longest when F coincides with D. Now $AF = 1$ and $AG = \frac{1}{2}$. Hence $FG = \frac{\sqrt{3}}{2}$. Since triangles AFE and GFA are similar, $\frac{EF}{AF} = \frac{AF}{FG}$. Hence $EF = \frac{2\sqrt{3}}{3}$ so that the

maximum area of triangle AEF is $\frac{\sqrt{3}}{6}$.

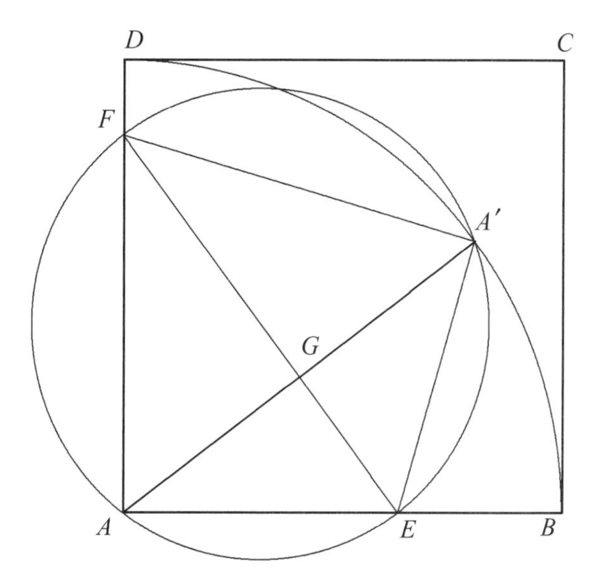

Second Solution: By symmetry, we may assume that $AE \leq AF$. Let A falls on the point A' on arc BD. Let G be the midpoint of AA'. Then $AG = \frac{1}{2}$. Now EF is the perpendicular biscctor of AA'. Hence it passes through G. Let $\angle EAG = \theta$. Then $0° < \theta < 45°$. Now the area of triangle AEF is given by $\frac{AE \cdot AF}{2} = \frac{(AG \sec\theta)(AG \csc\theta)}{2} = \frac{1}{4\sin 2\theta}$. Now this is minimum when $\sin 2\theta = 1$ or $\theta = 45°$, yielding $\frac{1}{4}$. Since $\sin 2\theta$ is an increasing function of θ from $0°$ to $45°$, the maximum area of triangle AEF occurs when θ is as small as possible. Now $1 \geq AF = AG \csc\theta$. Hence $2 \leq \csc\theta$, so that $\theta \geq 30°$. Since $\sin 2(30°) = \frac{\sqrt{3}}{2}$, the maximum area is $\frac{\sqrt{3}}{6}$.

Third Solution: By symmetry, we may assume that $AE \leq AF$. Let A falls on the point A' on arc BD. Let G be the midpoint of AA'. Then $AG = \frac{1}{2}$. Now EF is the perpendicular bisector of AA'. Hence it passes through G. Set up a coordinate system so that A is at (0,0), B at (1,0), C at (1,1) and D at (0,1). Let A' be at (u, v). Then $u \geq v$ and $u^2 + v^2 = 1$. Note that G is at $(\frac{u}{2}, \frac{v}{2})$. The slope of AA' is $\frac{v}{u}$, so that the slope of EF is $-\frac{u}{v}$. Hence the equation of EF is $y - \frac{v}{2} = -\frac{u}{v}(x - \frac{u}{2})$, which simplifies to $2ux + 2vy = 1$. Hence E is at $(\frac{1}{2u}, 0)$ and F is at $(0, \frac{1}{2v})$. Now the area of triangle AEF is given by $\frac{AE \cdot AF}{2} = \frac{1}{8uv}$. Since $0 \leq (u - v)^2 = 1 - 2uv$, we have $uv \geq \frac{1}{2}$, with equality if and only if $u = v$. This yields the minimum area $\frac{1}{4}$ of triangle AEF. Since $u^2 + v^2 = 1$ is constant, the value of u^2v^2 decreases as the difference between u^2 and v^2 increases. This follows from $(u^2 + d)(v^2 - d) = u^2v^2 - d^2 < u^2v^2$ for any real number d. Hence $\frac{1}{uv}$ is maximum if v is as small as possible. Now $\frac{1}{2v} = AF \leq AD = 1$. Hence the minimum value of v is $\frac{1}{2}$, and the corresponding value of u is $\frac{\sqrt{3}}{2}$. It follows that the maximum area of triangle AEF is $\frac{\sqrt{3}}{6}$.

February 9, 1993

1. Let $x = 1992$. Then $1992^{1993} - 1992 = x(x^{1992} - 1)$ while $1992^2 + 1993 = x^2 + x + 1$. Since $x^{1992} - 1$ is divisible by $x^3 - 1 = (x-1)(x^2+x+1)$, $x(x^{1992}-1)$ is divisible by $x^2 + x + 1$.

2. (a) If the first term is H, HTT must appear before TTH. This is also the case if the first term is T and the second term is H. If the first two terms are T's, then TTH must appear before HTT. Hence the desired probability is $\frac{3}{4}$.

   (b) Nothing happens until the first T appears. If the next term is T, then TTH must appear before THT. If the next two terms are H and T, then we have THT. If they are H's instead, we are back to square one. Hence the desired probability is $\frac{2}{3}$.

3. Let O_1, O_2 and O_3 be the centers of the spheres tangent to the plane at A, B and C respectively, where $AB = c$, $BC = a$ and $CA = b$. Let $O_1A = r_1$, $O_2B = r_2$ and $O_3C = r_3$. Let D be the point on O_2B such that O_1D is parallel to AB. Then $O_1D = c$, $O_2D = |r_1 - r_2|$ and $O_1O_2 = r_1 + r_2$. By Pythagoras' Theorem, $c^2 + (r_1 - r_2)^2 = (r_1 + r_2)^2$, which simplifies to $c^2 = 4r_1r_2$. Similarly, we have $a^2 = 4r_2r_3$ and $b^2 = 4r_3r_1$. Multiplying these equations together, we have $a^2b^2c^2 = 64r_1^2r_2^2r_3^2$ or $abc = 8r_1r_2r_3 = 2r_1a^2$. Hence $r_1 = \frac{bc}{2a}$. Similarly, $r_2 = \frac{ca}{2b}$ and $r_3 = \frac{ab}{2c}$.

4. We can replace $S = \{1, 2, \ldots, n\}$ with a set of n 0's and 1's by reducing each number modulo 2. We consider congruence classes modulo 8.

   (i) If $n = 1$, $S = \{1\}$ and Steven must lose.

   (ii) If $n = 2$, $S = \{0, 1\}$ and Steven wins by taking the 0.

   (iii) If $n = 3$, $S = \{0, 1, 1\}$. Todd can always take a 1 in his turn and win.

   (iv) If $n = 4$, $S = \{0, 0, 1, 1\}$. Todd can win by matching Steven's moves.

   (v) If $n = 5$, $S = \{0, 0, 1, 1, 1\}$. Steven can win by taking a 1 and then matching Todd's moves.

   (vi) If $n = 6$, $S = \{0, 0, 0, 1, 1, 1\}$. Steven can win by taking a 1 and then matching Todd's moves if possible. This is not possible only when Todd takes the last 0. In that case, Steven has only one 0 and must have two 1's, so that he wins also.

   (vii) If $n = 7$, $S = \{0, 0, 0, 1, 1, 1, 1\}$. Steven can win by taking a 0 and then matching Todd's moves.

   (viii) If $n = 8$, $S = \{0, 0, 0, 0, 1, 1, 1, 1\}$. Steven can win by taking a 0 and then matching Todd's moves if possible. This is not possible only when Todd takes the last 0. In that case, Steven has two 0's and must have two 1's, so that he wins also.

   Suppose $n = 8q + r$ where $q \geq 1$ and r is one of 2, 5, 6, 7 and 8. Then S contains at least $4q + 1$ 0's and $4q + 1$ 1's. Steven makes the first move as in the case $q = 0$. Then he matches Todd's 0-moves $2q$ times and Todd's 1-moves $2q$ times. These matching moves should be as early as possible but not necessarily consecutive. This reduces the game to the case $q =$, and Steven wins. Suppose $n = 8q + r$ where $q \geq 1$ and r is one of 1, 3 and 4. Then S contains at least $4q$ 0's and $4q$ 1's. Todd can win by matching Steven's 0-moves $2q$ times and Steven's 1-moves $2q$ times.

5. Suppose a_1 is the largest among a_i, $1 \le i \le n$. That these are the lengths of the sides of an n-gon means that $a_1 < a_2 + a_3 + \cdots + a_n$. Note that $\frac{a_1}{1+a_1} = 1 - \frac{1}{1+a_1}$ is the largest among $\frac{a_i}{1+a_i}$, $1 \le i \le n$. Now

$$\begin{aligned}
&\frac{a_2}{1+a_2} + \frac{a_3}{1+a_3} + \cdots + \frac{a_n}{1+a_n} \\
&\ge \frac{a_2}{1+a_1} + \frac{a_3}{1+a_1} + \cdots + \frac{a_n}{1+a_1} \\
&= \frac{a_2 + a_3 + \cdots + a_n}{1+a_1} \\
&\ge \frac{a_1}{1+a_1}.
\end{aligned}$$

Hence $\frac{a_i}{1+a_i}$, $1 \le i \le n$, are possible lengths of the sides of another n-gon.

February 8, 1994

1. $P(x)$ cannot be a constant polynomial. Otherwise, the left side of the equation is a quadratic polynomial and the right side linear. Let it be of degree $n > 1$. Then the left side is of degree $2n$ and the rightside degree $n+1$. We must have $2n = n+1$ or $n = 1$. Hence $P(x)$ is of the form $ax+b$. Now $ax^2+b+2x^2+10x = 2x(a(x+1)+b)+3$. Comparing the constant terms, we have $b = 3$. Comparing the linear terms, we have $10 = 2a+2b$ or $a = 2$. Finally we check that $a+2 = 2a$ holds, so that the quadratic terms also agree. Hence the only polynomial that satisfies the functional equation is $P(x) = 2x + 3$.

2. First observe that in order to get two triangles, the cut must pass through one of the vertices of the original triangle. When the cut reaches the opposite side, two angles will be formed, at least one of which will not be acute. It follows that the cut cannot be through the vertex of the angle of original triangle with smallest measure. In particular, the equalateral triangle is not an amoeba. In triangle ABC, let $AB = AC \ne BC$. Suppose $AB = AC < BC$. Then the cut is along AD for some point D on BC. Note that $AB = AC > AD$, so that there are only two ways for BAD to be isosceles. The first is to have $AB = BD$. Since $AB + AC > BC$, $AC > CD$. We must have $AD = CD$. Let $\angle ACD = \theta$. Then $\angle ABD = \angle CAD = \theta$, and $\angle ADB = \angle BAD = 2\theta$. From $5\theta = 180°$, we have $\theta = 36°$. Thus our first amoeba is the $(108°, 36°, 36°)$ triangle. The second is way is to have $AD = BD$. We may assume that $AC = CD$ does not hold. Otherwise we have the first amoeba again by symmetry. Hence $AD = CD$. Let $\angle ACD = \theta$. Then $\angle ABD = \angle CAD = \angle BAD = \theta$. From $4\theta = 180°$, we have $\theta = 45°$. Thus our second amoeba is the $(90°, 45°, 45°)$ triangle. Suppose instead $AB = AC > BC$. By symmetry we may assume that the cut is along BD for some point D on AC. Note that $AB > AD$. If $AB = BD$, then they must be shorter than BC, which is not the case. Hence $AD = BD$. Since $\angle BCD = \angle CBD = \angle CBD$, $BD > CD$, so that there are only two ways for BCD to be isosceles. The first is to have $BC = BD$. Let $\angle BAD = \theta$. Then $\angle ABD = \theta, \angle BDC = \angle BCD = 2\theta$ and $\angle CBD = \theta$. From $5\theta = 180°$, we

have $\theta = 36°$. Thus our third amoeba is the $(36°, 72°, 72°)$ triangle. The second way is to have $BC = CD$. Let $\angle BAD = \theta$. Then $\angle ABD = \theta$, $\angle BDC = \angle CBD = 2\theta$ and $\angle BCD = 3\theta$. From $7\theta = 180°$, we have $\theta = \frac{180}{7}°$. Thus our fourth and last amoeba is the $(\frac{180}{7}°, \frac{540}{7}°, \frac{540}{7}°)$ triangle.

3. (a) The length of the interval is

$$\left(n + \frac{1}{1993}\right)^2 - \left(n + \frac{1}{1994}\right)^2 = \left(2n + \frac{1}{1994} + \frac{1}{1993}\right)\left(\frac{1}{1993} - \frac{1}{1994}\right).$$

When n is sufficiently large, this length will be greater than 1, and the interval must contain some integer N.

(b) Although they eventually overlap, such intervals are disjoint initially, so that N grows with n. We want

$$n^2 + \frac{2n}{1994} + \frac{1}{1994^2} < N < n^2 + \frac{2n}{1993} + \frac{1}{1993^2}.$$

Let $N = n^2 + k$. Then $k > 0$ and

$$\frac{2}{1994} + \frac{1}{1994^2 n} < \frac{k}{n} < \frac{2}{1993} + \frac{1}{1993^2 n}.$$

It follows that n grows with k. For $k = 1$, $\frac{2}{1994} < \frac{1}{n}$ implies that $n < 997$ while $\frac{1}{n} - \frac{1}{1993^2 n} < \frac{2}{1993}$ implies that $1993 - \frac{1}{1993} < 2n$ or $n > 996$. There is no such integer n. For $k = 2$, $\frac{2}{1994} < \frac{2}{n}$ implies that $n < 1994$ while $\frac{2}{n} - \frac{1}{1993^2 n} < \frac{2}{1993}$ implies that $2 \cdot 1993 - \frac{1}{1993} < 2n$ or $n \geq 1993$. It follows that we should take $k = 2$ and $n = 1993$. Then $N = 1993^2 + 2$ is the smallest positive integer contained in such an interval.

4. Suppose there is a pentagon $ABCDE$ such that every area bisector through a vertex misses the opposite side. By symmetry, we may assume that the bisector through A hits BC. Then the area of ABC is at least half the area of the pentagon. If the bisector through E hits CD, then the area of CDE is also at least half that of the pentagon. Since ABC and CDE do not overlap, this is impossible. Hence the bisector through E hits AB and the area of ABE is at least half that of the pentagon. Now the bisector through D must hit AE or BC. In the first case, the total area of ADE and ABC will not be less than that of the pentagon. The same contradiction arises in the second case with BCD and ABE. Hence, no such pentagons exist, and the desired result follows.

5. (a) From $a^2 = 2a$, we have $a = 0$ or 2. Hence (0,0,0) and (2,2,2) are the only tadpoles of the form (a, a, a).

(b) We seek a tadpole (a, b, c) not of the form (a, a, a), satisfying $a + b = ca$, $b + c = ab$ and $c + a = bc$. Eliminating b from the first two equations, we have $c = \frac{a^2-a}{a^2-a-1}$. This in turn yields $b = \frac{a}{a^2-a-1}$ and

$$\frac{a(a-2)(a^3 + a^2 - 2a - 1)}{(a^2 - a - 1)^2} = 0.$$

We do not want $a = 0$ or 2 as these values yield the tadpoles in (a). We also do not want $a^2 - a - 1 = 0$. The roots of this equation are $a_1 = \frac{1+\sqrt{5}}{2}$ and $a_2 = \frac{1-\sqrt{5}}{2}$. Let $f(x) = x^3 + x^2 - 2x - 1$. Then $f(a_1) = (a_1 + 2)(a_1^2 - a_1 - 1) + a_1 + 1 > 0$ while $f(1) < 0$. Hence $f(x)$ has a root a where $1 < a < a_1$. For this value of a, (a, b, c) is a tadpole satisfying the given system of equations, with b and c defined in terms of a above. Moreover, $b \neq c$ as otherwise $a = 0$ or 2. Hence (a, b, c) is not of the form (a, a, a).

February 14, 1995

1. Let A and B be the first two students to leave. Place C 120 meters from the school in the other direction, D another 120 meters further, and E yet another 120 meters further. Let them all start at noon, and we have essentially the same situation as in the problem. Now C takes $\frac{120}{0.5} = 240$ seconds to catch up to B, D takes a further $\frac{120}{1} = 120$ seconds to catch up to C, and E takes a further $\frac{120}{1.5} = 80$ seconds to catch up to D. Hence the five friends will be together 440 seconds past noon. E has covered 880 meters during this period. Take away the 360 meters on the other side of the school, the five friends are 520 meters from the school.

2. Adding the first and the third equations, we have $a + c = b(a + c)$. Since $a + c > 0$, we have $b = 1$. Substituting into the third equation, we have $c = a + 1$. Substituting both into the second equation, we have $1 = a(a + 1) - 2$ or $a^2 + a - 3 = 0$. By the Quadratic Formula, $a = \frac{-1+\sqrt{13}}{2}$ since $a > 0$. We have $c = \frac{1+\sqrt{13}}{2}$. This is the only solution.

3. We have

4.
$$1 > \frac{a}{a+1} = \frac{b}{b+2} + \frac{c}{c+3} = \frac{bc + 3b + 2c + bc}{bc + 3b + 2c + 6}.$$

Hence $bc < 6$. There are only 10 cases to consider, namely, $(b, c) = (1, 1)$, $(1, 2)$, $(2, 1)$, $(1, 3)$, $(3, 1)$, $(1, 4)$, $(2, 2)$, $(4, 1)$, $(1, 5)$ and $(5, 1)$. They lead respectively to

$$\frac{a}{a+1} = \frac{7}{12}, \frac{11}{15}, \frac{3}{4}, \frac{5}{6}, \frac{17}{20}, \frac{9}{10}, \frac{11}{12}, \frac{19}{21}, \frac{23}{24}, \frac{27}{28}.$$

The only integral solutions are $(a, b, c) = (3, 2, 1)$, (5,1,3), (9,2,2), (11,4,1), (23,1,5) and (27,5,1).

5. We use boldface for vectors. We have **AB** = ±**DE**, **BC** = ±**EF** and **CD** = ±**FA**. Suppose **AB** = **DE**. If **BC** = **EF** also, then **AC** = **DF** and we cannot have **CD** = ±**FA**. It follows that **BC** = −**EF**. Similarly, **CD** = −**FA**, so that **BD** = −**EA**. Since **AB** + **BD** + **DE** + **EA** = **0**, we have **AB** = −**DE**, a contradiction. It follows that **AB** = −**DE** and **BC** = −**EF**, so that $ABDE$ and $BCEF$ are parallelograms. Hence AD, BE and CF are concurrent at their common midpoint O. Let U, V, W, X, Y and Z be the midpoints of AB, BC, CD, DE, EF and FA respectively. UX is halfway between the parallel lines CE and BF, so that its midpoint is O. Similarly, the midpoint of VY is also O. Hence $UVXY$ is a parallelogram. Now WX is a line

through O parallel to UV and XY. Hence it lies on the plane of $UVXY$. It follows that U, V, W, X, Y and Z are coplanar.

6. First, note that the degree of $P(x)$ is at least that of $Q(x)$. Otherwise, the absolute value of $\frac{P(x)}{Q(x)}$ can be made arbitrarily small by choosing sufficiently large values of x. Thus it will not be possible for $\frac{P(x)}{Q(x)}$ to be an integer if x is a sufficiently large integer. This is a contradiction. Let $R(x)$ be the quotient and $S(x)$ be the remainder when $P(x)$ is divided by $Q(x)$. Then $P(x) = Q(x)R(x) + S(x)$ and the degree of $S(x)$ is less than that of $Q(x)$. Since both $P(x)$ and $Q(x)$ have integral coefficients, $R(x)$ has rational coefficients. We claim that $S(x)$ is identically zero. Suppose it is a polynomial of degree $n \geq 0$. Let m be the least common denominator of the coefficients of $R(x)$. For any integer value of x, $R(x)$ is a rational number. In its lowest terms, the denominator is a positive integer less than or equal to m. Now take any integer value of x sufficiently large so that we have

$$-\frac{1}{m} < \frac{S(x)}{Q(x)} < \frac{1}{m}.$$

Since $R(x) + \frac{S(x)}{Q(x)} = \frac{P(x)}{Q(x)}$ is an integer, we must have $S(x) = 0$. Since x only has to be integral and sufficiently large, there are more than n such values for which $S(x) = 0$. This is a contradiction since a polynomial of degree n can have at most n roots. Hence $P(x) = Q(x)R(x)$.

February 13, 1996

1. Let a_k be the number consisting of k 4's. Suppose the largest of our numbers is a_n. We claim that their sum S has exactly n digits, and its first digit is 4. Now $a_n \leq S \leq a_1 + a_2 + \cdots + a_n$. Clearly, $a_n > 4 \cdot 10^{n-1}$. We have

$$\begin{aligned} a_1 + a_2 + \cdots + a_n &= \frac{4}{9}(10 - 1 + 10^2 - 1 + \cdots + 10^n - 1) \\ &= \frac{4}{9}(\frac{10^{n+1} - 1}{9} - n - 1) \\ &< \frac{4}{80}10^{n+1} \\ &= 5 \cdot 10^{n-1}. \end{aligned}$$

Thus the claim is justified.

2. Let the pentagon be $ABCDE$, with $CD = 3$ and $\angle BCD = \angle CDE = 135°$. We may assume that $BC \leq DE$. Hence there are six possible choices for their lengths. In each case, we need to know between which two integers BE lies, so that we can choose AB and EA to satisfy the Triangle Inequality. Let the extensions of BC and ED meet at H. Then CHD is a right triangle with $CH = DH = \frac{3}{\sqrt{2}}$. Hence $BE^2 = BH^2 + EH^2 = BC^2 + DE^2 + 3\sqrt{2}(BC + DE) + 9$. In the first of six cases, $BC = DE = 2$. Then $BE^2 = 17 + 12\sqrt{2}$. This is an irrational number, so that there exists an integer n such that $n < 17 + 12\sqrt{2} < n + 1$. From $(n - 17)^2 < 288$ and

$16^2 = 256 < 288 < 289 = 17^2$, we see that $33 < BE^2 < 34$. Hence $5 < BE < 6$, so that $EA + AB = 2 + 4, 3 + 3, 3 + 4$ or $4 + 4$. The remaining five cases are summarized, along with the first, in the following table:

BC	DE	BE^2	n	$[BE]$	$EA + AB$
2	2	$17 + 12\sqrt{2}$	33	5	6,7,8
2	3	$22 + 15\sqrt{2}$	43	6	7,8
3	3	$27 + 18\sqrt{2}$	52	7	8
2	4	$29 + 18\sqrt{2}$	54	7	8
3	4	$34 + 21\sqrt{2}$	63	7	8
4	4	$41 + 24\sqrt{2}$	64	8	none

We have ten mutually non-congruent pentagons, with $(BC, CD, DE, EA, AB) = (2, 3, 2, 3, 3)$, (2,3,2,2,4), (2,3,2,3,4), (2,3,2,4,4), (2,3,3,3,4), (2,3,3,4,3), (2,3,3,4,4), (3,3,3,4,4), (2,3,4,4,4) and (3,3,4,4,4) respectively.

3. (a) Take the first progression to be $0 < 1 < 2$ and the second $2n > n > 0$. Then

$$0 \cdot 2n + 1 \cdot n + 2 \cdot 0 = n.$$

(b) Let the first progression be $a < a + d < a + 2d < a + 3d$ and the second be either $b < b + e < b + 2e < b + 3e$ or $b > b + e > b + 2e > b + 3e$. Then $n = 4ab + 6ae + 6bd + 14de$, so that no odd integers can be so represented.

(c) For $k = 5$, let the first progression be $a < a + d < a + 2d < a + 3d < a + 4d$ and the second one be either $b < b + e < b + 2e < b + 3e < b + 4e$ or $b > b + e > b + 2e > b + 3e > b + 4e$. Then $n = 5ab + 10ae + 10bd + 30de$, so that only integers divisible by 5 may have such a representation. For $k = 6$, take the first progression to be $-3 < -2 < -1 < 0 < 1 < 2$ and the second $3n < 4n < 5n < 6n < 7n < 8n$. Then

$$-3 \cdot 3n - 2 \cdot 4n - 1 \cdot 5n + 0 \cdot 6n + 1 \cdot 7n + 2 \cdot 8n = n.$$

4. By symmetry, we may assume that $a = \min\{a, b, c, d\}$. Then $4a^4 \le a^2bc + b^2cd + c^2da + d^2ab$. Hence

$$4(a^4 + b^4 + c^4 + d^4) - (a^2bc + b^2cd + c^2da + d^2ab) \le 4(b^4 + c^4 + d^4) \le 12.$$

5. (a) We have $\angle BAC = \angle FAB$. By the Isosceles Triangle Theorem, $\angle ABF = \angle ADB$. By Thales' Theorem, $\angle ADB = \angle ACB$. Hence triangles ABC and AFB are similar. It follows that $\frac{AF}{AB} = \frac{AB}{AC} = \frac{AE}{AD}$. Since $AB = AD$, we have $AF = AE$.

(b) Join CD. Then

$$\angle FDA = \angle ABD = \angle ACD = \angle ACE + \angle ECD = \angle ACE + \angle EAD.$$

It follows from the Exterior Angle Theorem that

$$\angle AFG = \angle FDC + \angle ACD = \angle FDC + \angle FDA = \angle ADC = \angle AEG.$$

We have $AF = AE$ from (a), and AG is a common side of triangles AGE and AGF. Hence these triangles are congruent unless $\angle AGF + \angle AGE = 180°$. However, this would mean that B, C, D and E are collinear, which is not the case. Thus AGE and AGF are indeed congruent, so that $\angle EAG = \angle FAG$.

February 11, 1997

1. If $x < -2$, the inequality becomes $7 - x > (-2 - x) + (2 - x)$ which simplifies to $x > -7$. Hence all x in $(-7, -2)$ also satisfy the inequality. If $-2 \leq x \leq 2$, then $|x - 2| + |x + 2| = 4$ while $|x - 7| \geq 5$, so all x in $[-2, 2]$ satisfy the inequality. If $2 < x < 7$, then the inequality becomes $7 - x > (x - 2) + (x + 2)$ which simplifies to $x < \frac{7}{3}$. Hence all x in $(2, \frac{7}{3})$ satisfy the inequality. If $x \geq 7$, then $x - 7 < x - 2 < (x - 2) + (x + 2)$, and the inequality is not satisfied. In summary, x satisfies the inequality if and only if $-7 < x < \frac{7}{3}$.

2. Clearly, triangles AB_nC_n are similar to each other. In a $(30°, 60°, 90°)$ triangle, the hypotenuse is twice as long as the shorter leg. Let $AB_1 = x$. Then $AC_1 = 2x$ and $AB_2 = 4x$. It follows that $\text{area}(AB_nC_n) = 16 \cdot \text{area}(AB_{n-1}C_{n-1})$, so that the desired total area is given by $T = 1 + 16 + 16^2 + \cdots + 16^{1996}$. Multiplying this by 16, we have $16T = 16 + 16^2 + 16^3 + \cdots + 16^{1997}$. Subtraction yields $15T = 16^{1997} - 1$ so that $T = \frac{1}{15}(16^{1997} - 1)$.

3. Complete the circle. Extend EA to cut it at C', and extend DB to cut it at F'. By symmetry, $AC = AC'$ so that $\angle AC'C = \angle ACC'$. Similarly, $\angle BF'F = \angle BFF'$. Now

$$\begin{aligned}\angle EC'C &= 180° - \angle CAC' = 180° - 2\angle CAM = 180° - 2\angle FBN \\ &= 180° - \angle FBF' = \angle DF'F.\end{aligned}$$

Since the arcs CE and DF subtend equal angles at the circle, they have equal measure. It follows that the chords CE and DF are equal.

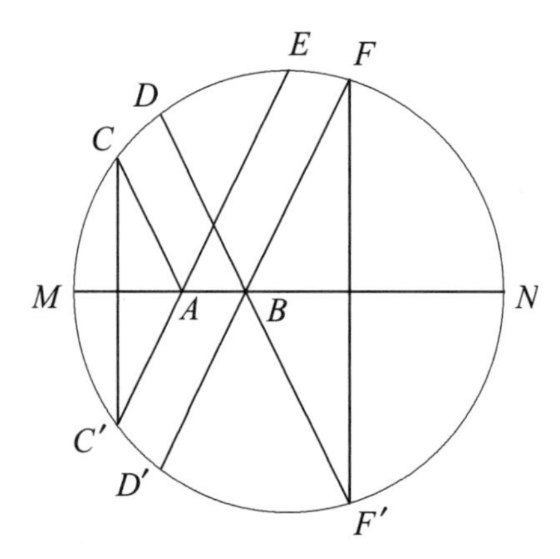

4. (a) Note that $a(a + b)^2 = a(a^2 + b^2) + 2a^2b$. Since p^4 divides both $a(a + b)^2$ and $a^2 + b^2$, it must also divide $2a^2b$. Since p is an odd prime, p^4 divides a^2b.

Suppose p^2 does not divide a. Then at most p^2 can divide a^2 so that p^2 must divide b. Hence p^4 divides b^2. However, this contradicts p^4 dividing a^2+b^2 but not a^2. It follows that we must have p^2 dividing a. Then p^4 divides a^2 so that it also divides b^2. Hence p^2 divides b, and it also divides $a+b$. It follows that p^4 divides $a(a+b)$.

(b) We look for a, b and p such that p^5 divides a^2+b^2, p^2 divides a and b, but p^3 does not divide $a+b$. Setting $a = p^2x$ and $b = p^2y$, these conditions become p divides x^2+y^2 and p does not divide $x+y$. We can pick $x=2$, $y=1$ and $p=5$. This gives $a=50$ and $b=25$ as an example.

5. The number of ways for $n=1$ is 2, the ways being B to A and B to C. For $n=2$, we have 4 ways, B via A to B, B via C to B, B via C to D and B via C to F. Note that the answer is the same had we started in D or F. We guess that the number of ways for any n is exactly 2^n. We now prove this by induction on k, our induction hypothesis being that the number of ways is exactly 2^{2k-1} for $n=2k-1$ and exactly 2^{2k} for $n=2k$. The basis $k=1$ has been settled. Suppose the result holds for a particular k. Then after $2k$ kilometers, we have arrived at B, D or F in any of the 2^{2k} ways. Since there are 2 ways to go an extra kilometer from any of B, D and F, the number of ways is exactly 2^{2k+1} for $n=2k+1$. Also, there are 4 ways to go another 2 kilometers from any of B, D or F. Hence the number of ways is exactly 2^{2k+2} for $n=2k+2$. This completes the inductive argument.

February 10, 1998

1. Let the first fraction be $\frac{1}{1}$ and the second be $\frac{a}{b}$. Then $100(1+\frac{a}{b}) = \frac{1}{b}+a$. Clearing denominators, we have $100a+100b = 1+ab$ or

$$(a-100)(b-100) = ab - 100a - 100b + 10000 = 9999.$$

We may take $a-100 = 99$ and $b-100 = 101$ so that the two fractions are 1 and $\frac{199}{201}$.

2. Suppose the task is possible. Except for possibly one horizontal segment, every other one connects a dot in the top row to a dot in the bottom row. If we have a horizontal segment, it will use up two dots in the same row, and this forces another horizontal segment on the other row. Since no two segments are parallel, there cannot be any horizontal segments at all. We may have a vertical segment, but the slopes of the other 19 segments must be one of $\pm\frac{1}{k}$ where $k = 1, 2, \ldots, 9$. Since there are 18 different slopes, it follows from the Pigeonhole Principle that two of the segments are parallel to each other. Thus the task is impossible.

3. Since $CE = EA$, area$(CDE) = \frac{1}{2}$ area(ADC). Since $CD = DB$, area$(ADC) = \frac{1}{2}$ area(ABC). Hence area$(CDE) = \frac{1}{4}$ area$(ABC) = 9$ so that area$(ABDE) = 27$. Suppose AD is not perpendicular to BE. Let G be the point of intersection of AD and BE. Then the altitude of triangle ABE from A to BE will be shorter than AG, so that area$(ABE) < \frac{1}{2}AG \cdot BE$. Similarly, area$(DBE) < \frac{1}{2}DG \cdot BE$, so that

area$(ABDE) < \frac{1}{2}(AG + DG)BE = 27$. This is a contradiction. It follows that AD must be perpendicular to BE.

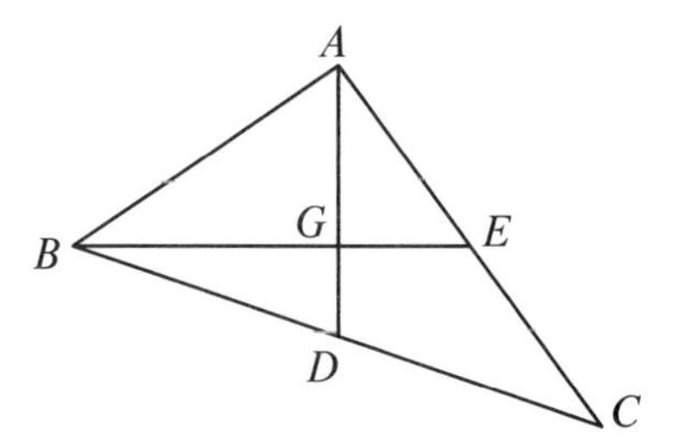

4. (a) A team has only two victories, and each of its vanquished foes beats only two other teams. Since a team must beat every other team either directly or indirectly via a third team, the total number of teams is at most 1+2+4=7.

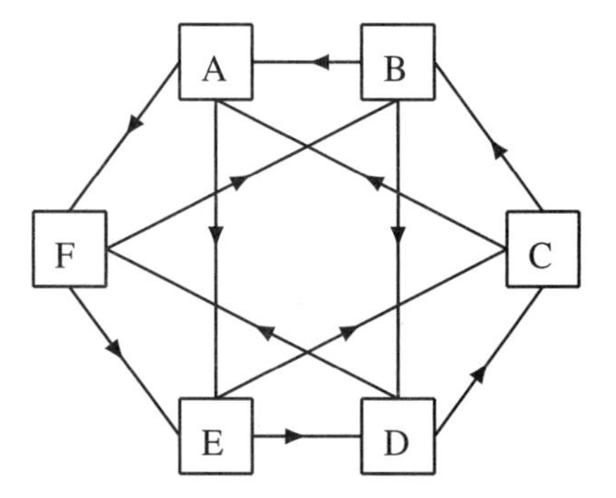

(b) The diagram above shows a tournament with six teams, each of which has played exactly four games so far. A beats E and F directly. E beats C and D while F beats B. By symmetry, C and E also beat every other team either directly or indirectly. Now B beats A and D. A beats E and F while D beats C. By symmetry again, D and F also beat every other team either directly or indirectly.

(c) Suppose it is possible to have 7 teams. We may assume that A has beaten B and C, B has beaten D and E, and C has beaten F and G. Now B has not beaten A, C, F or G. Hence each of D and E has to take care of two of them. We may assume that D has beaten A. Since D has not beaten E, it must have beaten some team which has beaten E, and this can only be F or G. We may assume that D has beaten F, so that E has beaten C and G. Now D has not beaten G either, and G has only lost to C and E. Since D's victories so far are over A and F, we have a contradiction.

5. The roots of the equation $(x + a)(x + b)(x + c)(x + d) = 0$ are $-a, -b, -c$ and $-d$. When expanded, it becomes

$$x^4 + (a + b + c + d)x^3 + (ab + ac + ad + bc + bd + cd)x^2 \\ + (abc + abd + acd + bcd)x + abcd = 0.$$

Since all the coefficients are given to be positive, its roots must all be negative. It follows that all of a, b, c and d are positive.

February 2, 1999

1. This is a relatively easy problem, and most contestants solved it. Since the last digit is the sum of the other four which are distinct, it must be at least $0 + 1 + 2 + 3 = 6$. If it is 6, the first four digits must be 0, 1, 2 and 3 in some order. Since 0 may not be the first digit, there are 3 choices for it. The numbers of choices for the other three digits are 3, 2 and 1 respectively, yielding $3 \times 3 \times 2 \times 1 = 18$ numbers.

   Sonny Chan of Western Canada High School obtained 18 in a slightly different way. He argued that there are 4! ways of arranging the first four digits. However, we have to exclude those with 0 as the first digit. In such a number, there are 3! ways of arranging the middle three digits. It follows that there are $4! - 3! = 18$ numbers with 6 as the last digit.

   Suppose the last digit is 7. Then the first four digits must be 0, 1, 2 and 4 in some order, and again we have 18 such numbers. Suppose the last digit is 8. Then the first four digits can either be 0, 1, 2 and 5, or 0, 1, 3 and 4. There are 36 such numbers. Finally, suppose the last digit is 9. Then the first four digits can either be 0, 1, 2 and 6, or 0, 1, 3 and 5, or 0, 2, 3 and 4. This time, we have 54 such numbers. The total is 126. Of course, we could have observed by counting that there are altogether 7 combinations of digits, so that the total is $7 \cdot 18 = 126$.

2. Many contestants were able to find the answers. If $A = 0°$, $B = 90°$ and $C = 0°$, then $\sin A = \cos B = \tan C = 0$ and $A + B + C = 90°$. This is the minimum. If $A = 90°$, $B = 0°$ and $C = 45°$, then $\sin A = \cos B = \tan C = 1$ and $A + B + C = 135°$. This is the maximum.

   However, far too many students left it at that, and did not bother justifying that these values are indeed extremal. The key observation is that since $\sin A = \cos B$ and $0° \le A, B \le 90°$, we must have $A + B = 90°$. It follows that the extremal values of $A + B + C$ coincide with those of C alone. Since $0° \le C \le 90°$, the minimum value of C is 0° and that of $A + B + C$ is 90° as noted above.

   Now $\tan 90°$ is undefined. For $0° \le C < 90°$, $\tan C$ increases with C. Since $\tan C = \sin A$ and $\sin A \le 1$, the maximum value of $\tan C$ is 1 when $C = 45°$. Hence the maximum value of C is 45° and that of $A + B + C$ is 135° as noted above.

   A number of students made the unwarranted assumption that A, B and C are angles of some triangle ABC, and claimed that $A + B + C = 180°$ all the time. They might have thought that this was just a "trick" question. However, when a ridiculously easy solution arises, read the question again carefully to see if you might have left out something that was there, or might have read into it something that was not there.

3. The performance on this problem was somewhat disappointing. It is what we call a *contestant-friendly* question. Everyone should be able to get at least some partial results, but a surprisingly large number of contestants avoided this problem altogether.

   Many contestants just tried various values. For instance, if $m = 1$ and $k = 2$, we may have x pairs of chickens from Minsk and y trios of chickens from Kiev. Then $2x + 3y = 18$ and $x + 2y = 11$. This yields $x = 3$ and $y = 4$. More generally, if $m = 1$, we may have x pairs of chickens from Minsk and y groups of chickens from Kiev with $k+1$ in each group. Then $2x + (k+1)y = 18$ and $x + ky = 11$. Subtraction yields $x + y = 7$, and subtracting this from the previous equation yields $(k-1)y = 4$. Hence $k - 1 = 1, 2$ or 4, so that $k = 2, 3$ or 5. This is as far as many contestants got.

Apparently, there are no further solutions with $m = 1$. Let us try $m = 2$ and $k = 3$. Suppose there are x trios of chickens from Minsk and y quartets of chickens from Kiev. Then $3x + 4y = 18$ and $2x + 3y = 11$. Subtraction yields $x + y = 7$ as always, but now $y = -3$, which is impossible.

Jack Chen, Kim Horel and **Andrew Lee** from Sir Winston Churchill High School, and **Filip Crnogorac, Roger Huang** and **Ryan Johnston** of Western Canada High School, were the only ones who observed that m must be equal to 1. Otherwise, each chicken lays on the average at least $\frac{2}{3} > \frac{11}{18}$ egg per day. Using this, we can approach the problem more systematically.

Suppose Cathy has n chickens from Minsk and $18 - n$ chickens from Kiev. Each chicken from Minsk lays on the average $\frac{1}{2}$ egg per day, while the average for the chickens from Kiev is $\frac{k}{k+1}$. Hence $\frac{n}{2} + \frac{(18-n)k}{k+1} = 11$. Note that if we had not taken $m = 1$, this would have been an equation in three unknowns. Several contestants got this far, but got tangled in the algebra.

Solving for n, we get $n = 14 - \frac{8}{k-1}$. Since n must be an integer, we must have $k - 1 = 1, 2, 4$ or 8, so that $k = 2, 3, 5$ or 9, when $n = 6, 10, 12$ or 13, respectively. We have already shown that each of $(m, k) = (1, 2)$, (1,3) and (1,5) is realizable. We now show that so is $(m, k) = (1, 9)$. Consider a cycle of 10 days. Each of the five chickens from Kiev takes a different odd-numbered day off, while one of the chickens from Minsk takes all even-numbered days off. These six chickens lay exactly 5 eggs per day. The remaining twelve chickens from Minsk are divided into six pairs, with each pair laying an egg per day.

Now we see why we missed the last case in our initial attempt. While the variables n and k are integers, this is not so with x or y. In the last case, we have a "mixed" group with chickens of both kinds. **Jack Chen** and **Ryan Johnston** were the only ones to find this case, but Ryan's reasoning was incomplete, and Jack inexplicably missed the first case $(m, k) = (1, 2)$.

4. Many contestants were intimidated by the statement of this problem, and quite a number got tangled in hopelessly messy algebra. What we are asked to do is to see if $2xy(x^6 + y^6)$ may exceed $(x^4 + y^4)^2$ for some real numbers x and y. A relatively large number of contestants tried $x = 2$ and $y = 3$. Then $2xy(x^6 + y^6) = 12(64 + 729) = 9516$ while $(x^4 + y^4)^2 = (16 + 81)^2 = 9409$. That is all it takes to solve this problem.

However, the presentations by a number of contestants left a lot to be desired. It went typically as follows:

$$(2^4 + 3^4)^2 < 2 \cdot 2 \cdot 3(2^6 + 3^6),$$
$$(16 + 81)^2 < 12(64 + 729),$$
$$9409 < 9516.$$

Implicitly, this is saying that since the last statement is true, so must the first one be. This is faulty logic. Suppose we wish to prove that $2 = 1$. If so, then certainly $1 = 2$. Addition now yields $2 + 1 = 1 + 2$, which is clearly correct. However, it does not follow that $2 = 1$.

Siuki Kwan and **Danny Ng**, both of Western Canada High School, used $x = 1$ and $y = \frac{3}{2}$, while **Jackie Chan** of McNally High School used $x = \frac{1}{3}$ and $y = \frac{1}{2}$. Both sets of numbers boil down to $9409 < 9516$ eventually. **Peter Dziegielewski** of Western Canada High School and **Irwin Tang** of Sir Winston Churchill High School had the simplest numerical task at hand. They tried $x = 1$ and $y = \sqrt{2}$. Now $2xy(x^6 + y^6) = 2\sqrt{2}(1 + 8) = 18\sqrt{2}$ while $(x^4 + y^4)^2 = (1 + 4)^2 = 25$. We have $18\sqrt{2} > 18(1.4) = 25.2 > 25$.

A large number of horrible blunders, both arithmetical and algebraic, were made by various contestants. Some who solved this problem inexplicably went on and made some wild and false statements. One claimed that the difference of x and y must be 1. Another claimed that one has to be rational and the other irrational. There were still other variations. This showed that they did not fully understand what the problem was about, but got lucky and hit upon the right numbers.

We are trying to determine if the inequality $(x^4 + y^4)^2 \geq 2xy(x^6 + y^6)$ is valid for all real numbers x and y. So we should not accept it as a statement. Instead, we should work with the expression $P(x, y) = (x^4 + y^4)^2 - 2xy(x^6 + y^6) = x^8 - 2x^7y + 2x^4y^4 - 2xy^7 + y^8$, and try to determine its sign.

Note that $P(x, y)$ is a polynomial of degree 8 in two variables. If $x = y$, it becomes identically 0. Hence the given inequality holds whenever $x = y$. On the other hand, if x is much larger than y, then the term x^8 dominates all the others, and again the original inequality holds. To seek counter-examples, x and y must be relatively close but unequal.

Another important observation is that every term of $P(x, y)$ is of degree 8. We call such a polynomial *homogeneous*. Let us see what happens if we replace x by λx and y by λy, where λ is some constant. We have

$$\begin{aligned}(\lambda x)^8 - 2(\lambda x)^7(\lambda y) + 2(\lambda x)^4(\lambda y)^4 - 2(\lambda x)(\lambda y)^7 + (\lambda y)^8 \\ = \lambda^8(x^8 - 2x^7y + 2x^4y^4 - 2xy^7 + y^8).\end{aligned}$$

In other words, $P(\lambda x, \lambda y) = \lambda^8 P(x, y)$.

It follows that $P(x, y)$ has the same sign as $P(\lambda x, \lambda y)$ for any nonzero constant λ. Recall that most of the contestants who solved this problem chose $x = 2$ and $y = 3$. If we take $\lambda = \frac{2}{3}$, we have $\lambda x = 1$ and $\lambda y = \frac{3}{2}$. If we take $\lambda = \frac{1}{6}$, we have $\lambda x = \frac{1}{3}$ and $\lambda y = \frac{1}{2}$.

We may as well choose y to be 1, and work with the polynomial $p(x) = x^8 - 2x^7 + 2x^4 - 2x + 1$. Let us do a bit of algebra and factor it as follows:

$$\begin{aligned}p(x) &= x^7(x - 1) - xy(x^6 - 2x^3 + 1) - (x - 1) \\ &= (x^7 - 1)(x - 1) - x(x^3 - 1)^2 \\ &= (x - 1)^2(x^6 + x^5 + x^4 + x^3 + x^2 + x + 1 - x(x^2 + x + 1)^2) \\ &= (x - 1)^2(x^6 - x^4 - 2x^3 - x^2 + 1).\end{aligned}$$

An alternative way of factoring is to observe that $p(1) = 0$ so that $p(x) = (x-1)q(x)$, and that $q(1) = 0$ also. Note that if $x = 1$, the last factor of $p(x)$ is equal to -2. If we now choose x to be very close to 1 but not equal to 1, then $x^6 - x^4 - 2x^3 - x^2 + 1$ is negative while $(x-1)^2$ is positive. Then the polynomial is also negative, and we have shown that counter-examples exist without exhibiting a specific pair of numbers.

There are many possible choices for counter-examples. All that is required is that $x \neq y$, and that $\frac{x}{y}$ lies roughly between 0.6 and 1.6. Now the *Golden Ratio* $\tau = \frac{1+\sqrt{5}}{2}$ is approximately 1.618 and $\frac{1}{\tau}$ is roughly 0.618. Does τ have anything to do with this problem? If we drop the constant term 1 from $x^6 - x^4 - 2x^3 - x^2 + 1$, then the remaining part factors into $x^2(x^2+x+1)(x^2-x-1)$, and τ is a root of $x^2 - x - 1$. So it has no direct bearing on this problem, but its value gives us an approximate range of where we can find counter-examples.

5. Most contestants found this problem from hard to inaccessible. The main difficulty lies in how to make use of the given condition $AD = BC$, when the two segments seem very far away from each other. There were only three successful attempts, though not without blemish.

Jack Chen made use of reflection to bring these two segments together. He first considered the case where $\angle BAC$ is a right angle. This is easily seen to be impossible since D would coincide with B, and $BC = \sqrt{2}AD \neq AD$.

Consider the case where $\angle BAC$ is acute, so that O lies inside triangle ABC. Let $\angle BAC = 2\theta$. Then $\angle DAO = \angle CAO = \angle ACO = \theta$. Let P be the reflection of D across the perpendicular bisector of AC. Then P lies on OA and $\angle PCO = \theta$. Now triangles ACD and CAP are congruent since $AC = CA, \angle DCA = \angle PAC = \theta$ and $\angle DAC = \angle PCA = 2\theta$. Hence $AD = CP$. Since we are given that $AD = BC$, and we have $BP = CP$ by symmetry, PCB is an equilateral triangle. With O inside triangle ABC, we have $\angle APC = 150°$, $\theta = 10°$ and $\angle BAC = 20°$.

Finally, consider the case where $\angle BAC$ is obtuse, so that O is outside triangle ABC. Jack was unable to handle this case, though it goes almost word for word the same as the previous one. It is only at the last stage when we make use of the fact that O is outside ABC to conclude that $\angle APC = 30°$, $\theta = 50°$ and $\angle BAC = 100°$.

Frank Chen of Harry Ainlay High School left the segments AD and BC where they were, and incorporated them into congruent triangles. He considered only the case where $\angle BAC$ is acute. As in Jack's solution, let $\angle BAC = 2\theta$. Then $\angle DAO = \theta$ and $\angle ADO = 2\theta$. Let Q be the point on the major arc BC such that $\angle CBQ = \theta$. Now $\angle BQC = \angle BAC = 2\theta = \angle AOD$ and $BC = AD$. Hence $BQ = OA$, so that OBQ is an equilateral triangle and $\angle OBQ = 60°$. Finally, $90° = \angle BAO + \angle ABC = \angle BAO + \angle ABO + \angle CBQ + \angle BOQ = 3\theta + 60°$. Hence $\theta = 10°$ and $\angle BAC = 20°$.

If $\angle BAC$ is obtuse, the point Q is on the minor arc BC and OBQ is still an equilateral triangle. Now $\angle OBC = \angle OBQ - \angle CBQ = 60° - \theta$ and $\angle ABC = \angle ABO - \angle OBC = 2\theta - 60°$. Finally, from $90° = \angle BAO + \angle ABC = 3\theta - 60°$, we have $\theta = 50°$ and $\angle BAC = 100°$.

The third approach, by **Michael James** of Sir Winston Churchill High School, is trigonometric. Like Frank, he also considered only the case where $\angle BAC$ is acute. As in Jack's solution, let $\angle BAC = 2\theta$. Then $\angle AOD = \angle COR = 2\theta$, where R is the point of intersection of BC and OA. Let r be the radius of the circle and let $AD = BC = 2x$. Applying the Law of Sines to triangle ADO, we have $\frac{\sin(180^\circ - 3\theta)}{r} = \frac{\sin 2\theta}{2x}$. From the right triangle COR, we have $x = r \sin 2\theta$. It follows that $\sin(180^\circ - 3\theta) = \frac{1}{2}$. Either $180^\circ - 3\theta = 30^\circ$ or 150°. These lead respectively to $\theta = 50^\circ$ and $\theta = 10^\circ$.

Since Michael took $\angle BAC$ to be acute, he rejected $\theta = 50^\circ$ and concluded that $\angle BAC = 20^\circ$. He should have realized that $\angle BAC$ could be obtuse. In that case, we have $\angle COR = 180^\circ - 2\theta$ instead, but that does not change the rest of the argument. This time, we reject $\theta = 10^\circ$ and conclude that $\angle BAC = 100^\circ$.

February 1, 2000

1. Slightly less than two-thirds of the contestants solved this problem, but most resorted to detailed case analysis with heavy computations, running in one case to six pages. Others displayed short-comings in the understanding of probabilities, adding them when they should be multiplied, and vice versa. In the heat of the contest, many arithmetical errors occurred.

   Part (a) of the problem can be solved with no computations at all. In order for the goblin to win, it must win an 8 versus 10 battle and another battle. The odds for the other battle are 8 versus 6 if the weaker elf fights first, and 10 versus 6 if the stronger elf fight first. It follows that the elves' better strategy is to have the weaker elf fight first. The actual value of the elves' winning probability is $\frac{13}{18}$ if the stronger elf fights first, and $\frac{47}{63}$ otherwise.

   Robert Barrington Leigh of Vernon Barford Junior High School, **Simon Yewchuk** of Archbishop MacDonald High School and **Teresa Leung** of Dr. E. P. Scarlett High School generalized part (b) of the problem as follows. Let the strength of the goblin be x and those of the elves be y_1 and y_2 respectively. The probability of the goblin winning both battles is $\frac{x}{x+y_1} \cdot \frac{x+y_1}{x+y_1+y_2} = \frac{x}{x+y_1+y_2}$. Since this expression is symmetric with respect to y_1 and y_2, it does not matter which elf fights first. The actual value of the goblin's winning probability is $\frac{1}{3}$.

   This problem is adapted from the paper *Games People Don't Play* by **Professor Peter Winkler** of Bell Telephones Laboratories. The paper was contributed to the fourth **Gathering for Gardner** in Atlanta, February, 2000. The gathering is in honor of **Martin Gardner**, the famous author of the column *Mathematical Games* in the magazine *Scientific American* from 1956 to 1986. The gathering is held in Atlanta once every two years.

2. The performance on this problem was very disappointing. Only one-sixth of the contestants succeeded in solving it. It is not a difficult problem, but many contestants misread the question and turned it into something totally trivial. They should have realized that their interpretation must be incorrect, and reread the question more

carefully. However, there is an unfortunate tendency to treat a task as an obstacle rather than a challenge, and to look for an easy way out.

In the question, a and b are given first, and A and B are to be found according to the given values of a and b. Many contestants went ahead to choose A and B, and then take their digit-sums as a and b. One went so far as to state that a and b are irrelevant.

Let us work through an example. The given values may be $a = 13$ and $b = 35$. We may take A to be 1111111111111 and B to be 77777. However, $A + B =$ 1111111188888, and its digit-sum is $a + b = 48 > 9$. How can this be made less than $a + b$? We have to take advantage of carrying.

Let $a = 9c + e$ and $b = 9d + f$, where c, d, e and f are integers such that $0 \le e, f \le 8$. First suppose $f = 0$. Choose B to be the integer consisting of d 9's. Then the digit sum of B is $9d = b$. Let the last d digits of A be $000\ldots01$. If $e = 0$, put $c - 1$ 9's in front and add a leading digit 8. If $e > 0$, put c 9's in front and add a leading digit $e - 1$ if $e > 1$. In each case, the digit sum of A is $9c + e = a$. Now only the leading digit of $A + B$ is non-zero, so that its digit sum is at most 9.

If $f > 0$, choose A as before and B to have the same number of digits as A. The leading digit of B is f and the last d digits are $999\cdots99$. All other digits of B are 0's. Then the digit sum of B is $9d + f = b$. It may still happen that only the leading digit of $A + B$ is non-zero. If that is not the case, then there is carrying at that position, so that the leading digit of $A + B$ is 1. The second digit is at most 8 since $f \le 8$. All other digits are 0. Hence the digit sum of $A + B$ is still at most 9.

3. By Pythagoras' Theorem, we have $b^2 + c^2 = 4bc$. This is as far as two-thirds of the contestants got. Most of the remaining one-third found that the tangent of one of the acute angles is $2 - \sqrt{3}$, but only a few recognized that the angle is 15°. However, the overall performance was quite encouraging since there were a couple of very nice solutions.

The best was that of **Jonathan Lau** of Sir Winston Churchill High School. He observed that $\sin 2\theta = 2\sin\theta\cos\theta = 2 \cdot \frac{c}{2\sqrt{bc}} \cdot \frac{b}{2\sqrt{bc}} = \frac{1}{2}$. Hence $2\theta = 30°$ or $150°$ so that the acute angles we seek are 15° and 75°.

Jeffrey Mo of the University Elementary School took $b = 1$ so that $c^2 - 4c + 1 = 0$. By the Quadratic Formula, $c = \frac{1}{2}(4 \pm \sqrt{12}) = 2 \pm \sqrt{3}$. Note that $\frac{1}{2+\sqrt{3}} = 2 - \sqrt{3}$. We may assume that $c < 1$ so that $c = 2 - \sqrt{3}$. Let θ be the acute angle between the side of length 1 and the hypotenuse, so that $\theta < 45°$. The hypotenuse has length $2\sqrt{2 - \sqrt{3}}$. Then $\sin^2\theta = \left(\frac{2 - \sqrt{3}}{2\sqrt{2 - \sqrt{3}}}\right)^2 = \frac{2 - \sqrt{3}}{4}$ while $\cos^2\theta = \left(\frac{1}{2\sqrt{2 - \sqrt{3}}}\right)^2 = \frac{2 + \sqrt{3}}{4}$. Hence $\cos 2\theta = \cos^2\theta - \sin^2\theta = \frac{\sqrt{3}}{2} = \cos 30°$. It follows that $\theta = 15°$. The other acute angle of the triangle is obviously 75°.

Gordon Kwok of John G. Diefenbaker High School and **Karim Versi** of Old Scona Academic High School also came up with $\sin 2\theta = \frac{1}{2}$, but in rather round-about ways.

Another variation of this solution goes as follows. Reflect the triangle about the side of length 1, so that it is an isosceles triangle with two sides of lengths $2\sqrt{2-\sqrt{3}}$ and an angle 2θ in between. The area of this triangle is given by $\frac{1}{2}(2\sqrt{2-\sqrt{3}})^2 \sin 2\theta = 2(2-\sqrt{3}) \sin 2\theta$. On the other hand, it has base $2(2-\sqrt{3})$ and height 1, so that its area is given by $2-\sqrt{3}$. Hence $\sin 2\theta = \frac{1}{2}$.

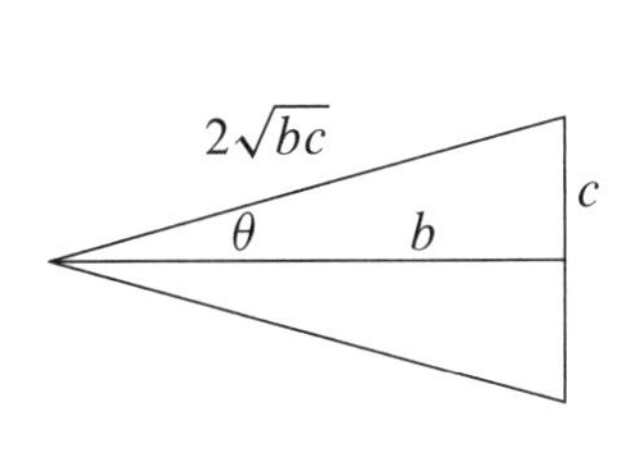

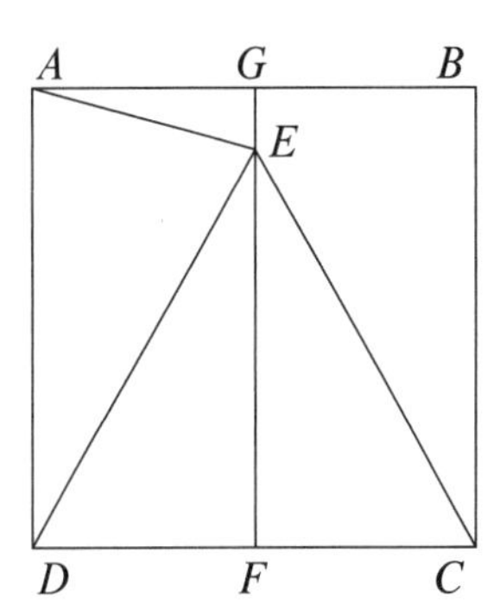

It is possible to solve this problem without trigonometry. Let $ABCD$ be a square of side length 2. Let CDE be an equilateral triangle inside $ABCD$. Draw the line through E perpendicular to CD, cutting it at F and AB at G. Then $EF = \sqrt{3}$ and $EG = 2-\sqrt{3}$. Hence AGE is congruent to the triangle in question. Now $\angle ADE = 30°$. Since $AD = DE$, we have $\angle DAE = 75°$ so that $\angle GAE = 15°$.

4. This is a fairly straight-forward problem in geometry, in that only straight lines are involved. Thus it can be solved routinely by analytical geometry. However, only Jeffrey Mo, **Jack Chen** and Jonathan Lau of Sir Winston Churchill High School and **Frédéric Dupuis** of Western Canada High School managed to do that. Jeffrey's presentation is the cleanest.

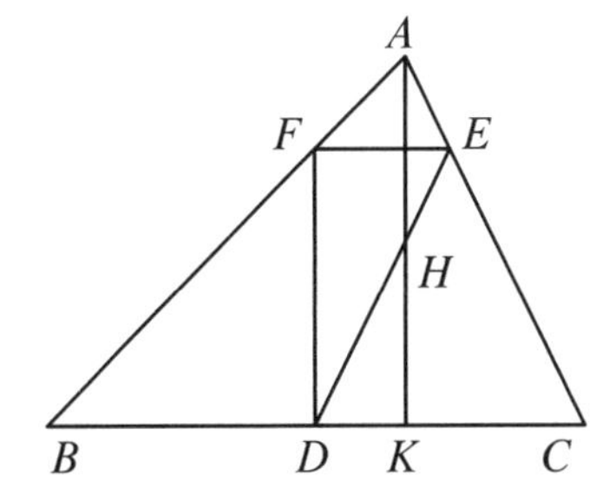

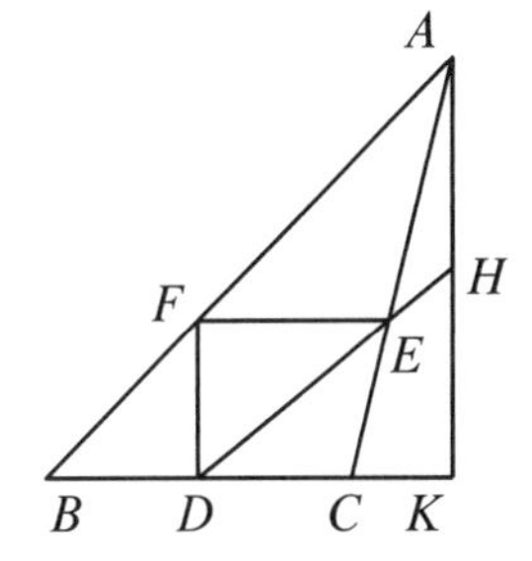

Choose A to be (a, b), B to be $(0, 0)$ and C to be (c,0). Then D is $(\frac{c}{2}, 0)$ and K is $(a, 0)$. The equation of AB is $y = \frac{bx}{a}$. Hence F is $(\frac{c}{2}, \frac{bc}{2a})$. The equation of AC is $y = \frac{bx-bc}{a-c}$. From $\frac{bc}{2a} = \frac{bx-bc}{a-c}$, E is $(\frac{3ac-c^2}{2a}, \frac{bc}{2a})$. Hence the equation of DE is $y = \frac{2bx-bc}{4a-2c}$. It crosses AK at the point $H(a, \frac{b}{2})$, which is indeed the midpoint of AK.

The computations may be simplified further by taking D to be (0,0), B to be $(-1, 0)$ and C to be (1,0). This problem can also be solved using synthetic geometry by considering various pairs of similar triangles.

Let DE meet AK at H. Since triangles ABC and AFE are similar, we have $\frac{AB}{AF} = \frac{BC}{EF}$. Since triangles BAK and BFD are similar, we have $\frac{AK}{BK} = \frac{DF}{BD}$ and $\frac{BK}{DK} = \frac{AB}{AF}$. Since triangles DHK and EDF are similar, we have $\frac{DF}{EF} = \frac{HK}{DK}$. Multiplying these four equations yields $AK = \frac{HK \cdot BC}{BD} = 2HK$. Hence H is indeed the midpoint of AK.

The above solution may be shortened somewhat if we appeal to a result known as **Menelaus' Theorem**. Let ABC be a triangle. A straight line which does not pass through any of A, B and C is called a transversal of triangle ABC. Let such a line cut BC at D, CA at E and AB at F. Now you may wonder how a line can cut all three sides of a triangle. What we must add is that one of D, E and F is on the extension of a side, or all three are. Menelaus' Theorem then states that $\frac{BD}{DC} \cdot \frac{CE}{EA} \cdot \frac{AF}{FB} = 1$.

We prove Menelaus' Theorem by area considerations. Denote by $[T]$ the area of triangle T. If two triangles have collinear bases and the same vertex, then their areas are in the same ratio as their bases. For instance, $\frac{AF}{FB} = \frac{[DAF]}{[DFB]} = \frac{[EAF]}{[EFB]}$. Let us denote this common ratio by r. Then we have $[ADE] = [DAF] - [EAF] = r([DFB] - [EFB]) = r[BDE]$, so that $\frac{AF}{FB} = \frac{[ADE]}{[BDE]}$. We also have $\frac{BD}{DC} = \frac{[BDE]}{[DCE]}$ and $\frac{CE}{EA} = \frac{[DCE]}{[ADE]}$. Multiplying these three equations yields Menelaus' Theorem.

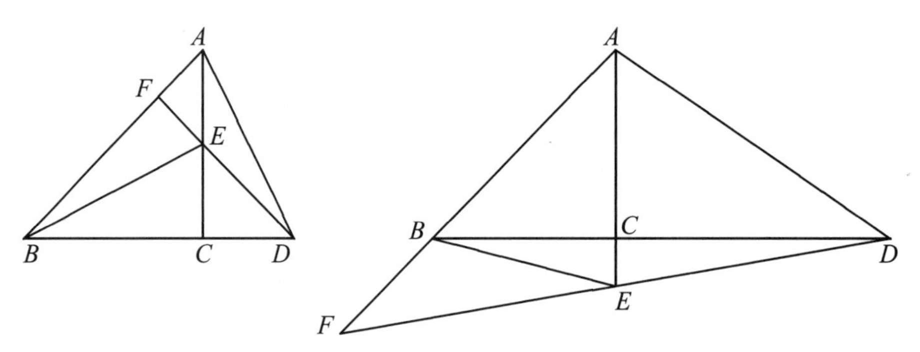

Note that the converse of Menelaus' Theorem also holds. Suppose D is a point on BC, E is a point on CA and F is a point on AB, with exactly one on the extension of a side or all three on extensions. If $\frac{BD}{DC} \cdot \frac{CE}{EA} \cdot \frac{AF}{FB} = 1$, then D, E and F are collinear.

We now give a third solution to Problem 4 in our competition. Let DE meet AK at H. Applying Menelaus' Theorem to triangle ACK with the transversal DHE, we have $\frac{AH}{HK} \cdot \frac{KD}{DC} \cdot \frac{CE}{EA} = 1$. Now $\frac{CE}{EA} = \frac{BF}{FA}$ since triangles AFE and ABC are similar, and $\frac{KD}{DC} = \frac{KD}{DB} = \frac{AF}{FB}$ since triangles ABK and FBD are similar. It follows that $\frac{KD}{DC} \cdot \frac{CE}{EA} = 1$, so that $AH = HK$.

5. This is a difficult problem. Most contestants merely substituted various values for x, y and z, and quite a few discovered that if one of them is equal to 1 and the other two equal to 0, the value 4 appears to be maximum. However, only one student managed to prove that this is indeed the case. Mercifully, nobody attempted to use calculus. Since the given expression involves several variables, the method of Lagrange multipliers would normally be applied. This is always rather messy. In this particular problem, the maximum occurs on the boundary and is easy to overlook.

The expression has cyclic symmetry in the three variables, and the given condition $x + y + z = 1$ is also symmetric. Making use of this, we transform one of the terms as follows:

$$\begin{aligned}4x + (y - z)^2 &= 4 - 4(y + z) + (y + z)^2 - 4yz \\ &= (2 - (y + z))^2 - (2\sqrt{yz})^2 \\ &= (2 - y - z + 2\sqrt{yz})(2 - y - z - 2\sqrt{yz}).\end{aligned}$$

At this point, we call upon the most important elementary inequality called the **Arithmetic-Mean Geometric-Mean Inequality**. Let a and b be positive real numbers. Their arithmetic mean is defined to be $\frac{a+b}{2}$ and their geometric mean is defined to be $\sqrt{ab}$. The Inequality states that the arithmetic mean is always greater than oe equal to the geometric mean, with equality if and only if $a = b$.

We know that the square of any real number is non-negative. Hence $(\sqrt{a} - \sqrt{b})^2 \geq 0$. This is equivalent to $a - 2\sqrt{ab} + b \geq 0$ or $\frac{a+b}{2} \geq \sqrt{ab}$. If equality holds, then $\sqrt{a} - \sqrt{b} = 0$ or $a = b$. This result can be generalized to three or more positive real numbers.

We now return to Problem 5 in our competition. By the Arithmetic-Mean Geometric-Mean Inequality, we have

$$\sqrt{4x + (y - z)^2} \leq \frac{1}{2}(2 - y - z + 2\sqrt{yz} + 2 - y - z - 2\sqrt{yz}) = 2 - y - z.$$

Similarly, $\sqrt{4y + (z - x)^2} \leq 2 - z - x$ and $\sqrt{4z + (x - y)^2} \leq 4 - x - y$, so that

$$\sqrt{4x + (y - z)^2} + \sqrt{4y + (z - x)^2} + \sqrt{4z + (x - y)^2} \leq 6 - 2(x + y + z) = 4.$$

The only contestant who solved this problem was Robert Barrington Leigh. His solution is more elegant and avoids using the Arithmetic-Mean Geometric-Mean Inequality. That the maximum value is 4 will follow if $\sqrt{4x + (y - z)^2} \leq 1 + x$, $\sqrt{4y + (z - x)^2} \leq 1 + y$ and $\sqrt{4z + (x - y)^2} \leq 1 + z$. Since $x + y + z = 1$, we have $|y - z| \leq 1 - x$ so that $(y - z)^2 \leq (1 - x)^2$. Now

$$\sqrt{4x + (y - z)^2} \leq \sqrt{4x + (1 - x)^2} = \sqrt{(1 + x)^2} = 1 + x.$$

The other two inequalities can be proved in an analogous manner. Hence the given expression is less than or equal to $1 + x + 1 + y + 1 + z = 4$.

February 6, 2001

1. This was meant to be an easy starter, but far too many contestants appeared to have spent hours on it, generating pages after pages of computations which in the end led nowhere. The most systematic approach was that of **Robert Barrington Leigh** of Old Scona Academic High School, Edmonton, and **Peter Du** of Sir Winston Churchill High School, Calgary.

We must have $b \leq 9$ as otherwise Fred would have covered at least 110 kilometers in 11 days. Similarly, $a \geq 12$ as otherwise Fred would have covered at most 99 kilometers in 9 days. If Fred spends x of the 9 days riding at a kilometers per day, then $ax + b(9 - x) = 100$ or $(a - b)x = 100 - 9b$. Similarly, if Fred spends y of the 11 days riding at a kilometers per day, then $(a - b)y = 100 - 11b$. It follows that $a - b$ is a common divisor of $100 - 9b$ and $100 - 11b$ so that it divides $2b$. Hence $2b \geq a - b$ so that $3b \geq a \geq 12$, yielding $b \geq 4$. On the other hand, $a - b$ cannot divide $100 - 10b$, so that it does not divide b. The following chart shows that only when $b = 4$ or $b = 8$ can we even have $a = 12$, and only when $b = 4$ do we have $a - b$ dividing $2b$ but not b. Hence $a = 12$ and $b = 4$ is the unique solution.

b	$100 - 9b$	$100 - 11b$	$a - b$	a
4	64	56	1,2,4,8	5,6,8,**12**
5	55	45	1,5	6,10
6	46	34	1,2	7,8
7	37	23	1	8
8	28	12	1,2,4	9,10,**12**
9	19	1	1	10

2. This was the problem in which the overall performance was the best. It also clearly identified those contestants who had acquired knowledge beyond the standard curriculum. The central idea behind this problem is to consider the remainder when an integer is divided by 7. Two integers a and b which leave the same remainder when so divided are said to be *congruent* to each other modulo 7. The standard notation is $a \equiv b \pmod 7$. Quite a few contestant used incorrect notations such as "a mod 7=b" or "a= mod 7 b".

The majority of the successful contestants approached the problem this way. We classify integers according to the remainders obtained when they are divided by 7. If the remainder is 0, the integer is put in class I. If it is 1 or 6, the integer is put in class II. If it is 2 or 5, the integer is put in class III. If it is 3 or 4, the integer is put in class IV. These classes are exhaustive and mutually exclusive. Among any five integers, two will be in the same class by the Pigeonhole Principle. If they have the same remainder when divided by 7, clearly their difference will be divisible by 7. If not, then their sum is is one of $(7a + 1) + (7b + 6)$, $(7a + 2) + (7b + 5)$ and $(7a + 3) + (7b + 4)$, each of which is divisible by 7.

Keith Chung of Western Canada High School, Calgary, assumed that all five numbers are non-congruent modulo 7, as otherwise the difference of two of them will be divisible by 7. Of the seven possible remainders, namely, 0, 1, 2, 3, 4, 5 and 6, exactly two are missing. Hence at least one of the pairs 1+6, 2+5 and 3+4 is intact. Hence the sum of two of the numbers is divisible by 7.

Robert Barrington Leigh used a slightly different approach. If at least two of the numbers are congruent to 0 modulo 7, clearly their sum as well as their difference will be divisible by 7. Hence we may assume that there is at most one such number, so that we have four numbers none of which is congruent to 0 modulo 7. Let them be

w, x, y and z. Throw in $-w$, $-x$, $-y$ and $-z$. By the Pigeonhole Principle, two of these numbers are congruent to each other. If both are original numbers, say x and y, then $x-y$ is divisible by 7. The same applies if neither is original. If one of them, say x, is an original number, while $-y$ is not, then $x+y$ is divisible by 7.

3. This problem revealed a disturbing trend that algebraic skills in general are on the decline. To cope with an increasingly technological society, we need more algebra, not less. Like Problem 1, many contestants went round and round without any sense of purpose, and made all sorts of elementary mistakes in algebraic computation along the way.

   Alex Kim of Glenmary School, Peace River, started by solving for y in terms of x. Subtracting $x(y^2-4x)=x$ from $x^3-4x^2=y$, we have $x(x^2-y^2)=y-x$. Since $x\neq y$, this reduces to $x(x+y)=-1$ so that $y=-\frac{x^2+1}{x}$. Now $x^3-4x^2=-\frac{x^2+1}{x}$ simplifies to $x^4-4x^3+x^2+1=0$. It follows that $x^3(y+4)=-x^2(x^2+1)+4x^3=-x^4+4x^3-x^2=1$.

   Jeffrey Mo of St. Paul's Academy, Okotoks, also got to $x(x+y)=-1=4x-y^2$. Adding $xy+y^2$ to both sides, he obtained $(x+y)^2=x(y+4)$. Hence $x^3(y+4)=(x(x+y))^2=(-1)^2=1$.

   Sonny Yue of Queen Elizabeth Jr/Sr High School, Calgary, got to $x^4-4x^3+x^2+1=0$ instead. Multiplying both sides of $x^4+x^2=4x^3-1$ by 4x and adding to both sides x^6-8x^5, he obtained $x^3(x^3-4x^2+4)=(x^3-4x^2)^2-4x$, which is equivalent to $x^3(y+4)=y^2-4x=1$.

4. Geometry continues to be the weak link in the mathematical background of our students. Only a handful got anywhere with this problem, and even less got complete solutions. There were essentially two approaches.

   The key step in the solution by **Charles Li** of Western Canada High School, is the similarity between triangles AMN and ACB, though he got to it in a more roundabout way. The most direct line of reasoning goes as follows.

   Since $\angle ABN=\angle ACM$ and $\angle BAN=\angle CAM$, triangles BAN and CAM are similar. It follows that $\frac{AN}{AM}=\frac{AB}{AC}$. Since $\angle MAN=\angle CAB$, triangles MAN and CAB are also similar, so that $\frac{MN}{BC}=\frac{AN}{AB}=\frac{AM}{AC}=\frac{AN+AM}{AB+AC}<1$.

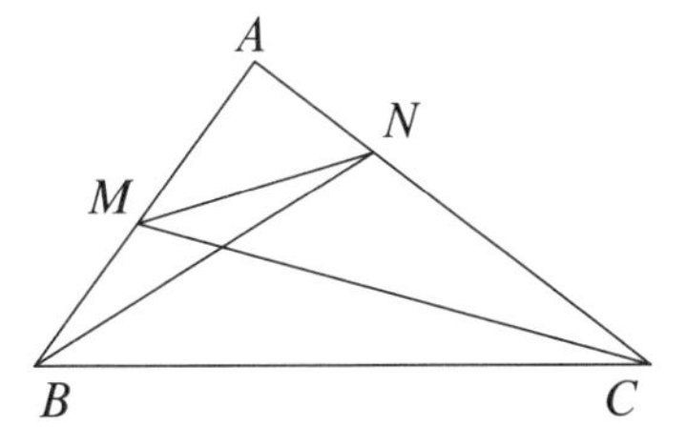

   The last step of the above solution uses a simple fact in algebra. If $\frac{a}{b}=\frac{c}{d}$, then they are also equal to $\frac{a+c}{b+d}$ and $\frac{a-c}{b-d}$, provided that $b\neq\pm d$. To see this, let $\frac{a}{b}=t$ so that $a=bt$. Then $c=dt$, $a+c=(b+d)t$ and $a-c=(b-d)t$. Hence all four ratios are equal. This useful result is apparently not well-known.

Charles assumed that $AC > AB$, since the result is trivial if $AB = AC$, and $AB > AC$ yields an analogous case. Since $AB > AM$, we have $\frac{MN}{BC} = \frac{AM}{AC} < 1$.

The key step in the solution by **Sarah Sun** of St. Mary's School, Okotoks, is that $BCNM$ is a cyclic quadrilateral. Actually, she proved it by first establishing the similarity of triangles AMN and ACB. What follows is a simplified approach.

Since $\angle MBN = \angle MCN$, B, C, N and M lie on a circle by the converse of the Circle-Angle Theorem. Rotate the arc MN along this circle so that N coincides with C and M with some point D. Then $\angle DBC = \angle MBN < \angle BNC = \angle BDC$ by the Exterior Angle Inequality and the Circle-Angle Theorem. It follows that $MN = BD < BC$ by the Angle-Side Inequality.

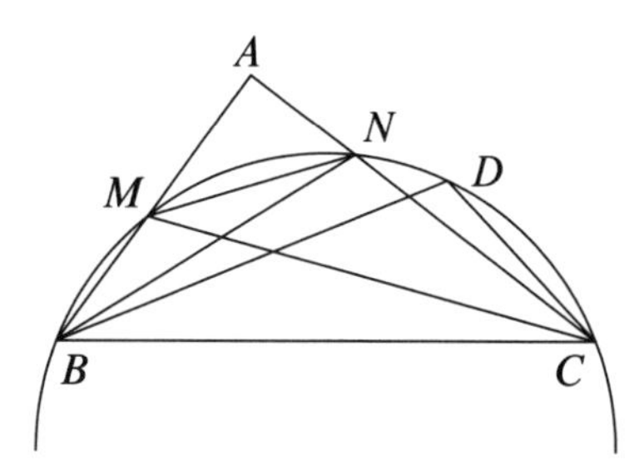

5. Most contestants found this problem hard, though they all made partial progress. The most popular answer was to ask each of (1) and (2) five times for a total of ten questions. Some reduced it to the correct answer of eight by ignoring the condition that all questions must be tabled in advance before any answers are given. By repeatedly asking (1) until we get three responses which agree, we may use up three, four or five questions. We then switch to asking (2). In the first case, one more question will do. In the second case, we may need three more. In the third case, we may need five more.

   The key step is the observation that A only needs to know the correct answers to any two of the three questions. Thus they are related symmetrically to one another, despite the apparent difference of (3) from the other two. Some contestants asked each question three times for a total of nine, but actually eight will suffice.

   Alex Fink of Queen Elizabeth Jr/Sr High School and **Alvin Tan** of McNally High School, Edmonton, asked each of (1) and (2) three times and (3) twice. By symmetry, we only need consider three cases:

   Case (a). *The three answers to (1) are the same, as are those to (2).*

   A can accept these answers.

   Case (b). *The three answers to (1) are the same, but not those to (2).*

   The answer to (1) can be accepted. B has lied at least once in answering (2). If the two answers to (3) are the same, they can be accepted. If not, A accepts the majority's answers to (2).

   Case (c). *The three answers to (1) are not the same, and neither are those to (2).*

   B has lied exactly once answering (1) and exactly once answering (2). A can accept the majority's answers to them.

   It is important to realize that we have not given a complete solution to the problem yet. In an extremal problem, not only do we have to show that the extremum can be attained, we have to establish that it is indeed an extremum, in that nothing more (or less, as in the present instance) will do.

Alex and Alvin then proved that A cannot get by with less than eight questions. If there is a functional list with seven questions, then two of (1), (2) and (3) will be asked a total of at most four times. We may assume that these are (1) and (2). We may as well assume that the answers to (3) are all the same and are to be accepted. By symmetry, we only need consider three cases:

Case (a). *A asks (1) four times.*

B may answer "Yes" twice and "No" twice, and A will be in trouble.

Case (b). *A asks (1) three times and (2) once.*

If the answers to (1) are not all the same, A will not be able to decide whether to accept the majority's answers or the minority answer.

Case (c). *A asks each of (1) and (2) twice.*

B may answer each question "Yes" once and "No" once, and A will be in trouble.

Robert Barrington Leigh used the language of error-correcting codes. He noted that there are four possible scenarios: both P and Q true, P true and Q false, P false and Q true, and both P and Q false. Representing the responses of B by a string of 0's and 1's, with 0 meaning "Yes" and 1 meaning "No", each scenario is assigned a code-word, or a response string in which B does not lie.

The fact that B may lie up to twice means that a response string can differ from a code-word by up to two digits. If two code-words differ in at most four digits, then there is a response string which sits between the two, and we will not be able to tell which is the correct code-word. Hence two code-words must differ in at least five digits.

Suppose we ask only seven questions. We may assume that one of the code-words is 0000000. Then each of the other three code-words must contain at least five 1's. However, any two of those three will differ in at most four digits. Hence seven questions are not sufficient. On the other hand, eight questions are, as we can take the code-words 00000000, 00011111, 11100011 and 11111100. This means that we ask each of (1) and (2) three times and (3) twice, as in the earlier solution.

February 5, 2002

1. This is a very easy question, but a large number of contestants failed to see their ways clearly, and ended up foundering with pages upon pages of case analysis. Among the successful contestants, most used the following approach by **Peter Du**, Sir Winston Churchill High School, Calgary.

   By (ii), each A must be flanked on both sides by D's. Hence we may replace each DAD with a single D, so that there are no more A's. By (iii), if there is a D next to a B, it must be flanked by two B's. It follows that if there is a D next to a C, it must be flanked by two C's. Hence we may replace each BDB with a single B and each CDC with a single C, so that there are no more D's. By (i), the modified molecule must consist alternately of B's and C's. Hence the total number of atoms is even. Since each replacement preserves the parity of the total number of atoms, this number is even before the molecule is modified. To complete the argument, we must proved that the modified molecule cannot consist of a single atom. Otherwise it will consist of three atoms before the last replacement, and the two end molecules which are identical will now be adjacent, contradicting (i).

A different argument was used by **Robert Barrington Leigh**, Old Scona Academic High School, Edmonton. He counted the number of bonds between two atoms along the circular molecule instead of the number of atoms.

By an XY bond is meant an X atom followed by a Y atom in clockwise order. With four kinds of atoms, there are potentially sixteen kinds of bonds. However, four of them, namely, AA, BB, CC and DD, are forbidden by (i) and four others, namely, AB, BA, AC and CA, are forbidden by (ii). Now an AD bond may potentially be followed by a DA, DB or DC bond, but the second is forbidden by (iii). Similarly, a CD bond cannot be followed by a DB bond, and a BD bond cannot be followed by either a DA or a DC bond. The following diagram represents all permissible pairs of bonds that can follow each other.

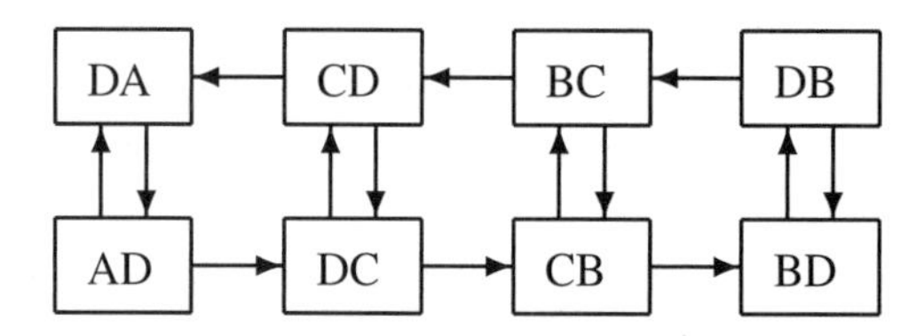

Color the bonds AD, CD, CB and DB white and the bonds DA, DC, BC and BD black. Then the bonds must alternate in color. Hence the number of bonds is even. It follows that the number of atoms is also even.

2. The performance on this problem is disappointing, since it can be worked out with relatively simple computations. However, some of the contestants who managed to come up with the correct answer did it rather laboriously. On the other hand, we do have an absolutely beautiful synthetic solution.

A relatively simple approach is as follows. We used a simplified version of the solution by **Keith Chung**, Western Canada High School, Calgary.

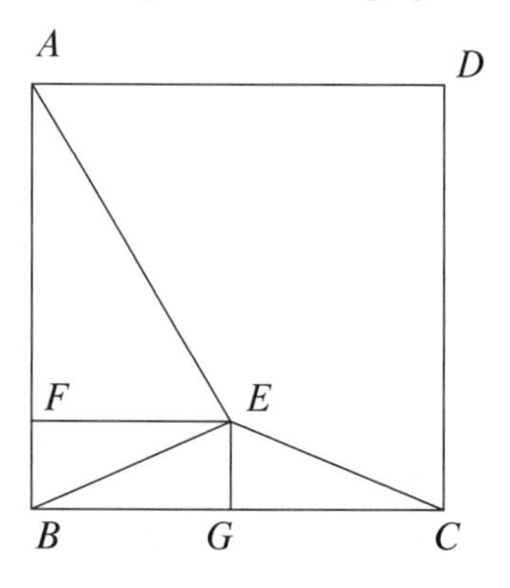

Drop perpendiculars EF and EG from E onto AB and BC respectively. Let $EF = x$, $EG = y$ and $AB = s$. Then

$$x^2 + y^2 = 8, \tag{6}$$
$$(s-x)^2 + y^2 = 9, \tag{7}$$
$$x^2 + (s-y)^2 = 25. \tag{8}$$

From (1) and (2). we have $x = \frac{s^2-1}{2s}$. From (1) and (3), we have $y = \frac{s^2-17}{2s}$. Substituting back into (1) and simplifying, we have $(s^2-5)(s^2-29) = 0$. Since E is inside $ABCD$ and $AE = 5$, we must reject $s^2 = 5$. It follows that the area of $ABCD$ is 29.

Xiao Lin, Western Canada High School, used a trigonometric approach. Again, let the side length of the square be s. By the Law of Cosines, $\cos ABE = \frac{s^2-17}{4\sqrt{2}s}$ and $\cos CBE = \frac{s^2-1}{4\sqrt{2}s}$. Since $\angle ABE + \angle CBE = 90°$, the sum of the squares of their cosines is 1. Hence $\frac{s^4-34s^2+289}{32s^2} + \frac{s^4-2s^2+1}{32s^2} = 1$ or $0 = s^4 - 34s^2 + 145 = (s^2-5)(s^2-29)$. As before, we reject $s^2 = 5$, so that the desired area is 29.

We now present the solution by **Robert Barrington Leigh**.

Rotate triangle ABE $90°$ about B to CBH. Then $CH = 5$, $BH = 2\sqrt{2}$ and $\angle HBE = 90°$, so that BEH is a right isosceles triangle. By Pythagoras' Theorem, $HE = \sqrt{BH^2 + BE^2} = 4$. By the converse of Pythagoras' Theorem, $\angle CEH = 90°$. Extend CE to K where BK is perpendicular to CK. Then BEK is also a right isosceles triangle. Hence $BH = EK = 2$, $CK = 5$ and $BC^2 = BK^2 + CK^2 = 29$.

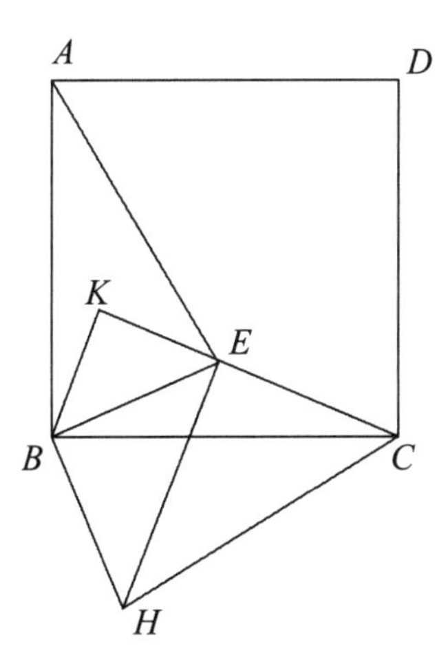

3. Most contestants approached this problem by graphical means, as typified by the solution from **Nathanael Wu**, Western Canada High School.

   Equating each of the three given quantities to $\frac{4}{9}$, we have $10x + y = 4$, $5y = 4x + 2$ and $x + 5 = 8y$. Solving the system of equations consisting of the first two, we have $20 - 50x = 4x + 2$ or $x = \frac{1}{3}$ and $y = 4 - 10x = \frac{2}{3}$. It is easy to verify that $(x, y) = (\frac{1}{3}, \frac{2}{3})$ also satisfies $x+5 = 8y$. It follows that the three lines are concurrent, dividing the Cartesian plane into six sextants. It is easy to verify that in each sextant, at least one of the three given quantities is less than $\frac{4}{9}$.

Alex Fink of Queen Elizabeth Jr/Sr High School, Calgary, along with **Alec Mills** and **Charles Li**, both of Western Canada High School, used the following indirect approach.

Assuming that each of the three given quantities is greater than $\frac{4}{9}$, we have $10x + y > 4$, $5y > 4x + 2$ and $x + 5 > 8y$. From the first and third, we have $x + 5 > 8y > 32 - 80x$ so that $x < \frac{1}{3}$. From the second and third, we have $10x + 50 > 80y > 8x + 4$ so that $x > \frac{1}{3}$. This is a contradiction.

Robert Barrington Leigh and **Sonny Yue**, Queen Elizabeth Jr/Sr High School, had the cleanest solution.

Let $a = \frac{2x+y}{2x+2y+1}$, $b = \frac{2y}{2x+2y+1}$ and $c = \frac{x+1}{2x+2y+1}$. Then

$$2a + 3b + 4c = \frac{4x + 2y + 6y + 4x + 4}{2x + 2y + 1} = 4.$$

It follows that not all of a, b and c can exceed $\frac{4}{9}$.

4. This turned out to be the hardest of the five problems. Very few contestants got the correct answer, and some who did failed to prove that it cannot be any larger. An alarming minority was handicapped right from the start by stating that 0 is neither odd nor even. This is blatantly false. Perhaps they were thinking of 1 being neither prime nor composite, or 0 being neither positive nor negative. However, 0 is definitely an even digit.

We follow the solution by **Sumudu Fernando**, Harry Ainlay High School, Edmonton.

Let $N = MK$. Since M divides N, K is a positive integer. Since N is even and M is odd, K is even. If we pick M and K accordingly to these rules but arbitrarily otherwise, one of the potential problems is that N may contain an odd digit. In order for M to be large, we expect K to be small. Thus it is likely to be a single-digit number. If in computing $N = MK$, there is no carrying in the multiplication, odd digits will not appear in N. So let us check the carrying when each odd digit is multiplied by a single-digit K.

K	1	3	5	7	9
2	0	0	1	1	1
4	0	1	2	2	3
6	0	1	3	4	5
8	0	2	4	5	7

Our attention is drawn immediately to the last column. The digit 9 is going to be troublesome. When multiplied by K, the carrying is an odd digit $K - 1$. Suppose we leave it out of M. Then the largest possible value of M is 7531. Fortunately, this works since if we take $K = 8$, then $N = 60248$.

Suppose a larger value of M works. Obviously $K \neq 10$. If $K \geq 12$, then $MK > 7531 \times 12 > 86420$. Hence K is indeed a single-digit number. Recall that the troublesome 9 must appear somewhere in M. Consider first an example where $M = 15973$.

Separate it into two parts at where the 9 is, with the 9 going to the second part. In other words, write $15973 = 15000 + 973$. Now $1000(K-1) < 973K < 1000K$. It follows that $973K$ is a four-digit number whose first digit is $K-1$, which is odd. In computing $N = 15973K$, we have to add this digit to the last digit of $15K$, which is even. Hence N will contain an odd digit, and we have a contradiction.

We now write out the proof in general terms. Let M_2 be the part of M after and including the digit 9, and M_1 be the part before the digit 9. Suppose M_2 is an n-digit number. Then $M = 10^n M_1 + M_2$. Note that $10^n(K-1) < M_2 K < 10^n K$. Hence the first digit of $M_2 K$ is $K-1$, which is odd. In computing $N = (10^n M_1 + M_2)K$, this digit will be added to the last digit of $M_1 K$, which is even. Hence N will contain an odd digit, and we have a contradiction. In conclusion, the largest possible value of M is 7531.

5. Many contestants were able to come up with the correct answer $f(x) = 2x^2 - 1$ for part (b). We have $f(3) = 17$. Now $f(x)$ is a parabola facing upward, with vertex at $(0, -1)$. Hence it is increasing for $1 \le x \le 3$, so that its maximum value in this range is indeed 17. However, many neglected to show that this function satisfies $|f(x)| \le 1$ for $-1 \le x \le x$. We have $f(\pm 1) = 1$ and $f(0) = -1$. Since $f(x)$ is decreasing for $-1 \le x \le 0$ and increasing for $0 \le x \le 1$, it has the desired property.

   We do not have any completely satisfactory solutions to part (a), so we turn to that provided by the proposer.

   Let $f(x) = ax^2 + bx + c$. Then $f(\pm 1) = a \pm b + c$ and $f(0) = c$. Hence $a = \frac{1}{2}f(1) + \frac{1}{2}f(-1) - f(0)$ and $b = \frac{1}{2}f(1) - \frac{1}{2}f(-1)$. Now

$$f(x+2) = a(x+2)^2 + b(x+2) + c = f(x) + 2(2ax+b) + 4a.$$

   Hence

$$|f(x+2)| \le |f(x)| + 2|2ax+b| + 4|a|.$$

   For $-1 \le x \le 1$, $|2ax+b| \le \max\{|2a+b|, |2a-b|\}$ where

$$|2a+b| \le \frac{3}{2}|f(1)| + \frac{1}{2}|f(-1)| + 2|f(0)| \le 4$$

   and

$$|2a-b| \le \frac{1}{2}|f(1)| + \frac{3}{2}|f(-1)| + 2|f(0)| \le 4.$$

   Combined with $4|a| \le 2|f(1)| + 2|f(-1)| + 4|f(0)| \le 2+2+4 = 8$ and $|f(x)| \le 1$, we have $|f(x+2)| \le 1 + 2 \times 4 + 8 = 17$ for $-1 \le x \le 1$, which means that $|f(x)| \le 17$ for $1 \le x \le 3$.

February 4, 2003

1. Quite a number of contestants know the test of divisibility by 11. Compute the two sums of alternate digits of a positive integer. Call the one which contains the last digit the even-placed sum and the other the odd-placed sum. Then the alternate digit-sum of the number is the difference obtained when the odd-placed sum is subtracted

from the even-placed sum. The number is congruent modulo 11 to this alternate digit-sum, so that one is divisible by 11 if and only if so is the other. For instance, if the number is 1234567, the the even-placed sum is 1+3+5+7=16 and the odd-place sum is 2+4+6=12, so that the alternate digit-sum is 4. Thus the number is not divisible by 11.

The approach used by **Sarah Sun**, Holy Trinity Academy, Okotoks, is representative of many contestants. Let the even-placed and odd-placed sums of a positive integer be x and y respectively, and let $x - y \equiv d \not\equiv 0 \pmod{11}$. If $d \neq 10$, simply add the single digit d at the end. The new even-placed sum is $y + d$ while the new odd-placed sum is x. Hence the new alternate digit-sum is $y + d - x \equiv 0 \pmod{11}$. If $d = 10$, let the last digit of the number be u. Let $v \equiv 2u + 1 \pmod{11}$, with $0 \leq v \leq 9$ since $u \neq 10$. Insert the digit v before u. Then the new even-placed sum is $y + u$ while the new odd-placed sum is $x - u + v$. Hence the new alternate digit-sum is $y + u - x + u - v \equiv 2u + 1 - v \equiv 0 \pmod{11}$.

It is possible to solve this problem without knowing this test of divisibility. We give the approach used by **Boris Braverman**, Branton Junior High School, Calgary. Divide the positive integer n by 11 and let the remainder be $r > 0$. If $r \neq 10$, just add the single-digit number r at the end, and we have a multiple of 11. If $r = 10$, let $n = 10m + u$. Note that $m \not\equiv 0 \pmod{11}$ as otherwise $u \equiv 10 \pmod{11}$, which is impossible since u is a single-digit number. Let $v \equiv m + u$, $0 \leq v \leq 10$. Since $10 \equiv 10m + u \equiv 9m + v$, we cannot have $v = 10$. Add the digit v just before u. Then $100m + 10v + u \equiv m - v + u \equiv 0 \pmod{11}$.

2. In the classical version of this problem, the bee flies at a constant speed of 75 meters per minute. Since it takes the birds exactly one minute to meet, the bee flies for exactly one minute, covering 75 meters. When the famous mathematician and computing scientist John von Neumann was told this problem, he gave immediately the correct answer. When asked whether he did it the easy way, he said, "Yes, I summed the infinite series." We have many budding von Neumanns among our contestants, as many tried to sum an infinite series. Actually, it is only necessary to work things out for one cycle, when the bee returns to the penguin. We give the approach used by **Boris Braverman**.

   Keep the penguin stationary, and let the pelican approach it at 90 meters per minutes. The bee will fly away from the penguin at 35 meters per minute while back to the penguin at 105 meters per minute. Replace the distance between the birds by a more general value of d meters. The first meeting between the bee and the pelican takes place after $\frac{d}{90+35} = \frac{d}{125}$ minutes, and the fly has ocverec a distance of $\frac{35d}{125}$ meters. Hence it will take the bee $\frac{35d}{125\times 105} = \frac{d}{125\times 3}$ minutes to return to the penguin. So the bee spends three times as much time flying at the higher speed, so that its average speed is $(3 \times 80 + 60) \div 4 = 75$ meters per minutes. Since it takes the birds 1 minutes to meet, the bee has covered 75 meters.

3. This is an easy problem for students with good backgrounds and quite hard for the others. Typically, the approach is that used by **Radoslav Marinov**, Harry Ainlay High School, Edmonton. Suppose $x^3 + ax^2 + bx + c = (x - p)(x - q)(x - r) = x^3 - (p + q + r)x^2 + (pq + qr + rp) - pqr$, where p, q and r are the integer roots.

Then $a = -(p+q+r)$, $b = pq+qr+rp$ and $c = -pqr$. If all three roots are odd, all of a, b and c are odd. If none are odd, all of a, b and c are even. If exactly one root is odd, then only a is odd. If exactly two roots are odd, then only b is odd. In none of the cases can exactly two of a, b and c be odd.

Robert Barrington Leigh, Old Scona Academic High School, Edmonton, gave the above approach a new twist. Putting $x = 1$ into $x^3 + ax^2 + bx + c = (x-p)(x-q)(x-r)$, we have $(1-p)(1-q)(1-r) = 1+a+b+c$. Since exactly two of a, b and c are odd, the right side of the last equation is odd. This means that all of $1-p$, $1-q$ and $1-r$ are odd, so that all of p, q and r are even. However, this means that all of a, b and c must be even, which is a contradiction.

4. In retrospect, this is not a very satisfactory problem, in that there is essentially one way to solve it, based on the fact that the only lines through the centroid which bisect the area of the triangle are precisely the medians. There is nothing much to the problem once this is assumed. It is very easy to show that the three medians divide the triangle into six small triangles with equal area, so that no two of them can quadrisect the overall area. Many students claimed the above result, but not too many provided justifications. We give below three different approaches.

   We first give the approach used by **Charles Li**, Western Canada High School, Calgary. Let G be the centroid and draw a line through G parallel to BC, cutting AB at P and AC at Q. It is easy to show that the area of APQ is $\frac{4}{9}$, and therefore less than $\frac{1}{2}$, that of ABC. Note that $PG = QG$, though we choose to record this as $PG \le QG$. Rotate the line PQ about G so that it now intersects AP at P' and CQ at Q'. Draw the line through Q parallel to AB, cutting GQ' at R. Then triangles GPP' and GQR are similar. Since $PG \le QG$, $P''G \le RG < Q'G$ and $AP'Q'$ has larger area than APQ. Continue the rotation until Q' lands on G. It is only at this point that the two parts of ABC have equal area, justifying the claim.

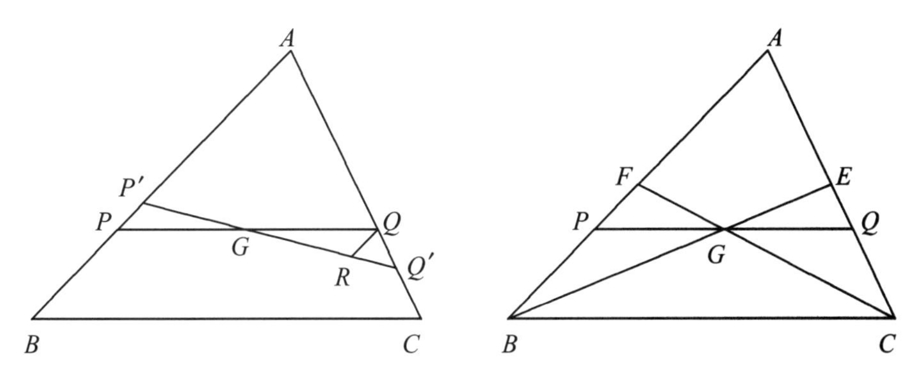

   We now give a composite approach used by **David Rhee**, Vernon Barford Junior High School, Edmonton, and **Robert Barrington Leigh**. Let G be the centroid and let BE and CF be two of the medians. Suppose there exists a line through G, cutting BF at P and CR at Q, such that it bisects the area of ABC. Then PFG and QCG have equal area, as do PBG and QEG. From the former, we have $PG \cdot FG = QG \cdot CG$. Since $CG = 2FG$, we have $PG = 2QG$. Similarly, from the lattice, we have $QG = 2PG$, which is a contradiction. Thus the claim is justified.

5. The contestants found this problem difficult, partly because of its abstract mathematical formulation. Very few succeeded in finding a complete solution. We give the approach of **Robert Barrington Leigh**.

   Set $f(x) = -4|x|^2 + 3|x| - 1$. Completion of the square yields $f(x) = (|x| - \frac{3}{8})^2 - \frac{7}{16} < -\frac{1}{4}$. Since $f(x) < 0$, $|f(x)| = -f(x)$ so that $|f(x) > |x|$ is equivalent to $0 < 4|x|^2 - 3|x| + 1 - |x| = (2|x| - 1)^2$, which is in turn equivalent to $|x| \neq \frac{1}{2}$. On the other hand, $f(-\frac{1}{2}) = -\frac{1}{2}$. Hence the sequence $x, f(x), f(f(x)), \ldots$ either becomes constant with value $-\frac{1}{2}$, or takes on larger and larger absolute values. Since the set of values is finite, the sequence must converge to $-\frac{1}{2}$. Suppose $f(x) = -\frac{1}{2}$. Then $0 = 8|x|^2 - 6|x| + 1 = (4|x| - 1)(2|x| - 1)$. Hence $x = \pm\frac{1}{2}$ or $x = \pm\frac{1}{4}$. We have proved earlier that $f(x) < -\frac{1}{4}$, so that $-\frac{1}{2}$ is the only one of these four values that $f(x)$ can take. Clearly, $-\frac{1}{2}$ must be in our set, to which we can add any subset of $\{\frac{1}{2}, \frac{1}{4}, -\frac{1}{4}\}$.

February 3, 2004

1. This is a straight-forward problem, and a large number of contestants successfully solved it. Below is a typical approach.

   Let the number of bricks the workman used be n. Since the length of Bill's yard is twice its width, the workman would have to use $2n$ bricks to carry out Bill's actual plan. Let x be the length of each brick. In Bill's plan, each brick after the first contributes $\frac{x}{13}$ to the total length. Hence $x + (2n - 1)\frac{x}{13} = 10$. Similarly, $x + (n-1)\frac{x}{3} = 20$. Dividing the first equation by the second, we have $\frac{13+2n-1}{3+n-1} = \frac{13}{6}$. Clearing denominators, we have $12n + 72 = 13n + 26$, which yields $n = 46$. It follows that the width of each brick is $\frac{10}{46} = \frac{5}{23}$. Finally, from $x + (92 - 1)\frac{x}{13} = 10$, we have $x = \frac{10}{8} = \frac{5}{4}$.

2. This "political" problem was composed before the Auditor General of Canada made headline news. Many contestants were confused by the last part of the question. Some simply gave up and wrote nothing down at all. Those who began solving the problem soon discovered that there were two possible solutions, and the last part of the question now makes perfect sense. The Auditor General announced the lower value, but the actual cost was the higher one. However, quite a few contestants failed to consider the various cases, while others committed careless arithmetical errors, so that they came up with only one solution. Some threw up their hands at this point. More amusingly, others began speculating on how extravagant the Upper House must have been or how corrupt the Lower House must have been, and came up with astounding values of λ via creative accounting! Below is a typical approach to this rather straight-forward problem.

   Suppose the Lower House hearing lasts x days. We cannot have $x \leq 15$ as otherwise the total bill from the Lower House will exceed that from the Upper House. Suppose $15 < x \leq 30$. Then the total bill from the Lower House is $\$600 \cdot 13x$ while that from the Upper House is $\$600 \cdot 6 \cdot 30 + \$900 \cdot 6(2x - 30)$. Equating these two amounts and solving for x, we have $x = 18$ so that each bill is \$140400. Suppose $x > 30$. Then the total bill from the Upper House has the same expression as before while that from the Lower House is $\$600 \cdot 13 \cdot 30 + \$900 \cdot 13(2x - 30)$. Equating these two amounts and solving for x, we have $x = 70$ so that each bill is \$702000, five times as large as \$140400. It follows that $\lambda = 5$.

3. This problem is also not hard, and was on the whole very well done. The most common error was overlooking the number $m^2 + 2m$, and obtaining the wrong answer $668 \times 2 = 1336$. Below is a typical approach.

   Let $m = \lfloor \sqrt{n} \rfloor$. Since $700^2 = 490000 > 447560$, m is less than 700 but not by much. Testing reveals that $669^2 = 447561$. Hence $m \le 668$. Note that $m \le \sqrt{n} < m + 1$. Hence $m^2 \le n < m^2 + 2m + 1$. Since n is an integer, $n \le m^2 + 2m$. The multiples of m in this range are m^2, $m^2 + m$ and $m^2 + 2m$, so that each value of m yields 3 values of n. Hence the number of n with the desired properties is $3 \times 668 = 2004$.

4. Although this is really a very easy geometry problem, it stood out as the major roadblock for most contestants. It highlights once again the sad neglect of geometry in our curriculum, especially geometric reasoning. Those contestants who solved this problem all used trigonometry.

   Everybody knows that $\sin 45° = \cos 45° = \frac{1}{\sqrt{2}}$, $\tan 45° = 1$, $\sin 30° = \frac{1}{2}$, $\cos 30° = \frac{\sqrt{3}}{2}$ and $\tan 30° = \frac{1}{\sqrt{3}}$. What are the values of the trigonometric ratios of a 15° angle? In the Compound Angle Formula $\cos(A + B) = \cos A \cos B - \sin A \sin B$, if we put $A = B = 15°$, we have $\frac{\sqrt{3}}{2} = \cos 30° = \cos^2 15° - \sin^2 15°$. We also know that $\cos^2 15° + \sin^2 15° = 1$. It follows that $\sin 15° = \sqrt{\frac{1-\cos 30°}{2}} = \frac{\sqrt{3}-1}{2\sqrt{2}}$. Similarly, $\cos 15° = \sqrt{\frac{1+\cos 30°}{2}} = \frac{\sqrt{3}+1}{2\sqrt{2}}$. From these, we obtain $\tan 15° = \frac{\sqrt{3}-1}{\sqrt{3}+1} = 2 - \sqrt{3}$.

   Peter Zhang of Sir Winston Churchill High School used a straight-forward approach using the Law of Sines. The majority of the successful solvers of this problem also followed the same path, though one of them appealed to the Law of Sinuses.

   Let $\angle ABC = \theta$ so that $\angle ACB = 135° - \theta$. Applying the Law of Sines to triangles BAD and CAD, we have $\frac{\sin 30°}{\sin\theta} = \frac{BD}{AD} = \frac{CD}{AD} = \frac{\sin 15°}{\sin(135°-\theta)}$. Hence

$$\sin 30°(\sin 45° \cos\theta + \cos 45° \sin\theta) = \sin 15° \sin\theta.$$

   Hence $\sin\theta + \cos\theta = (\sqrt{3} - 1)\sin\theta$. This simplifies to $\tan\theta = -(2 + \sqrt{3})$, so that $\angle ABC = 105°$.

   A few contestants stopped at the point $\tan\theta = -(2 + \sqrt{3})$. Since the question did not ask for the answer to be given in any specific form, this was accepted.

   If we draw a reasonably accurate diagram, it is not hard to see that the answer is $\angle ABC = 105°$, so that $\angle BCA = 30°$. This allows us to give an alternative version of the above solution, but without the difficulty at the end.

   Let $\angle BCA = \theta$ so that $\angle ABC = 135° - \theta$. Applying the Law of Sines to triangles BAD and CAD, we have $\frac{\sin 30°}{\sin(135°-\theta)} = \frac{BD}{AD} = \frac{CD}{AD} = \frac{\sin 15°}{\sin\theta}$. Hence

$$\sin 30° \sin\theta = \sin 15°(\sin 45° \cos\theta + \cos 45° \sin\theta).$$

   It follows that $2\sin\theta = (\sqrt{3} - 1)(\sin\theta + \cos\theta)$. This simplifies to $\tan\theta = \frac{1}{\sqrt{3}}$. Hence $\angle BCA = 30°$ and $\angle ABC = 105°$.

David Rhee of McNally High School, offered the following variation.

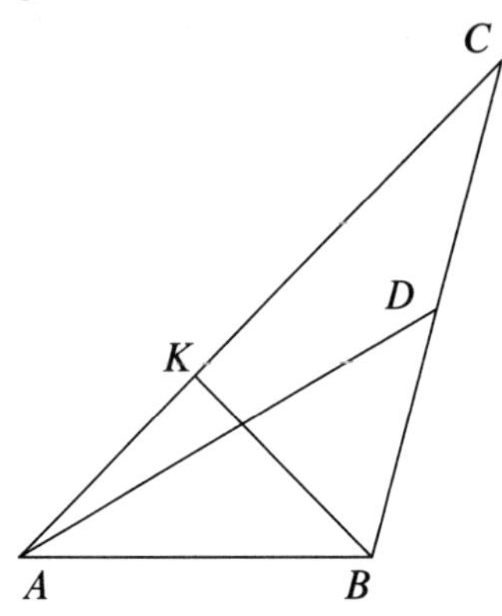

Let K be the foot of the perpendicular from B to AC. Let $BK = 1$. Then $AK = 1$ and $AB = \sqrt{2}$. Applying the Law of Sines to triangles BAD and CAD, we have $\frac{\sin 30^\circ}{\sin ADB} = \frac{BD}{AB}$ and $\frac{CD}{AC} = \frac{\sin 15^\circ}{\sin ADC}$. Now $AC = \frac{\sqrt{2}\sin 30^\circ}{\sin 15^\circ} = \sqrt{3}+1$. Hence $CK = \sqrt{3}$ and $\angle KBC = 60^\circ$, so that $\angle ABC = 105^\circ$.

Ken Zhang of Western Canada High School used even less trigonometry.

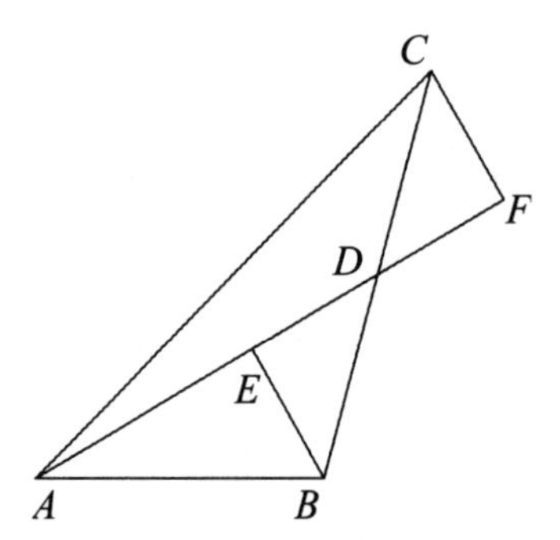

Let E and F be the respective feet of the perpendiculars from B and C onto AD. Then BED and CFD are congruent triangles. We may take $BE = 1$. Then $CF = 1$ also. Now $AE = \frac{1}{\tan 30^\circ} = \sqrt{3}$ while $AF = \frac{1}{\tan 15^\circ} = 2 + \sqrt{3}$. Hence $DE = \frac{1}{2}(AF - AE) = 1 = BE$. It follows that $\angle EBD = 45^\circ$ so that $\angle ABC = 105^\circ$.

Jerry Lo of Vernon Barford Junior High School used trigonometry at the very end instead of at the beginning.

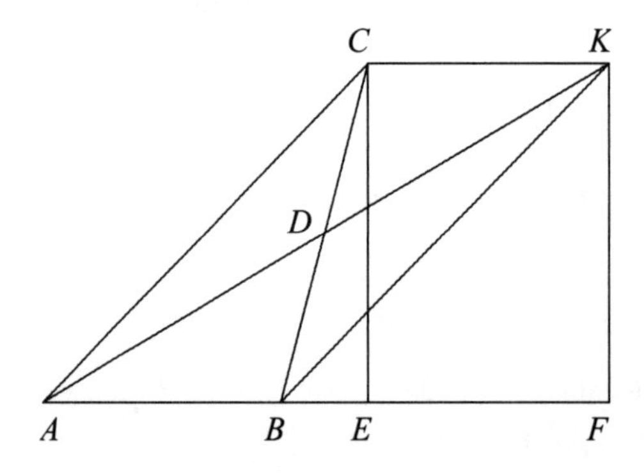

Extend AD to K so that $DK = AD$. Then $ABKC$ is a parallelogram. Let E and F be the respective feet of the perpendiculars from C and K onto AB. Let $KF = 1$. Then $AE = BF = CE = 1$ while $AF = \sqrt{3}$. Hence $BE = AE + BF - AF = 2 - \sqrt{3}$. Now $\tan CBE = \frac{1}{2-\sqrt{3}} = 2 + \sqrt{3}$, so that $\angle CBE = 75°$. It follows that $\angle ABC = 105°$.

The solution by **Dennis Chuang** of Strathcona Tweedsmuir School is essentially the same as the preceding one, at one half the scale.

Let E and F be the respective feet of the perpendiculars from C and D to AB. Let $DF = 1$. Then $AF = \sqrt{3}$. Since D is the midpoint of BC and DF is parallel to CE, we have $CE = 2$. Hence $AE = 2$ and $EF = 2 - \sqrt{3}$. Now $\tan CBE = \frac{1}{2-\sqrt{3}} = 2 + \sqrt{3}$, so that $\angle CBE = 75°$. It follows that $\angle ABC = 105°$.

The preceding solution is one step away from a purely geometric argument.

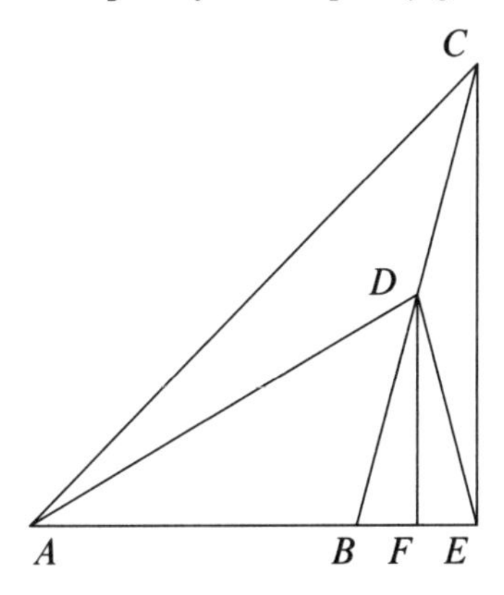

Let E and F be the respective feet of perpendicular from C and D to AB. Let $DF = 1$. Then $AD = 2$. Since D is the midpoint of BC and DF is parallel to CE, we have $CE = 2$. Hence $AE = 2 = AD$. Moreover, $DE = DB$ since D is the circumcenter of triangle CBE. Hence $\angle CBE = \angle AED = \angle ADE = \frac{1}{2}(180° - \angle DAE) = 75°$, so that $\angle ABC = 105°$.

5. (a) This can be done in a straight-forward manner, as typified by the approach of **Qu Chen** of Harry Ainlay High School.

   The desired inequality is equivalent to $\frac{1}{a+1} + \frac{b}{b+1} - \frac{c}{c+1} \geq 0$. We can take common denominator and then discard this positive quantity. The numerator simplifies to $abc + 2ab + a + b - c$. Since $a + b \geq c$, the desired inequality follows.

   Zheng Guo of Western Canada High School provided the following beautiful solution.

   If $c \geq a$ and $c \geq b$, then $\frac{a}{a+1} + \frac{b}{b+1} \geq \frac{a+b}{c+1} \geq \frac{c}{c+1}$. Otherwise, we may assume that $a > c$. Then $\frac{c}{c+1} = 1 - \frac{1}{c+1} < 1 - \frac{1}{a+1} = \frac{a}{a+1} \leq \frac{a}{a+1} + \frac{b}{b+1}$.

   (b) Zheng's argument works in exactly the same way here, but the approach used by **Brian Yu** of Old Scona Academic High School is typical of those of most of the successful solvers.

   Divide the polygon into $n - 2$ triangles using the diagonals from the vertex where the sides of lengths a_1 and a_n meet, and let the diagonals have lengths $b_2, b_3, \ldots, b_{n-2}$ respectively. Applying the result in (a) to each triangle, we have

$\frac{a_1}{a_1+1} + \frac{a_2}{a_2+1} \geq \frac{b_2}{b_2+1}$, $\frac{b_2}{b_2+1} + \frac{a_3}{a_3+1} \geq \frac{b_3}{b_3+1}, \ldots, \frac{b_{n-2}}{b_{n-2}+1} + \frac{a_{n-1}}{a_{n-1}+1} \geq \frac{a_n}{a_n+1}$.
Adding these yields the desired inequality.

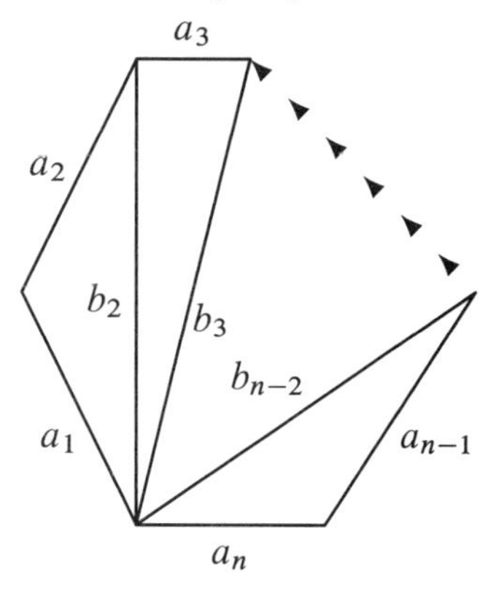

February 1, 2005

1. Although this is a very straight-forward problem, only one-third of the contestants solved it completely. The typical approach was like the one below, by **Graham Hill** of Bev Facey High School, Sherwood Park.

   Note that $70.\overline{7} = 70\frac{7}{9}$ while $70.\overline{67} = 70\frac{67}{99}$. Let the number of students in the two classes be x and y respectively. Then the total score is $\frac{353x}{5} + \frac{637y}{9} = \frac{6997(x+y)}{99}$. Clearing the denominators and simplifying, we have $19x = 25y$. Since 19 and 25 are relatively prime to each other, we have $x = 25t$ and $y = 19t$ for some positive integer t. Note that x is already divisible by 5. We need y to be divisible by 9 and $x + y$ to be divisible by 99. Since $x + y = 44t$, any t which is a multiple of 9 will work. However, we must keep $x + y$ under 500, so that the only possible choice is $t = 9$. Hence the two classes have 225 and 171 students respectively.

2. This is a very annoying problem. On the surface, it looks ridiculously easy. Yet, once we sink our teeth into it, we discover that there seem to be an endless array of possibilities. It calls for management skills.

   Boris Braverman of Sir Winston Churchill High School, Calgary, and **David Rhee** of McNally High School, Edmonton, hit upon the idea of using mathematical induction, without explicitly saying so.

   Consider the simple case where there are just two pearls. The daughter can do no worse than a draw in the inheritance game trivially, by making the only cut possible and taking the pearl of greater value, unless they are of the same value. When there are four pearls, they may be strung together in one of two ways, as shown in the diagram below.

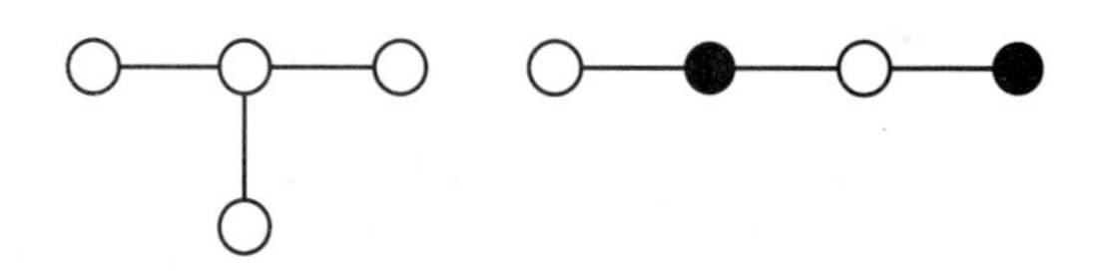

In the first case, the daughter has an easy task. She cuts off and takes the exposed pearl of the highest value. The son can at best match that. Now they are left with two pearls, a case already settled. However, if the daughter adopts the same strategy in the second case, she may lose as she may expose a pearl of enormous value.

Instead of merely looking at the boundary conditions, we should examine the global conditions. Mark the pearls black and white as shown. We may assume by symmetry that the total value of the black pearls is at least that of the white pearls. The daughter should take the exposed black pearl. Then she can get both black pearls.

With this background, we can now tackle the actual problem in the competition. Mark the pearls black and white as shown in the diagram below. Suppose the total value of the three black pearls is at least that of the three white pearls. The daughter cuts off the only black pearl available initially. The son must cut off the white pearl from either end, exposing a black pearl for the daughter to take. Thus she can get all three black pearls.

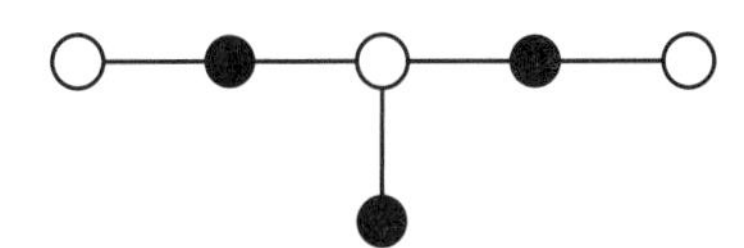

Suppose the total value of the three white pearls exceeds the total value of the three black pearls. The daughter cuts off the white pearl at an end that has a higher value than the one at the other end. If the son cuts off either black pearl now available, the daughter can get all three white pearls. Thus the son must take the white pearl at the other end, at best matching what the daughter has. Now they are left with four pearls, a case already settled.

3. This problem looks formidable and scared off many contestants. However, with two rules at our disposal, it should not be too hard to make some headway. A typical approach is like the one below by **Gary Huang** of Sir Winston Churchill High School, Calgary.

If x belongs to S, then $(x+2)^2+1=x^2+4x+5$ belongs to S, so that $x+2$ also belongs to S. Now $0^2+1=1$ belongs to S. Hence 0 belongs to S. Therefore every non-negative integer belongs to S. Let n be any integer. Then $2n^2+1$ is a positive integer and belongs to S, so that $\sqrt{2}n$ also belongs to S. Now $\sqrt{2}n+2$, $(\sqrt{2}n+2)+2$, $\dots$ all belong to S. In other words, $2m+\sqrt{2}n$ belongs to S for any nonnegative integer m. In particular, $2(n^2+1)+\sqrt{2}(-2n)$ belongs to S. Since this may be rewritten as $(\sqrt{2}n-1)^2+1$, $\sqrt{2}n-1$ belongs to S. Then $\sqrt{2}n+1$, $\sqrt{2}n+3$, $\dots$ all belong to S. In other words, $m+\sqrt{2}n$ belongs to S for any integer $m\geq -1$. Finally, if x belongs to S, then $(-x-2)^2+1=x^2+4x+5$ belongs to S, so that $-x-2$ belongs to S. For any integer $m\leq -2$, $(-m-2)+\sqrt{2}(-n)$ belongs to S. Hence $-((-m-2)+\sqrt{2}(-n))-2=m+\sqrt{2}n$ belongs to S.

4. This is a problem in computational geometry, and quite a number of contestants were able to show off their technical skills.

Boris Braverman used the most popular approach.

Let O be the center of the circle. Let $\angle BOC = 2\theta$ and $\angle EOF = 2\phi$. We have $6\theta + 6\phi = 360°$ so that $\theta + \phi = 60°$. Let r be the radius of the circle. Then $\sin\theta = \frac{1}{2r}$ while $\sin\phi = \frac{1}{r}$, so that $\cos\phi = \frac{\sqrt{r^2-1}}{r}$. Now $\frac{1}{2r} = \sin(60° - \phi) = \frac{\sqrt{3}}{2}\cdot\frac{\sqrt{r^2-1}}{r} - \frac{1}{2}\cdot\frac{1}{r}$, which yields $r = \sqrt{\frac{7}{3}} = \frac{\sqrt{21}}{3}$.

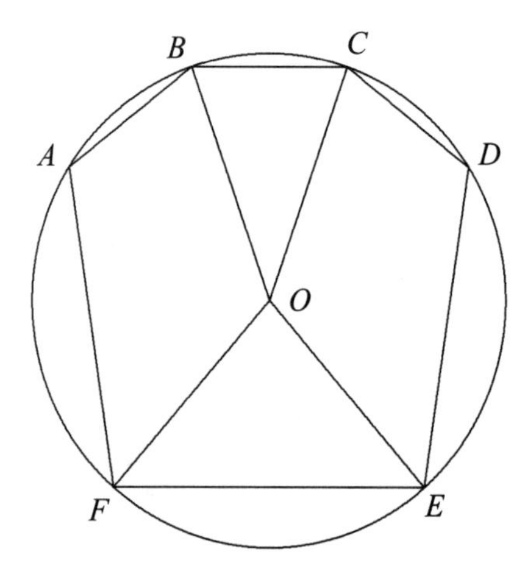

Malka Wrigley of Old Scona Academic High School, Edmonton, ended up with an eighth degree equation by using the identity $\sin 60° = \sin\theta\cos\phi + \cos\theta\sin\phi$ instead. Fortunately, it factored easily into $3r^4(r^2 - 1)(3r^2 - 7) = 0$, and yields the correct answer after the extraneous roots have been rejected. **Sarah Sun** of Holy Trinity Academy, Okotoks, worked with the base angles of triangles BOC and EOF instead. After a longer string of calculations, she arrived at the equation $(r^2 - 1)(3r^2 - 7) = 0$.

Zheng Guo of Western Canada High School, Calgary, also worked with base angles, but in a totally different way.

Let O be the center of the circle. Let $\angle OAB = \theta$ and $\angle OAF = \phi$. Then $6\theta + 6\phi = 720°$ so that $\angle\theta + \angle\phi = 120°$. It follows that $\angle BDF = 180° - \angle BAF = 60°$ and $\angle BOF = 2\angle BDF = 120°$. Let r be the radius of the circle. Then $BF = \sqrt{3}r$. Applying the Cosine Law to triangle BAF, we have $3r^2 = BF^2 = BA^2 + AF^2 - 2BA \cdot AF \cos BAF = 7$, so that $r = \sqrt{\frac{7}{3}} = \frac{\sqrt{21}}{3}$.

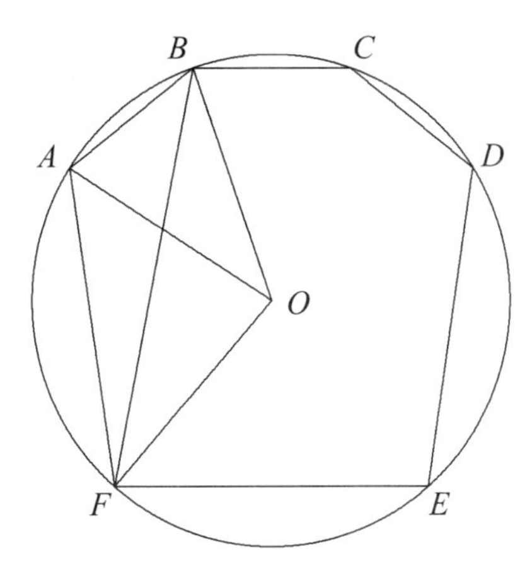

We now give the approach by **David Rhee**.

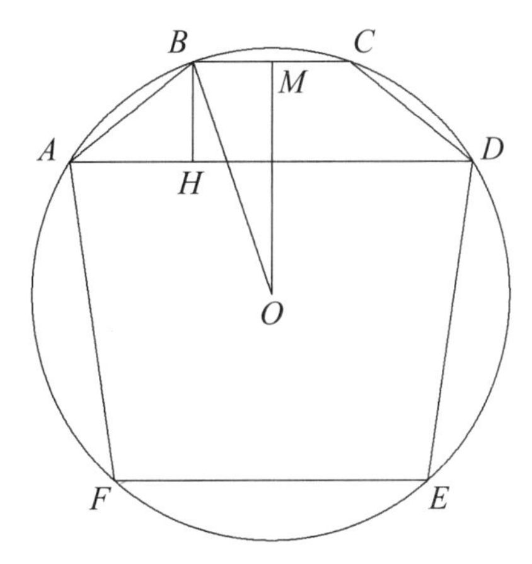

Let O be the center of the circle. Let H and M be the respective feet of the perpendiculars from B and O onto AD and BC. Let r be the radius and $AD = x$. Then $AH = \frac{x-1}{2}$. Note that $\angle MBA = 180° - \angle CDA = 180° - \angle BAH$. Hence $\angle MBO = \frac{1}{2}\angle MBA = 90° - \frac{1}{2}\angle BAH$. It follows that

$$\begin{aligned}\frac{1}{2r} &= \frac{BM}{BO}\\ &= \cos MBO\\ &= \sin\frac{BAH}{2}\\ &= \sqrt{\frac{1-\cos BAH}{2}}\\ &= \sqrt{\frac{1}{2}\left(1-\frac{AH}{AB}\right)}\\ &= \frac{\sqrt{3-x}}{2},\end{aligned}$$

so that $r = \frac{1}{\sqrt{3-x}}$. Similar calculations applied to the quadrilateral $DEFA$ instead of $ABCD$ yields $r = \frac{2\sqrt{2}}{\sqrt{6-x}}$. Equating the two expressions for r and solving for x, we have $x = \frac{18}{7}$, which yields $r = \frac{\sqrt{21}}{3}$.

Actually, this problem can be solved with no trigonometry, in fact, almost no calculations at all.

Suppose the center of the circle lies outside the hexagon. Then all six sides are inside the same semi-circle, so that the side closest to the center must be longer than each of the others, which is not the case. Hence the center of the circle is inside the hexagon. The lines joining the center to the six vertices divide the hexagon into six isosceles triangles, which can be rearranged into a new hexagon with sides of lengths 1 and 2 alternately. Since its vertices still lie on the circle, all its angles are equal, and the common measure is $120°$. By adding three equilateral triangles of side 1, we obtain an equilateral triangle of side 4. Now the center of the circle coincides with the center of this triangle. Hence its distance to a side is $\frac{2\sqrt{3}}{3}$. By Pythagoras' Theorem, its distance to a vertex of the new hexagon is $\sqrt{(\frac{2\sqrt{3}}{3})^2 + 1^2} = \frac{\sqrt{21}}{3}$.

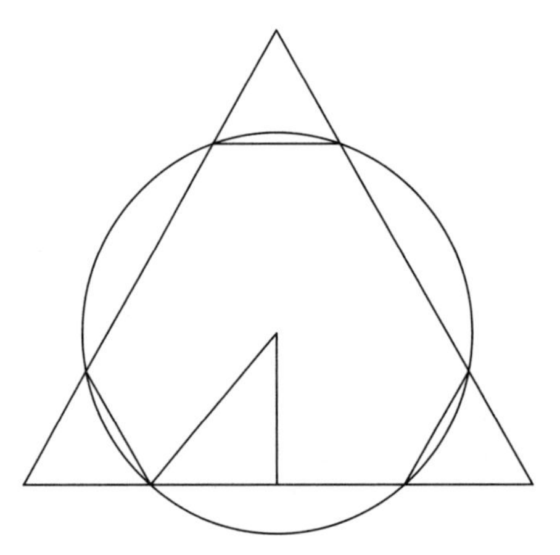

5. Synthetic geometry continues to be the bane for the vast majority of our students. A surprisingly large number of them actually got the correct answer, which was perhaps not hard to guess. However, most arrived at their conclusion via faulty reasoning.

   In this problem, we have plenty of angle relations. **Yiyi Yang** of Western Canada High School exploited this as follows.

   Let G be the point on AC such that $\angle CGD = \angle BAD = \angle BFC$. It follows from Thales' Theorem that $\angle GCD = \angle ABD = \angle ADB = \angle FCB$. Hence ABD, FBC and GCD are isosceles triangles similar to one another. Since $\angle CGD = 2\angle GFD$, we have $\angle GFD = \angle GDF$ so that $CG = DG = FG$. Since AC bisects $\angle BCD$, we have $\frac{BE}{DE} = \frac{BC}{DC} = \frac{CF}{CG} = 2$.

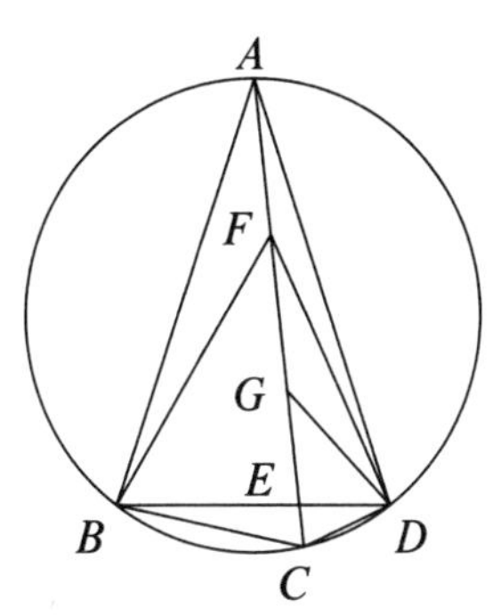

David Rhee and **Sarah Sun** used the following approach.

Since $AB = AD$, we have $\angle ABD = \angle ADB = \theta$. By Thales' Theorem, $\angle ACB = \theta = \angle ACD$ also. Let $\angle DFC = \phi$. Then $\angle BAD = \angle BFC = 2\phi$. Summing the angles of triangle BAD, we see that $\theta + \phi = 90°$, so that $\angle CDF = 90°$. Moreover, $\angle FBC = 180° - \theta - 2\phi = \theta = \angle FCB$. Hence $FB = FC$. Now the median FM divides triangle FBC into two parts, each of which is easily seen to be congruent to triangle FCD. It follows that $BC = 2CD$. Finally, since AC bisects $\angle BCD$, $\frac{BE}{DE} = \frac{BC}{DC} = 2$.

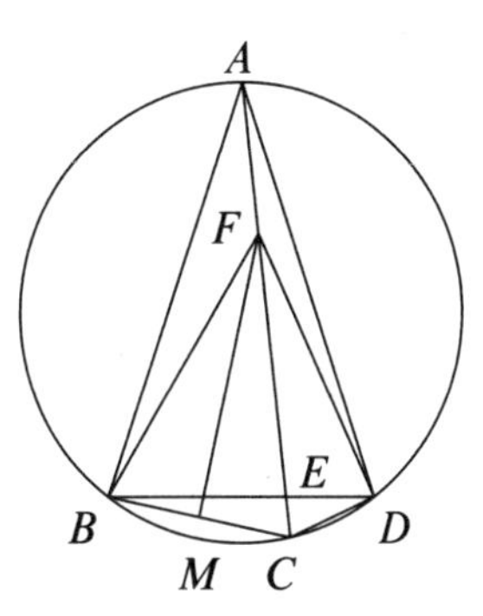

February 7, 2006

1. This is a straight-forward problem, though many contestants still resorted to the machinery of analytic geometry. The following is a simple approach used by **Brian Yu** of Old Scona Academic High School, Edmonton.

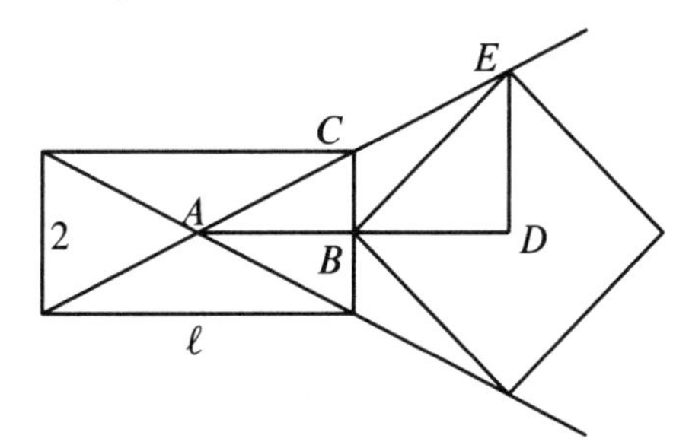

Join the center A of the rectangle to the center D of the square, passing through the vertex B of the square which lies on the perimeter of the rectangle. Join D to a vertex

E of the square adjacent to B. The line AE passes through a vertex C of the rectangle. Let $BD = DE = k$. Then the area of the square is $2k^2$. Since this is equal to the area of the rectangle, which is 2ℓ, we have $\ell = k^2$. Now triangles ABC and ADE are similar. Hence

$$\frac{\frac{k^2}{2}}{1} = \frac{AB}{BC} = \frac{AD}{DE} = \frac{\frac{k^2}{2} + k}{k}.$$

Since $k \neq 0$, this simplifies to $k^2 = k + 2$ or $(k + 1)(k - 2) = 0$. Since $k \neq -1$, we must have $k = 2$. It follows that $\ell = 4$.

2. This is a very easy problem, and almost every contestant solved it. Let the discounts be $m\%$ and $n\%$ respectively. Then $100(1 - \frac{m}{100})(1 - \frac{n}{100}) = 79.17$ so that $(100 - m)(100 - n) = 7917$. All that is required now is to express the number 7917 as a product of two factors each less than 100. The usual test of divisibility shows that 7917 is divisible by 3 since 7+9+1+7=24 is divisible by 3. We have $7917 = 3 \times 2639$. Now 2639 is clearly a multiple of 13, and we have $2639 = 13 \times 203$. Also, since 91 is a multiple of 7, so is 7917, and therefore so is 203. We have $203 = 7 \times 29$. Now 3, 7, 13 and 29 are the prime factors of 7917. They must be combined into two products each less than 100, and the only possible combination is $7 \times 13 = 91$ and $3 \times 29 = 87$. It follows that $m = 100 - 91 = 9$ and $n = 100 - 87 = 13$. Since multiplication is commutative, we may also have $m = 13$ and $n = 9$.

   A number of contestants expanded the equation $(100 - m)(100 - n) = 7917$ and obtained $2083 + mn = 100(m + n)$. Since $7917 + mn$ is divisible by 100, we have $mn = 100k + 17$ for some non-negative integer k, whereupon $m + n = 21 + k$. For $k = 0$, there are no solutions. The case $k = 1$ yields the correction solution of $\{m, n\} = \{9, 13\}$. However, they failed to check that there are no other solutions.

3. This is a straight-forward problem. Most contestants found all three answers. However, quite a few just conducted an exhaustive search, and then concluded at some point that the search would yield no further answers, without giving adequate justification. The following approach is used by **Graham Hill** of Sir Winston Churchill High School of Calgary.

   Let $S(n)$ denote the sum of the digits of n in base ten. We know that 9 divides $n - S(n) = 12S(n)$. Hence 3 divides $S(n)$, so that 3 divides n. Since 13 also divides n, n is a multiple of 39. Since 39 and 78 do not have the desired properties, n has at least three digits. Suppose it has exactly three digits. Then $S(n) \leq 9 + 9 + 9 = 27$ and $n \leq 13 \times 27 = 351$. Now $S(n) \leq 2 + 9 + 9 = 20$, so that $n \leq 13 \times 20 = 260$. The only multiples of 39 in range are 117, 156, 195 and 234. All but the last one has the desired properties. Suppose n has d digits where $d \geq 4$. We claim that $2d < 10^{d-3}$. This certainly holds for $d = 4$. Suppose it holds for some $d \geq 4$. Then $2(d + 1) < 10^{d-3} + 2 < 10^{(d+1)-3}$. This justifies the claim. Now $S(n) \leq 9d$ so that $n \leq 117d < 200d < 10^{d-1}$. This contradicts the assumption that n has d digits. It follows that the only numbers with the desired properties are 117, 156 and 195.

4. This is a very technical problem. A few contestants left it off completely, probably because they were overwhelmed by the details. Most of the others handled it routinely

and competently. The following approach by **David Rhee** of McNally High School, Edmonton stands out in its simplicity.

The slope of AB is $\frac{b^2-0}{b-0} = b$ and the slope of AD is $\frac{d^2-0}{d-0} = d$. Since AB is perpendicular to AD, we have $bd = -1$.

(a) Let **b**, **c** and **d** denote the vectors from A to B, C and D respectively. Then

$$\mathbf{c} = \mathbf{b}+\mathbf{d} = (b, b^2)+(d, d^2) = (b+d, b^2+d^2+2bd+2) = (b+d, (b+d)^2+2).$$

Hence C lies on the parabola $y = x^2 + 2$.

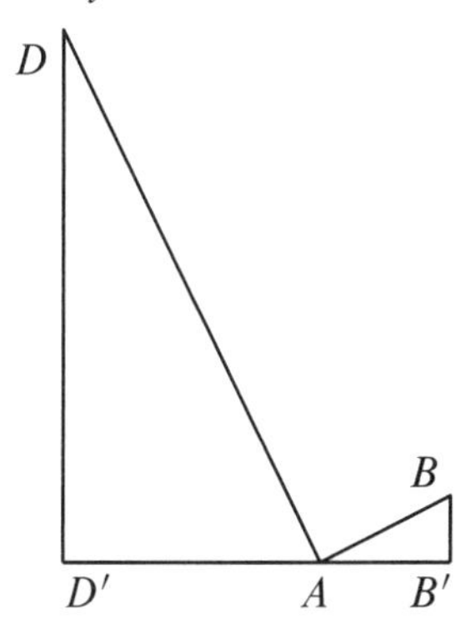

(b) Let B' and D' be the respective feet of perpendiculars from B and D to the x-axis. Then $\angle BAB' + \angle DAD' = 90°$ since $\angle BAD = 90°$. Also, $\angle ADD' + \angle DAD' = 90°$ since $\angle AD'D = 90°$. Hence $\angle BAB' = \angle ADD'$, so that triangles BAB' and ADD' are similar. It follows that $\frac{AB}{AD} = \frac{BB'}{D'A} = \frac{b^2}{-d} = b^3$.

5. Most the contestants found this problem challenging. Many just skipped it, or ran out of time before getting to it. Quite a few came to the erroneous conclusion that the set is infinite. The key observation is that there can be at most one large number in the set. Below is the approached used by **Michael Zhou** of Western Canada High School, Calgary.

Consider a set S obtained from the original set by reducing every number by 1. Denote $m+1$ by a and $n+1$ by b. By symmetry, we may take $a > b$. Then the given condition becomes $20(a-b) \geq ab$ or equivalently $(a+20)(b-20) = ab-20a+20b-400 \leq -400$. This means that b is at most 19, so that there can only be one large number in S. We now pack as many small numbers into it. Note that a increases as b increases. Hence if we put in the smallest available number at every stage, our construction will yield a set of maximum size. We begin by putting 0 into S. For $b = 0$, we have $a \geq \frac{400}{20} - 20 = 0$. Since $a > b$, we have $a \geq 1$. So we put 1 into S. For $b = 1$, we have $a \geq \frac{400}{19} - 20 = \frac{20}{19}$. Since a is an integer, we have $a \geq 2$. So we put 2 into S. Similarly, for $b = 2$, $a \geq 3$; for $b = 3$, $a \geq 5$; for $b = 5$, $a \geq 9$; and for $b = 9$, $a \geq 23$. So we put 2, 3, 5, 9 and 23 into S. Since we cannot have $b = 23$, no more elements can be added. Thus S has at most 7 elements, and so does the original set. An example would be $\{1, 2, 3, 4, 6, 10, 24\}$.

Winners 1983–2006

First Round Winners

Individual Prizes

First Prizes—W. H. Freeman Scholars:

1983/1984	**Ron Miller**	Old Scona Academic High School, Edmonton.
1984/1985	**Russell Schultz**	Lorne Jenken High School, Barrhead.
	Alan Williams	Harry Ainlay Composite High School, Edmonton.
1985/1986	**Russell Schulz**	Lorne Jenken High School, Barrhead.
1986/1987	**Mike Moser**	Ecole J. H. Picard, Edmonton.
1987/1988	**Chris Nichols**	Ross Sheppard Composite High School, Edmonton.
1988/1989	**Mischa Hooker**	Sir Winston Churchill High School, Calgary.
	John Yoon	Sir Winston Churchill High School, Calgary.
1989/1990	**Ozzie Gelbord**	Western Canada High School, Calgary.
1990/1991	**Rahim Hirji**	Western Canada High School, Calgary.
	Jason Colwell	Old Scona Academic High School, Edmonton
1991/1992	**Robert Kry**	Western Canada High School, Calgary.
1992/1993	**Peter Hwang**	Sir Winston Churchill High School, Calgary.
1993/1994	**Byung-Kyu Chun**	Edwin Parr High School, Athabasca.
	Michael Forbes	Strathcona-Tweedsmuir School, Okotoks.
1994/1995	**Derek Kisman**	Queen Elizabeth High School, Calgary.
1995/1996	**Byung-Kyu Chun**	Harry Ainlay High School, Edmonton.
1996/1997	**Byung-Kyu Chun**	Harry Ainlay High School, Edmonton.
1997/1998	**Eddie Ng**	Bishop Carroll High School, Calgary.
	Brian Tsai	Western Canada High School, Calgary.
	Gilbert Lee	McNally High School, Edmonton.
1998/1999	**Sonny Chan**	Western Canada High School, Calgary.
	Peter Dziegielewski	Western Canada High School, Calgary.
1999/2000	**Rowena Luk**	Old Scona Academic High School, Edmonton.
2000/2001	**Alex Fink**	Queen Elizabeth Jr/Sr High School, Calgary.
	Peter Du	Sir Winston Churchill High School, Calgary.
2001/2002	**Alex Fink**	Queen Elizabeth Jr/Sr High School, Calgary.

2002/2003	**R. Barrington Leigh**	Old Scona Academic High School, Edmonton.
2003/2004	**Dennis Chuang**	Strathcona Tweedsmuir School, Okotoks.
2004/2005	**Malka Wrigley**	Old Scona Academic High School, Edmonton.
2005/2006	**Boris Braverman**	Sir Winston Churchill High School, Calgary.
	Jeffrey Mo	William Aberhart High School, Calgary.

Second Prizes:

1983/1984	**Peter Newnham**	Old Scon Academic High School, Edmonton.
1985/1986	**Anant Murthy**	Old Scona Academic High School, Edmonton.
1986/1987	**David Joffe**	Western Canada High School, Calgary.
	Kevin Kornelson	Medicine Hat High School, Medicine Hat.
1987/1988	**Bartek Muszynski**	William Aberhart High School, Calgary.
1989/1990	**Dean Anderson**	Western Canada High School, Calgary.
1991/1992	**Joshua Lagrange**	Ecole Secondaire Beaumont, Beaumont.
1992/1993	**Anthony Fok**	St. Francis Xavier High School, Edmonton.
1994/1995	**Byung-Kyu Chun**	Harry Ainlay High School, Edmonton.
1995/1996	**Daniel Robbins**	Ecole Secondaire Beaumont, Beaumont.
1996/1997	**Benjamin Mirus**	Strathcona High School, Edmonton.
1999/2000	**James Penske**	Western Canada High School, Calgary.
	Chris Samuel	Harry Ainlay High School, Edmonton.
2001/2002	**R. Barrington Leigh**	Old Scona Academic High School, Edmonton.
2002/2003	**Boris Braverman**	Branton Junior High School, Calgary.
2003/2004	**Boris Braverman**	Simon Fraser Junior High School, Calgary.
	Jerry Lo	Vernon Barford Junior High School, Edmonton.
	David Rhee	McNally High School, Edmonton.
	Ken Zhang	Western Canada High School, Calgary.
	Hongyi Li	Sir Winston Churchill High School, Calgary.
	Peter Zhang	Sir Winston Churchill High School, Calgary.
2004/2005	**Ken Zhang**	Western Canada High School, Calgary.

Third Prizes:

1983/1984	**Andy Jenkins**	Harry Ainlay Composite High School, Edmonton.
1984/1985	**Roy Maltby**	Bishop Carroll High School, Calgary.
	Richard Rush	Western Canada High School, Calgary.
1985/1986	**Najib Ayas**	Old Scona Academic High School, Edmonton.
1987/1988	**Sam Maltby**	Bishop Carroll High School, Calgary.
1988/1989	**Michael Roy**	Salisbury Composite High School, Sherwood Park.
	David Koch	Strathcona Composite High School, Edmonton.
	Marc Mulligan	Salisbury Composite High School, Sherwood Park.
1989/1990	**David Adams**	Strathcona Composite High School, Edmonton.
	Miroslav Oballa	Western Canada High School, Calgary.
	Michael Roy	Salisbury Composite High School, Sherwood Park.
	James Lee	Wetaskiwin Composite High School, Wetaskiwin.
	Joni Walker	Sir Winston Churchill High School, Calgary.

1990/1991	**Matt Fenwick**	Sir Winston Churchill High School, Calgary.
1991/1992	**Jason Colwell**	Old Scona Academic High School, Edmonton.
1992/1993	**Cyrus Master**	Western Canada High School, Calgary.
1993/1994	**Alaister Savage**	Western Canada High School, Calgary.
1994/1995	**Daniel Robbins**	Ecole Secondaire Beaumont, Beaumont.
	Mathew Wong	Old Scona Academic High School, Edmonton.
1995/1996	**Derek Kisman**	Queen Elizabeth High School, Calgary.
1996/1997	**Shelby Haque**	Sir Winston Churchill High School, Calgary.
1998/1999	**Michael Busheikin**	Western Canada High School, Calgary.
2000/2001	**Sumudo Fernando**	Harry Ainlay High School, Edmonton.
	Kunyoung Kim	Glenmary High School, Peace River.
2001/2002	**Peter Du**	Sir Winston Churchill High School, Calgary.
	Sumudo Fernando	Harry Ainlay High School, Edmonton.
2002/2003	**Nathanael Wu**	Western Canada High School, Calgary.
2004/2005	**Zheng Guo**	Western Canada High School, Calgary.
	Yakov Shklarov	Henry Wise Wood High School, Calgary.
2005/2006	**David Rhee**	McNally Composite High School, Edmonton.

Grade 11 Prizes:

1983/1984	**Grant Parks**	Harry Ainlay Composite High School, Edmonton.
1984/1985	**Geoffrey Haynes**	Tempo School, Edmonton.
	David Peters	Western Canada High School, Calgary.
	Vladimir Zhivov	Western Canada High School, Calgary.
1985/1986	**Kenneth Ng**	Bishop Carroll High School, Calgary.
1986/1987	**Chris Nichols**	Ross Sheppard Composite High School, Edmonton.
1987/1988	**Linda Zhao**	Western Canada High School, Calgary.
1988/1989	**Teviet Creighton**	Western Canada High School, Calgary.
	Adon Crook	Harry Ainlay Composite High School, Edmonton.
1989/1990	**Keith Silva**	Archbishop MacDonald High School, Edmonton.
1990/1991	**Ozzie Gelbord**	Western Canada High School, Calgary.
1991/1992	**Bryant Swanson**	Western Canada High School, Calgary.
1992/1993	**John Roy**	Salisbury Composite High School, Sherwood Park.
1993/1994	**Andrew Klug**	Western Canada High School, Calgary.
1994/1995	**Jennifer Gordon**	Western Canada High School, Calgary.
1995/1996	**Cynthia Luk**	Old Scona Academic High School, Edmonton.
1996/1997	**Simon Chan**	Sir Winston Churchill High School, Calgary.
1997/1998	**Filip Crnogorac**	Western Canada High School, Calgary.
	Roger Huang	Western Canada High School, Calgary.
1998/1999	**Jack Chen**	Sir Winston Churchill High School, Calgary.
	Steven Lyster	Lorne Jenken High School, Barrhead.
1999/2000	**Jarett Prouse**	Old Scona Academic High School, Edmonton.
2000/2001	**Alvin Tan**	McNally High School, Edmonton.
2001/2002	**Charles Li**	Western Canada High School, Calgary.
2002/2003	**Dennis Chuang**	Strathcona-Tweedsmuir School, Okotoks.

2003/2004	**Polly Han**	Western Canada High School, Calgary.
2004/2005	**Gary Huang**	Sir Winston Churchill High School, Calgary.
2005/2006	**Graham Hill**	Sir Winston Churchill High School, Calgary.

Grade 10 Prizes:

1983/1984	**Robert Simard**	Bishop Grandin High School, Calgary.
1984/1985	**Rodney Gitzel**	Spruce Grove Composite High School, Spruce Grove.
1985/1986	**Ritchie Annand**	Western Canada High School, Calgary.
1986/1987	**Jamie Penney**	Ecole J. H. Picard, Edmonton.
1987/1988	**Siku Adam**	Harry Ainlay Composite High School, Edmonton.
	Jennifer Brockway	James Fowler Senior High School, Calgary.
1988/1989	**Michael Coward**	Old Scona Academic High School, Edmonton.
1989/1990	**Calvin Li**	Archbishop MacDonald High School, Edmonton.
1990/1991	**Robert Kry**	Western Canada High School, Calgary.
1991/1992	**Anthony Fok**	St. Francis Xavier High School, Edmonton.
1992/1993	**Paula Ayer**	Western Canada High School, Calgary.
1993/1994	**Scott Ireland**	Old Scona Academic High School, Edmonton.
	Brian Yeh	Western Canada High School, Calgary.
1994/1995	**Chi Hoang**	Bishop Carroll High School, Calgary.
	Talib Rajwani	Old Scona Academic High School, Edmonton.
1995/1996	**Joel Runka**	Western Canada High School, Calgary.
1996/1997	**Sonny Chan**	Western Canada High School, Calgary.
1997/1998	**Titus Yeung**	Harry Ainlay High School, Edmonton.
1998/1999	**Ryan Wall**	Bishop Carroll High School, Calgary.
1999/2000	**Peter Du**	Sir Winston Churchill High School, Calgary.
2000/2001	**Jeffrey Mo**	St. Paul's Academy, Okotoks.
2001/2002	**Nathanael Wu**	Western Canada High School, Calgary.
2002/2003	**Steven Melenchuk**	Henry Wise Wood High School, Calgary.
	Eric Tran	Western Canada High School, Calgary.
2003/2004	**Adrian Keet**	Westmount Charter School, Calgary.
2004/2005	**Boris Braverman**	Sir Winston Churchill High School, Calgary.
2005/2006	**Michael Wong**	Tempo School, Edmonton.

Pacific Institute for the Mathematical Sciences Special Prizes:

1992/1993	**Steven Laffin**	Ecole J. H. Picard, Edmonton.
1998/1999	**R. Barrington Leigh**	Vernon Barford Junior High School, Edmonton.
	Alex Fink	Queen Elizabeth Jr/Sr High School, Calgary.
1999/2000	**Jeffrey Mo**	University Elementary School, Calgary, (Grade 5) .
2002/2003	**Sarah Sun**	Holy Trinity Academy, Okotoks.

Zone I First Prizes:

1983/1984	**John Abraham**	John G. Diefenbaker High School, Calgary.
1984/1985	**Thomas Quinn**	Bishop Carroll High School, Calgary.

1985/1986	**Peter Gibson**	Western Canada High School, Calgary.
1986/1987	**Ritchie Annand**	Western Canada High School, Calgary.
1987/1988	**Stephen Chen**	Sir Winston Churchill High School, Calgary.
1988/1989	**Jason Nicholson**	Sir Winston Churchill High School, Calgary.
1989/1990	**Aaron Pollock**	Western Canada High School, Calgary.
	Tova Tanenbaum	Western Canada High School, Calgary.
1990/1991	**Reed Ball**	Lord Beaverbrook High School, Calgary.
1991/1992	**Jan Rubak**	Sir Winston Churchill High School, Calgary.
1992/1993	**Bryant Swanson**	Western Canada High School, Calgary.
1993/1994	**William Hu**	Sir Winston Churchill High School, Calgary.
1994/1995	**Daniel Glin**	Western Canada High School, Calgary.
1995/1996	**Chi Hoang**	Bishop Carroll High School, Calgary.
1996/1997	**David Cameron**	Western Canada High School, Calgary.
1997/1998	**Joshua Grosse**	Western Canada High School, Calgary.
1998/1999	**Danny Ng**	Western Canada High School, Calgary.
1999/2000	**Jack Chen**	Sir Winston Churchill High School, Calgary.
2000/2001	**Gary Seto**	Western Canada High School, Calgary.
2001/2002	**Alec Mills**	Western Canada High School, Calgary.
2002/2003	**Jan Owoc**	St. Francis High School, Calgary.
2003/2004	**Yiyi Yang**	Western Canada High School, Calgary.
2004/2005	**Eugene Sit**	Queen Elizabeth Jr/Sr High School, Calgary.
	Pang Song	Western Canada High School, Calgary.
2005/2006	**Yiyi Yang**	Western Canada High School, Calgary.

Zone I Second Prizes:

1983/1984	**Michelle Milne**	Western Canada High School, Calgary.
	Andreas Vollmerhaus	Sir Winston Churchill High School, Calgary.
1984/1985	**Richard Kadonaga**	Sir Winston Churchill High School, Calgary.
	Lynn Kondo	Sir Winston Churchill High School, Calgary.
1985/1986	**David Peters**	Western Canada High School, Calgary.
1986/1987	**Bartek Muszynski**	William Aberhart Senior High School, Calgary.
	Kevin Foltinek	Sir Winston Churchill Senior High School, Calgary.
1987/1988	**Albert Yoon**	Sir Winston Churchill High School, Calgary.
1988/1989	**Heidi Petersen**	Sir Winston Churchill High School, Calgary.
	Craig Story	D. E. P. Scarlett High School, Calgary.
1990/1991	**Aaron Pollack**	Western Canada High School, Calgary.
1991/1992	**James Baughan**	Western Canada High School, Calgary.
	Rahim Hirji	Western Canada High School, Calgary.
1992/1993	**Robin Damm**	Western Canada High School, Calgary.
1993/1994	**Peter Hwang**	Sir Winston Churchill High School, Calgary.
	Eugene Shih	Western Canada High School, Calgary.
1994/1995	**Gavin Duggan**	Western Canada High School, Calgary.
	Ammon Piepgrass	Western Canada High School, Calgary.
1995/1996	**Chad To**	Western Canada High School, Calgary.

1996/1997	**Matthew Ford**	Dr. E. P. Scarlett High School, Calgary.
	Scott Ure	Western Canada High School, Calgary.
	Stephen Yuen	Sir Winston Churchill High School, Calgary.
	Thomas Halford	Western Canada High School, Calgary.
	Mark Kwan	Western Canada High School, Calgary.
1997/1998	**Sonny Chan**	Western Canada High School, Calgary.
	Wilson Chan	Western Canada High School, Calgary.
1998/1999	**Dennis Bang**	Western Canada High School, Calgary.
1999/2000	**Jason Chang**	Western Canada High School, Calgary.
2000/2001	**Bob Cao**	Sir Winston Churchill High School, Calgary.
	Keith Chung	Western Canada High School, Calgary.
	Alex Kominek	Henry Wise Wood High School, Calgary.
	Samir Pradhan	Sir Winston Churchill High School, Calgary.
2001/2002	**Eric Nause**	Dr. E. P. Scarlett High School, Calgary.
2002/2003	**Bob Cao**	Sir Winston Churchill High School, Calgary.
2003/2004	**Selena Huang**	Sir Winston Churchill High School, Calgary.
	Nathanael Wu	Western Canada High School, Calgary.
2005/2006	**Pang Song**	Western Canada High School, Calgary.

Zone II First Prizes:

1983/1984	**Lisa Briosi**	Crescent Heights High School, Medicine Hat.
1984/1985	**Montgomery Simus**	Springbank Community High School, Calgary.
1985/1986	**Tracy Sutela**	Camille J. Lerouge Collegiate, Red Deer.
1986/1987	**Norichika Okada**	Lethbridge Collegiate Institute, Lethbridge.
1987/1988	**Jennifer Spanbauer**	St. Mary's School, Taber.
1988/1989	**Gregory Letal**	Olds Junior-Senior High School, Olds.
	Steven Nygard	Carbon School, Carbon.
1989/1990	**Travis Smith**	Strathcona Tweedsmuir School, Okotoks.
1990/1991	**Marc Lim**	St. Mary's School, Taber.
1991/1992	**Neil Kennedy**	Didsbury High School, Didsbury.
1992/1993	**Chris Nissen**	Spruce View High School, Spruce View.
1993/1994	**Rayhan Behin**	Cochrane High School, Cochrane.
	Andrew Thomson	Lacombe High School, Lacombe.
1994/1995	**Michael Rix**	Camille J. Lerouge Collegiate, Red Deer.
1995/1996	**Keng In Yu**	Winston Churchill High School, Lethbridge.
1996/1997	**Steve Dubbelboer**	Winston Churchill High School, Lethbridge.
1997/1998	**Ryan Kerner**	J. T. Foster High School, Nanton.
1998/1999	**Ryan Bolinger**	Strathmore High School, Strathmore.
	Sean Borchert	Crowsnest Consolidated School, Coleman.
	Graham Nelson	Olds Koinonia Christian School, Olds.
1999/2000	**Landis Stankievech**	Trochu Valley School, Trochu.
2000/2001	**Sarah Sun**	St. Mary's School, Okotoks.
2001/2002	**Meghan Shan**	Winston Churchill High School, Lethbridge.
2002/2003	**Lily Liu**	Winston Churchill High School, Lethbridge.

2003/2004	**Sarah Sun**	Holy Trinity Academy, Okotoks.
2004/2005	**Sarah Sun**	Holy Trinity Academy, Okotoks.
2005/2006	**Sarah Sun**	Holy Trinity Academy, Okotoks.

Zone II Second Prizes:

1983/1984	**Gwen Alison**	Delburne Centralized School, Delburne.
	Paul Thomsen	Olds Junior-Senior High School, Olds.
1984/1985	**Lisa Hunter**	Springbank Community High School, Calgary.
1985/1986	**Claude Daigle**	Lethbridge Collegiate Institute, Lethbridge.
1986/1987	**David Stewart**	Lethbridge Collegiate Institute, Lethbridge).
1987/1988	**Lori Polanchek**	Senator Riley High School, High River.
1989/1990	**Michelle Hof**	McCoy High School, Medicine Hat.
	Kim Rubak	Cochrane High School, Cochrane.
1990/1991	**Craig Sellars**	Winston Churchill High School, Lethbridge.
	Cory Pregoda	Eckville Junior-Senior High School, Eckville.
	Murray Robinson	Acme School, Acme.
1991/1992	**Karen Widish**	Didsbury High School, Didsbury.
1992/1993	**Heather Cseke**	St. Mary's School, Taber.
1994/1995	**David Good**	Lacombe High School, Lacombe.
1995/1996	**Anthony Mills**	Prairie High School, Three Hills.
1996/1997	**Samuel Conard**	Winston Churchill High School, Lethbridge.
1997/1998	**Jalilian Ehsan**	Strathcona-Tweedsmuir School, Okotoks.
1999/2000	**Paul Tarjan**	Springbank School, Calgary.
2000/2001	**Landis Stankievech**	Trochu Valley School, Trochu.
	Christina Johnston	Canmore School, Canmore.
2001/2002	**David Cunningham**	Oilfields Jr/Sr High School, Black Diamond.
2002/2003	**Sunimal Fernando**	St. Paul's Academy, Okotoks.
	Ben Stevens	Prairie High School, Three Hills.
2003/2004	**Lily Liu**	Winston Churchill High School, Lethbridge.
2004/2005	**Brad Kruse**	New Norway School, New Norway.
2005/2006	**David Liu**	Winston Churchill High School, Lethbridge.

Zone III First Prizes:

1983/1984	**Rick Neuls**	Archbishop MacDonald High School, Edmonton.
1984/1985	**Naomi Makins**	Tempo School, Edmonton.
1985/1986	**Lola Sim**	Harry Ainlay Composite High School, Edmonton.
1986/1987	**Jean Duteau**	Archbishop MacDonald High School, Edmonton.
1987/1988	**Graham Denham**	Old Scona Academic High School, Edmonton.
1988/1989	**Walter Lai**	St. Joseph Composite High School, Edmonton.
	Jennifer Zou	Harry Ainlay Composite High School, Edmonton.
1989/1990	**Mark Fokema**	Harry Ainlay Composite High School, Edmonton.
1990/1991	**Alan Hughes**	Old Scona Academic High School, Edmonton.
	William Lee	Harry Ainlay Composite High School, Edmonton.
1991/1992	**Calvin Li**	Archbish-op MacDonald High School, Edmonton.

1992/1993	**Joseph Modayil**	Ross Sheppard Composite High School, Edmonton.
1993/1994	**Eric Finley**	Harry Ainlay High School, Edmonton.
	Anthony Fok	St. Francis Xavier High School, Edmonton.
1994/1995	**Lei Jia**	Harry Ainlay High School, Edmonton.
1995/1996	**Zhiqi Zhong**	St. Luke's College, Edmonton.
1996/1997	**Robert Lutz**	Harry Ainlay High School, Edmonton.
1997/1998	**Frank Chen**	Harry Ainlay High School, Edmonton.
	Shane McNalley	Strathcona High School, Edmonton.
1998/1999	**Frank Chen**	Harry Ainlay High School, Edmonton.
1999/2000	**Ryan Vogt**	Old Scona Academic High School, Edmonton.
2000/2001	**Tze Luck Chia**	M. E. Lazerte High School, Edmonton.
2001/2002	**Richard Ng**	Archbishop MacDonald High School, Edmonton.
2002/2003	**Geoff Lywood**	Archbishop MacDonald High School, Edmonton.
2003/2004	**Qu Chen**	Harry Ainlay High School, Edmonton.
2004/2005	**Michael Wong**	Tempo School, Edmonton.
2005/2006	**Brian Yu**	Old Scona Academic High School, Edmonton.

Zone III Second Prizes:

1983/1984	**Terry Liu**	Old Scona Academic High School, Edmonton.
	Kenneth Melax,	Harry Ainlay Composite High School, Edmonton.
	Andrew Stephenson	Old Scona Academic High School, Edmonton.
1984/1985	**Anant Murthy**	Old Scona Academic High School, Edmonton.
1985/1986	**Geoffrey Haynes**	Tempo Secondary School, Edmonton.
1986/1987	**Lianne Durocher**	Ecole J. H. Picard, Edmonton.
	Andrew Burton	Harry Ainlay Composite High School, Edmonton.
1987/1988	**Xiaochang Cheng**	Harry Ainlay Composite High School, Edmonton.
1989/1990	**Amina Danial**	Old Scona Academic High School, Edmonton.
	Dennis Wong	Ross Sheppard Composite High School, Edmonton.
1991/1992	**Gilbert Lai**	Archbishop MacDonald High School, Edmonton.
	Steven Taschuk	Victoria Composite High School, Edmonton.
1992/1993	**Theresa Winski**	St. Francis Xavier High School, Edmonton.
	Tin-Yau Kwan	St. Luke's High School, Edmonton.
	Merwin Siu	Old Scona Academic High School, Edmonton.
1994/1995	**Dean Sohnle**	McNally High School, Edmonton.
1995/1996	**Hubert Chan**	Archbishop MacDonald High School, Edmonton.
	Dean Sohnle	McNally High School, Edmonton.
1996/1997	**Philip Stein**	Ross Sheppard High School, Edmonton.
1998/1999	**Anton Cherney**	Old Scona Academic High School, Edmonton.
1999/2000	**Alvin Ho**	Old Scona Academic High School, Edmonton.
2000/2001	**Robert Barrington Leigh**	Old Scona Academic High School, Edmonton.
2001/2002	**Alvin Tan**	McNally High School, Edmonton.
2002/2003	**Paul Atkins**	Jasper Place High School, Edmonton.

2003/2004	**Malka Wrigley**	Old Scona Academic High School, Edmonton.
2004/2005	**Brent Thompson**	Tempo School, Edmonton.
2005/2006	**Xi Chen**	Hary Ainley High School, Edmonton.

Zone IV First Prizes:

1983/1984	**Sylvia Haener**	St. Patrick's High School, Yellowknife, N.W.T..
1984/1985	**Emma Barnes**	Grand center High School, Grand center.
1985/1986	**Bruce Johnson**	Grand center High School, Grand center.
1986/1987	**Aaron Humphrey**	Grande Prairie Composite High School, Grande Prairie.
1987/1988	**Keriley Romanufa**	Archbishop Jordan High School, Sherwood Park.
1988/1989	**Tony Bayduza**	F. G. Miller Junior-Senior High School, Elk Point.
1989/1990	**David Bullas**	Leduc Composite High School, Leduc.
1990/1991	**Michael Roy**	Salisbury Composite High School, Sherwood Park.
1991/1992	**Toby Heinrichs**	Grand Trunk High School, Grand Trunk.
1992/1993	**David Dilworth**	Salisbury Composite High School, Sherwood Park.
	Richard Yeomans	Salisbury Composite High School, Sherwood Park.
1993/1994	**Doug Rae**	Lloydminster Comprehensive High School, Lloydminster.
1994/1995	**Colette Fluet**	Lorne Jenken High School, Barrhead.
1995/1996	**Jonathan Backer**	Central Peace School, Spirit River.
1996/1997	**Laura Harms**	Lorne Jenken High School, Barrhead.
1997/1998	**Kevin Geddert**	Father Patrick Mercredi School, Fort McMurray.
1998/1999	**Michael Smith**	Paul Kane High School, St. Albert.
1999/2000	**Matt Larocque**	St. Albert High School, St. Albert.
2000/2001	**Neil Pandya**	Salisbury High School, Sherwood Park.
2001/2002	**Colin Wilkbur**	Paul Kane High School, St. Albert.
2002/2003	**Colin Wilkbur**	Paul Kane High School, St. Albert.
2003/2004	**Ruetz Nathen**	St. Johns School of Alberta, Stony Plain.
2004/2005	**Graham Hill**	Bev Facey Community High School, Sherwood Park.
2005/2006	**Eliot Buchanan**	Salisbury Composite High School, Sherwood Park.

Zone IV Second Prizes:

1983/1984	**Randy Saunders**	Sturgeon Composite High School, Namao.
1984/1985	**Rhonda Metrunec**	New Myrnam High School, Myrnam.
1985/1986	**Michael Roth**	Fort McMurray Composite High School, Fort McMurray.
1986/1987	**Greg Hackman**	Sturgeon Composite High School, Namao.
1987/1988	**Michael Fisher**	Salisbury Composite High School, Sherwood Park.
1988/1989	**Chris Harrison**	Salisbury Composite High School, Sherwood Park.
	Ian Harrison	St. Albert High School, St. Albert.
	Kendal Seaton	Sturgeon Composite High School, Namao.
	Christopher Sudyk	Lamont Junior-Senior High School, Lamont.

1989/1990	**Charlie Kim**	Salisbury Composite High School, Sherwood Park.
	Angela McCormick	Lorne Jenken High School, Barrhead.
	Kevin Romanchuk	Lorne Jenken High School, Barrhead.
	Basil Vandegriend	Salisbury Composite High School, Sherwood Park.
1990/1991	**Karin Lu**	Salisbury Composite High School, Sherwood Park.
1991/1992	**Shawn Loewen**	Salisbury Composite High School, Sherwood Park.
	Greg Ritter	Lorne Jenken High School, Barrhead.
1993/1994	**Daniel Robbins**	Ecole Secondaire Beaumont, Beaumont.
	Dylan George	Paul Kane High School, St. Albert.
1994/1995	**Rich Kreuger**	Salisbury High School, Sherwood Park.
1995/1996	**Maurice Bujold**	Lorne Jenken High School, Barrhead.
1996/1997	**Ian Dmytrash**	Lamont Secondary School, Lamont.
1997/1998	**Cameron Barr**	Paul Kane High School, St. Albert.
	Erik Poelzer	Harry Collinge High School, Hinton.
	Derek Williams	Harry Collinge High School, Hinton.
1998/1999	**Spencer Giffin**	Calmar School, Calmar.
1999/2000	**Kunyoung Kim**	Glenmary High School, Peace River.
2000/2001	**Christopher Lerohl**	Fox Creek School, Fox Creek.
2001/2002	**Stephen Arnason**	Salisbury High School, Sherwood Park.
2002/2003	**Andrew Hulleman**	Father Patrick Mercredi Community School, Fort McMurray.
	Veronica MacInnis	Archbishop Jordan High School, Sherwood Park.
	Matthew McKeown	Assumption Jr/Sr High School, Grand center.
2003/2004	**Sean Normandeau**	Archbishop Jordan School, Sherwood Park.
2004/2005	**Matthew Reynolds**	Paul Kane High School, St. Albert.
2005/2006	**Megan Chartrand**	Archbishop Jordon High School, Sherwood Park.
	Michael Schaub	J.A. Williams High School, Lac La Biche.

Team Prizes

First Prizes—Peter H. Denham Memorial Plaques:

1983/1984 **Archbishop MacDonald High School**, Edmonton, with **Rick Neuls, Jeff Candy** and **Derek Brenneis**.

1984/1985 **Western Canada High School**, Calgary, with **Richard Rush, David Peters** and **Vladimir Zhivov**.

1985/1986 **Old Scona Academic High School**, Edmonton, with **Anant Murthy, Najib Ayas** and **Jean Rochet**, managed by **Mr. R. Fizzell**.

1986/1987 **Ecole J. H. Picard**, Edmonton, with **Mike Moser, Jamie Penney** and **Lianne Durocher**, managed by **Mr. V. Bratu**.

1987/1988 **Sir Winston Churchill High School**, Calgary, with **Stephen Chen, Albert Yoon** and **David Hwang**, managed by **Mr. D. Cantrill**.

1988/1989 **Sir Winston Churchill High School**, Calgary, with
Mischa Hooker, John Yoon and **Jason Nicholson**,
managed by **Mr. D. Cantrill**.

1989/1990 **Western Canada High School**, Calgary, with
Ozzie Gelbord, Dean Anderson and **Miroslav Oballa**,
managed by **Mr. M. Milner**.

1990/1991 **Western Canada High School**, Calgary, with
Rahim Hirji, Aaron Pollack and **Rhys Yarranton**,
managed by **Mr. M. Milner**.

1991/1992 **Western Canada High School**, Calgary, with
Robert Kry, Bryant Swanson and **J. Baughan/R. Hirji**,
managed by **Mr. M. Milner**.

1992/1993 **Sir Winston Churchill High School**, Calgary, with
Peter Hwang, Douglas Puzzie and **Jennifer Winnitoy**,
managed by **Mr. D. Cantrill**.

1993/1994 **Western Canada High School**, Calgary, with
Alaister Savage, Eugene Shih and **Mari Sampei**,
managed by **Ms. Diane Barry**.

1994/1995 **Queen Elizabeth High School**, Calgary, with
Derek Kisman, Chris Haines and **Vienna Ng**,
managed by **Mr. Jim Byrne**.

1995/1996 **Harry Ainlay High School**, Edmonton, with
Byung-Kyu Chun, Vivian Yu and **Sudhakar Sivapalan**,
managed by **Mr. Lorne Lindenberg**.

1996/1997 **Harry Ainlay High School**, Edmonton, with
Byung-Kyu Chun, Robert Lutz and **Gary Lai**,
managed by **Mr. Lorne Lindenberg**.

1997/1998 **Western Canada High School**, Calgary, with
Brian Tsai and **Filip Crnogorac/Roger Huang/Joshua Grosse**,
managed by **Ms. Evelyn Grosse**.

1998/1999 **Western Canada High School**, Calgary,
with **Sonny Chan, Peter Dziegielewski** and **Michael Busheikin**,
managed by **Mrs. Hazel Williams**.

1999/2000 **Old Scona Academic High School**, Edmonton,
with **Rowena Luk, Jarett Prouse** and **Ryan Vogt**,
managed by **Mr. Lorne Pascoe**.

2000/2001 **Sir Winston Churchill High School**, Calgary,
with **Peter Du, Bob Cao** and **Samir Pradhan**,
managed by **Mr. Dennis Cantrill**.

2001/2002 **Queen Elizabeth Jr/Sr High School**, Calgary,
with **Alex Fink, Sonny Yue** and **Andy Liu**,
managed by **Ms. Sharon Reid**.

2002/2003 **Western Canada High School**, Calgary,
with **Nathanael Wu, Gordon Tam** and **Ye Ly Lin**,
managed by **Ms. Renata Fitzner**.

2003/2004 **Sir Winston Churchill High School**, Calgary,
with **Hongyi Li, Peter Zhang** and **Selena Huang**,
managed by **Mr. Patrick Ancelin**.

2004/2005 **Western Canada High School**, Calgary,
with **Ken Zhang, Zheng Guo** and **Pang Song**,
managed by **Mrs. Renata Delisle**.

2005/2006 **Sir Winston Churchill High School**, Calgary,
with **Boris Braverman, Graham Hill** and **Linzhou Fang**,
managed by **Mr. Patrick Ancelin**.

Second Prizes:

1983/1984 **Harry Ainlay Composite High School**, Edmonton, with
Andy Jenkins, Randy Pawluk and **Alan Williams**.

1984/1985 **Tempo School**, Edmonton, with
Naomi Makins, Geoffrey Haynes and **Alexander Shetsen**.

1985/1986 **Western Canada High School**, Calgary, with
Peter Gibson, David Peters and **Ritchie Annand**,
managed by **Mrs. L. MacRae**.

1986/1987 **Western Canada High School**, Calgary, with
David Joffe, Ritchie Annand and **Nora Sleumer**,
managed by **Mrs. L. MacRae**.

1987/1988 **Western Canada High School**, Calgary, with
Ritchie Annand, Ian Wright and **Linda Zhao**,
managed by **Mr. M. Milner**.

1988/1989 **Salisbury Composite High School**, Sherwood Park, with
Michael Roy, Marc Mulligan and **Chris Harrison**,
managed by **Mr. R. Broemling**.

1989/1990 **Sir Winston Churchill High School**, Calgary, with
Joni Walker, Deidre Sorensen and **Roger Carbol**,
managed by **Mr. D. Cantrill**.

1990/1991 **Old Scona Academic High School**, Edmonton, with
Jason Colwell, Alan Hughes and **Taha Taher**,
managed by **Mr. L. Pascoe**.

1991/1992 **Sir Winston Churchill High School**, Calgary, with
Jan Rubak, Essam Metwally and **Sean Corbett**,
managed by **Mr. D. Cantrill**.

1992/1993 **Western Canada High School**, Calgary, with
Cyrus Master, Bryant Swanson and **Robin Damm**,
managed by **Mr. M. Milner**.

1993/1994 **Sir Winston Churchill High School**, Calgary, with
William Wu, Peter Hwang and **Christopher Davies**,
managed by **Mr. Dennis Cantrill**.

1994/1995 **Harry Ainlay High School**, Edmonton, with
Byung-Kyu Chun, Lei Jia and **Michael Lin**,

	managed by **Mr. Lorne Lindenberg**.
	Western Canada High School, Calgary, with
	Jennifer Gordon, Gavin Duggan and **Daniel Glin**,
	managed by **Ms. Diane Barry**.
1995/1996	**Western Canada High School**, Calgary, with
	Chad To, Jaisel Vadgama and **Mark Kwan**,
	managed by **Mr. Mark Milner**.
1996/1997	**Sir Winston Churchill High School**, Calgary, with
	Shelby Haque, Simon Chan and **Paul Chen**,
	managed by **Mr. Dennis Cantrill**.
1997/1998	**Harry Ainlay High School**, Edmonton, with
	Frank Chen, Titus Yeung and **Alex Fung**,
	managed by **Mr. Lorne Lindenberg**.
1998/1999	**Sir Winston Churchill High School**, Calgary,
	with **Michael James, Jamie Batuwantudawe** and **Jack Chen**,
	managed by **Dennis Cantrill**.
1999/2000	**Western Canada High School**, Calgary,
	with **James Fenske, Jason Chang** and **Joel Jackson**,
	managed by **Mrs. Hazel Williams**.
2000/2001	**Queen Elizabeth Jr/Sr High School**, Calgary,
	with **Alex Fink, Sonny Yue** and **Erika Harrison**,
	managed by **Ms. Sharon Reid**.
2001/2002	**Western Canada High School**, Calgary,
	with **Chalres Li, Alec Mills** and **Keith Chung**,
	managed by **Mrs. Hazel Williams**.
2002/2003	**Sir Winston Churchill High School**, Calgary,
	with **Bob Cao, Peter Zhang** and **Hongyi Li**,
	managed by **Mr. Patrick Ancelin**.
2003/2004	**Western Canada High School**, Calgary,
	with **Ken Zhang, Polly Han** and **Yiyi Yang**,
	managed by **Mrs. Renata Delisle**.
2004/2005	**Old Scona Academic High School**, Edmonton,
	with **Malka Wrigley, Jennifer Lo** and **Brian Yu**,
	managed by **Mr. Lorne Pascoe**.
2005/2006	**Western Canada High School**, Calgary,
	with **Yiyi Yang, Song Pang** and **J. Liu/E. Cui/L. Yu/K. Boone**,
	managed by **Mrs. Renata Delisle**.

Third Prizes:

1983/1984	**Old Scona Academi High School**, Edmonton, with
	Andrew Stephenson, Jun Kawashima and **Atul Malhotra**.
1984/1985	**Sir Winston Churchill High School**, Calgary, with
	Richard Kadonaga, Lynn Kondo and **Sherry Weaver**.

1985/1986 **Lorne Jenken High School**, Barrhead, with
Russell Schulz, Kelly Stratford and **F. Diekmann/H. Wemekamp**,
managed by **Mr. H. Schabert**.

1986/1987 **Sir Winston Churchill Senior High School**, Calgary, with
Kevin Foltinek, Stephen Chen and **David Hwang**,
managed by **Mr. D. Cantrill**.

1987/1988 **Ross Sheppard Composite High School**, Edmonton, with
Chris Nichols, Kurt Tober and **Offir Spanglet**,
managed by **Mr. E. Wasylyk**.

1988/1989 **Strathcona Composite High School**, Edmontonr with
David Koch, Richard Wan and **Carl Kovithavongs**,
managed by **Ms. J. Frost**.

1989/1990 **Strathcona Composite High School**, Edmonton, with
David Adams, Gordon Lipford and **Carl Kovithavongs**,
managed by **Ms. J. Forst**.

1990/1991 **Harry Ainlay Composite High School**, Edmonton, with
William Lee, Patrick Chan and **Simon Wong**,
managed by **Mr. L. Lindenberg**.

1991/1992 **Archbishop MacDonald High School**, Edmonton, with
Calvin Li, Gilbert Lai and **Gregory Charrois**,
managed by **Mr. H. Marcuk**.

1992/1993 **Salisbury Composite High School**, Sherwood Park, with
John Roy, David Dilworth and **Richard Yeomans**,
managed by **Mrs. T. Baumgartner**.

1993/1994 **Old Scona Academic High School**, Edmonton, with
Rock Leung, Jeremy Sit and **Matthew Wong**,
managed by **Mr. Lorne Pascoe**.

1995/1996 **Bishop Carroll High School**, Calgary, with
Chi Hoang, Denise Garvey and **Robert Behm**,
managed by **Mrs. Pat McManus**.

1996/1997 **Western Canada High School**, Calgary, with
David Cameron, Scott Ure and **T. Halford/M. Kwan**,
managed by **Mrs. Hazel Williams**.

1997/1998 **Bishop Carroll High School**, Calgary, with
Eddie Ng, Russell Ford and **F. Ng/L. Maier**,
managed by **Ms. Susan Osterkampf**.
Dr. E. P. Scarlett High School, Calgary, with
Matthew Ford, Tim Mark and **Jason Vanderzwaag**,
managed by **Mr. Mark Milner**.

1998/1999 **Harry Ainlay High School**, Edmonton,
with **Frank Chen, Titus Yeung** and **Alex Fung**,
managed by **Mr. Lorne Lindenberg**.

1999/2000 **Sir Winston Churchill High School**, Calgary,
with **Peter Du, Jack Chen** and **Yibin Guo**,
managed by **Mr. Dennis Cantrill**.

2000/2001 **Western Canada High School**, Calgary,
with **Gary Seto, Keith Chung** and **Shannon Long**,
managed by **Mrs. Hazel Williams**.

2001/2002 **Old Scona Academic High School**, Edmonton,
with **Robert Barrington Leigh, Vu Ly** and **Darren Lau**,
managed by **Mr. Lorne Pascoe**.

2002/2003 **Old Scona Academic High School**, Edmonton,
with **Robert Barrington Leigh, Jonathan Chi** and **Maria Lee**,
managed by **Mr. Lorne Pascoe**.

2003/2004 **Harry Ainlay High School**, Edmonton,
with **Qu Chen, Boyan Marinov** and **Radoslav Marinov**,
managed by **Mr. Steve Hardy**.

2004/2005 **Sir Winston Churchill High School**, Calgary,
with **Boris Braverman, Gary Huang** and **S. Li/A. Liu**,
managed by **Mr. Patrick Ancelin**.

2005/2006 **Harry Ainlay High School**, Edmonton,
with **Xi Chen, Kristen Climenhaga** and **Shu Kan**,
managed by **Ms. Jacqueline Coulus.**

Zone I Prizes:

1983/1984 **Bishop Carroll High School**, Calgary, with
David Krebes, Arun Lakra and **Marc Kelly**.

1984/1985 **Bishop Carroll High School**, Calgary, with
Roy Maltby, Thomas Quinn and **Kenneth Ng**.

1985/1986 **Bishop Carroll High School**, Calgary, with
Kenneth Ng, Samuel Maltby and **Michael Born**,
managed by **Mrs. P. McManus**.

1986/1987 **Henry Wise Wood High School**, Calgary, with
Kosta Vasilakos, Thomas Yoon and **Daniel Ross**,
managed by **Mr. J. Rogers**.

1987/1988 **Bishop Carroll High School**, Calgary, with
Samuel Maltby, Michael Born and **Nicole Richer**,
managed by **Mrs. P. McManus**.

1988/1989 **Western Canada High School**, Calgary, with
Suresh Pillai, Teviet Creighton and **Paul Malik**,
managed by **Mr. M. Milner**.

1989/1990 **Bishop Carroll High School**, Calgary, with
Sean Monkman, Jocelyn Donnelly and **Michael Papsdorf**,
managed by **Mrs. P. McManus**.

1990/1991 **Sir Winston Churchill High School**, Calgary, with
Matt Fenwick, Shinichi Nakane and **James Kao**,
managed by **Mr. D. Cantrill**.

1991/1992 **St. Mary's Community School**, Calgary, with
Quyen Lam, Kristine Schaus and **C. Braun/G. Leroux/A. Tankard**,
managed by **Mr. G. Dorscher**.

1992/1993 **Henry Wise Wood High School**, Calgary, with
Blake Schroeder, Bernie Sattin and **Cheryl Ning**,
managed by **Mr. J. Rogers**.

1993/1994 **St. Mary's Community High School**, Calgary, with
Ramona Corbiell, Bich Hoang and **David Ng**,
managed by **Ms. Charlotte White**.

1994/1995 **Sir Winston Churchill High School**, Calgary, with
Aditya Bharatha, Oliver Chen and **E. Fung/J. Secord**,
managed by **Mr. Dennis Cantrill**.

1995/1996 **Queen Elizabeth High School**, Calgary, with
Derek Kisman, Mark Peterson and **Erik Fjeldstrom**,
managed by **Mr. Curt Gullacher**.

1996/1997 **St. Mary's Community High School**, Calgary, with
Stephen Yuen, Derek Law and **Erika Dempsey**,
managed by **Ms. Charlotte White**.

1997/1998 **St. Mary's Community High School**, Calgary,
with **Alexander Poda, Robert Chan** and **Kathryn Schrage**,
managed by **Ms. Charlotte White**.

1998/1999 **Henry Wise Wood High School**, Calgary,
with **Erin Feldman** and **Emmi Driedger/Cassandra Johnston/Jeremy Vink**,
managed by **Mr. Stan Bold**.

1999/2000 **Dr. E. P. Scarlett High School**, Calgary,
with **Nadine Danard, Teresa Leung** and **Mike Thompson**,
managed by **Ms. Maureen Mills**.

2000/2001 **Henry Wise Wood High School**, Calgary,
with **Alex Kominek, David Sohn** and **Jeff Gonis**,
managed by **Mr. Stan Bold**.

2001/2002 **Sir Winston Churchill High School**, Calgary,
with **Peter Du, Lisa Chen** and **Nicholas Tam**,
managed by **Mr. Dennis Cantrill**.

2002/2003 **Henry Wise Wood High School**, Calgary,
with **Stephen Melenchuk, Audrey Kertesz** and **Stephan Badragan**,
managed by **Mr. Stan Bold**.

2003/2004 **Henry Wise Wood High School**, Calgary,
with **Stephan Badragan, Audrey Kertesz** and **Stephen Melenchuk**,
managed by **Mr. Stan Bold**.

2004/2005 **Henry Wise Wood High School**, Calgary,
with **Yakov Shklarov, Nathan Adolph** and **Elly Im**,
managed by **Mr. Court Bedford**.

2005/2006 **William Aberhart High School**, Calgary,
with **Jeffrey Mo, Gary Wu** and **S. Miller/D. Johns**,
managed by **Mr. Jim Kotow**.

Zone II Prizes:

1983/1984 **Olds Junior-Senior High School**, Olds, with
Michael Johnson, Wes Petersen and **Annette Woodruff**.

1984/1985 **Springbank Community High School**, Calgary, with
Montgomery Simus, Lisa Hunter and **Paul Fatovich**.

1985/1986 **Camille J. Lerouge Collegiate**, Red Deer, with
Tracy Sutela, Michael Szeto and **Ken Farion**,
managed by **Mr. G. Keogh**.

1986/1987 **Lethbridge Collegiate Institute**, Lethbridge, with
Norichika Okada, David Stewart and **Ka Yin Leung**,
managed by **Mr. B. Haig**.

1987/1988 **Kate Andrews High School**, Coaldale, with
Walter Blank, Mike Piekema and **Kathryn Koliaska**,
managed by **Mr. R. Teramura**.

1988/1989 **Olds Junior-Senior High School**, Olds, with
Gregory Letal, Virginia Whitehair and **Andrea Coupal**,
managed by **Mr. D. Remillard**.

1989/1990 **Cochrane High School**, Cochrane, with
Kim Rubak, Lorinda Moore and **Dale Rossetti**,
managed by **Mr. B. O'Neil**.

1990/1991 **Acme School**, Acme, with
Murray Robinson, Tam Harder and **Jonathan Stade**,
managed by **Mr. D. Denham**.

1991/1992 **Didsbury High School**, Didsbury, with
Neil Kennedy, Karen Widish and **Andrew Whittaker**,
managed by **Ms. M. Barkley**.

1992/1993 **St. Mary's School**, Taber, with
Heather Cseke, Diana Wever and **Michael Wagner**,
managed by **Mr. B. Wagner**.

1993/1994 **Strathcona-Tweedsmuir School**, Okotoks, with
Michael Forbes, Jai Jacob and **B. Fong/C. Kent**,
managed by **Ms. Tina Lerakidis**.

1994/1995 **Camille J. Lerouge Collegiate**, Red Deer, with
Michael Rix, Darrin Lindout and **Thom Harmon**,
managed by **Mr. A. Quinn**.
Lacombe High School, Lacombe, with
David Good, Michael King and **Elisabeth Specht**,
managed by **Mrs. Linda Jaffray**.

1995/1996 **Brooks High School**, Brooks, with
Jeremy Wojtowicz, Justin MacCallum and **Matthew Gillett**,
managed by **Dr. Santokh Singh**.
Winston Churchill High School, Lethbridge, with

Keng In Yu, Steven Dubbleboer and **Blair Chandler**, managed by **Mr. Neil Gilbert**.

1996/1997 **Winston Churchill High School**, Lethbridge, with **Steve Dubbelboer, Samuel Conard** and **Russell Goodman**, managed by **Mr. Ken May**.

1997/1998 **Strathcona-Tweedsmuir School**, Okotoks, with **Ehsan Jalilian, David Klein** and **Nicholas Koning**, managed by **Ms. Tina Ierakidis**.

1998/1999 **Winston Churchill High School**, Lethbridge, with **Jenna Goodrich, Jill Joevanazzo** and **O. J. Wagontall**, managed by **Mr. Ken May**.

1999/2000 **Winston Churchill High School**, Lethbridge, with **Donald Howard, Peter Kim** and **Nathan Chronik**, managed by **Mr. Ken May**.

2000/2001 **Springbank School**, Calgary, with **Josh Seeley, Kyla Selk** and **Jan Shuites**, managed by **Ms. Stephanie Smith**.

2002/2002 **Oilfields Jr/Sr High School**, Black Diamond, with **David Cunningham, Amy Wescott** and **Danielle Paron**, managed by **Mr. C. Hughes**.

2002/2003 **Strathcona-Tweedsmuir School**, Okotoks, with **Dennis Chuang, Philip Woodard** and **Kevin Lemke**, managed by **Ms. Tina Ierakidis**.

2003/2004 **Strathcona Tweedsmuir School**, Okotoks, with **Dennis Chuang, Katherine Olsen** and **C. Gibson/M. Haslett**, managed by **Mr. Henk Koning**.

2004/2005 **New Norway School**, New Norway, with **Brad Kruse, Nolan Sand** and **Tamara van Lieshout**, managed by **Mrs. Denise Chromik**.

2005/2006 **Winston Churchill High School**, Lethbridge, with **David Liu, Julie Xu** and **Alexis Kaminski**, managed by **Ms. Terri Yamagishi**.

Zone III Prizes:

1983/1984 **McNally Composite High School**, Edmonton, with **Tom Harke, Heather Winitoy** and **Richard Soluk**.

1984/1985 **Harry Ainlay Composite High School**, Edmonton, with **Alan Williams, Todd Lee** and **Grant Parks**.

1985/1986 **Harry Ainlay composite High School**, Edmonton, with **Lola Sim, Victoria Filanovsky** and **E./G. Feltham**, managed by **Mr. D. Hunka**.

1986/1987 **Harry Ainley Composite High School**, Edmonton, with **Andrew Burton, Graeme Feltham** and **Lola Sim**, managed by **Mr. D. Hunka**.

1987/1988 **Old Scona Academic High School**, Edmonton, with
Graham Denham, Yiu Liu and **Patrick Chuang**,
managed by **Mr. R. Fizzell.**

1988/1989 **Harry Ainlay Composite High School**, Edmonton, with
Jennifer Zou, Adon Crook and **Mark Fokema**,
managed by **Mr. L. Lindenberg.**

1989/1990 **Archbishop MacDonald High School**, Edmonton, with
Keith Silva, Calvin Li and **Anthony Arendt**,
managed by **Mr. H. Marcuk.**
Old Scona Academic High School, Edmonton, with
Amina Danial, Brent Ellingson and **Chris Barrington Leigh**,
managed by **Mr. R. Frizzell.**

1990/1991 **Ross Sheppard Composite High School**, Edmonton, with
Albert Lee, Shui-Yeung Lam and **Hussein Waljee**,
managed by **Ms. K. Skrypnek.**

1991/1992 **Old Scon Academic High School**, Edmonton, with
Jason Colwell, Olga Cherney and **Merwin Siu**,
managed by **Mr. L. Pascoe.**

1992/1993 **St. Francis Xavier High School**, Edmonton, with
Anthony Fok, Theresa Winski and **Angie Debenedetto**,
managed by **Mr. L. Tomko.**
Ecole J. H. Picard, Edmonton, with
Steven Laffin, Chris Chandra and **Mark Chandra**,
managed by **M R. Mercier.**
Archbishop MacDonald High School, Edmonton, with
Patricia Lau, Sinisa Urban and **Ronnie Cheung**,
managed by **Mr. H. Marcuk.**

1993/1994 **Harry Ainlay High School**, Edmonton, with
Eric Finley, Peter Tai and **Lance Doherty**,
managed by **Mr. Lorne Lindenberg.**

1994/1995 **Old Scona Academic High School**, Edmonton, with
Matthew Wong, Talib Rajwani and **Beatrice Wu**,
managed by **Mr. Lorne Pascoe.**

1995/1996 **Archbishop MacDonald High School**, Edmonton, with
Hubert Chan, Jason Ding and **Tullia Dymarz**,
managed by **Mr. John Campbell**,
Old Scona Academic High School, Edmonton, with
Cynthia Luk, Tim Poon and **Peter Wei**,
managed by **Mr. Lorne Pascoe.**

1996/1997 **Archbishop MacDonald High School**, Edmonton, with
Tullia Dymarz, Stephen Somogyi and **Jason Ding**,
managed by **Mr. John Campbell.**

1997/1998 **McNally High School**, Edmonton, with
Gilbert Lee, Darryl Schneider and **Jackie Chan**,
managed by **Mr. Ed Heilman.**

1998/1999 **Old Scona Academic High School**, Edmonton,
with **Anton Cherney, Aden Grue** and **S. Olausen/M. Hirji**,
managed by **Lorne Pascoe**.

1999/2000 **Harry Ainlay High School**, Edmonton,
with **Chris Samuel, Kenman Gan** and **J. Wong/T. Yeung**,
managed by **Mr. Lorne Lindenberg**.

2000/2001 **Harry Ainlay High School**, Edmonton,
with **Sumudo Fernando, John Hu** and **Joseph Cheung**,
managed by **Mr. Lorne Lindenberg**.

2001/2002 **Archbishop MacDonald High School**, Edmonton,
with **Richard Ng, Maya Kumar** and **Laurence Yang**,
managed by **Mr. John Campbell**.

2002/2003 **Harry Ainlay High School**, Edmonton,
with **Radoslav Marinov, Jia Hu** and **D. Li/K. Szynkarczuk**,
managed by **Mr. Steven Hardy**.

2003/2004 **Old Scona Academic High School**, Edmonton,
with **Malka Wrigley, Brian Yu** and **Christopher Cheung**,
managed by **Mr. Lorne Pascoe**.

2004/2005 **Temp School**, Edmonton,
with **Michael Wong, Brent Thompson** and **Nicholas Horeczky**,
managed by **Mr. Lorne Rusnell**.

2005/2006 **Old Scona Academic High School**, Edmonton,
with **Brian Yu, Marshall Yuan** and **Douglas Cheung**,
managed by **Mr. Lorne Pascoe**.

Zone IV Prizes:

1983/1984 **Grand center High School**, Grand center, with
Emma Barnes, Chris Goss and **Kathy Pratt**.

1984/1985 **Lorne Jenken High School**, Barrhead, with
Russell Schultz, Neil McKellar and **Paul Gilbertson**.

1985/1986 **Fort McMurray Composite High School**, Fort McMurray, with
Michael Roth, Stephen Jones and **Kimberley Rogers**,
managed by **Mr. S. Kizior**.

1986/1987 **Sturgeon Composite High School**, Namao, with
Greg Hackman, Steve Prokopchuk and **Stuart Kane**,
managed by **Mr. R. Dolinski**.

1987/1988 **Salisbury Composite High School**, Sherwood Park, with
Michael Fisher, Vikram Karvat and **Charles Morse**,
managed by **Ms. J. Swainson**.

1988/1989 **St. Albert High School**, St. Albert, with
Ian Harrison, Rachel Harrison and **Allan McDonald**,
managed by **Mr. H. Reinbold**.

1989/1990 **Salisbury Composite High School**, Sherwood Park, with
Michael Roy, Charlie Kim and **Basil Vandegriend**,
managed by **Mr. R. Broemling**.

1990/1991 **Salisbury Composite High School**, Sherwood Park, with
Michael Roy, Karin Lu and **R. Godwaldt/A. Karvat/J. Tsang**,
managed by **Mr. R. Broemling**.

1991/1992 **Lorne Jenken High School**, Barrhead, with
Greg Ritter, Gina Harty and **Henrik Eskilsson**,
managed by **Mr. H. Schabert**.

1992/1993 **Paul Kane High School**, St. Albert, with
Spencer Ling, Adam Beacham and **Colin Reynolds**,
managed by **Mr. P. Zalasky**.

1993/1994 **Edwin Parr High School**, Athabasca, with
with **Byung-Kyu Chun, Su-Jin Chun** and **Joel Stobee**,
managed by **Mr. S. McMillan**.

1994/1995 **Ecole Secondaire Beaumont**, Beaumont, with
Daniel Robbins, Patrick Kobly and **Jean Fortin**,
managed by **Ms. Corlene Balding**.

1995/1996 **Ecole Secondaire Ste. Marguerite d'Youville**, St. Albert, with
Jeffrey Greeniaus, Terri-Lynn MacIntosh and **Tamara Cracknell**,
managed by **M Marcel Ouellette**.

1996/1997 **Lorne Jenken High School**, Barrhead, with
Laura Harms, Christine DeVries and **Robert Lopetinsky**,
managed by **Mr. Herbert Schabert**.

1997/1998 **Harry Collinge High School**, Hinton, with
Erik Poelzer, Derek Williams and **Amy Newman**,
managed by **Mr. Barry Davidoff**.

1998/1999 **Lorne Jenken High School**, Barrhead,
with **Steven Lyster, Tammy Setterington** and **Larisa Long**,
managed by **Mr. Herbert Schabert**.

1999/2000 **St. Albert High School**, St. Albert,
with **Matt Larocque, Adrian Pegoraro** and **Rob Lane**,
managed by **Mr. Hank Reinbold**.

2000/2001 **Salisbury High School**, Sherwood Park,
with **Neil Pandya, Andrew Pearson** and **Andrew Crowe**,
managed by **Ms. Sharon Sereda**.

2001/2002 **Salisbury High School**, Sherwood Park,
with **Stephen Arnason** and **O. Amer/S. McGladdery/K. Smith/H. Xue**,
managed by **Ms. Sharon Sereda**.

2002/2003 **Archbishop Jordan High School**, Sherwood Park,
with **Veroinica MacInnis, Daniel D'Lima** and **Sean Normandeau**,
managed by **Ms. Marge Hallonquist**.
Paul Kane High School, St. Albert,
with **Colin Wilbur, Matthew Sloboda** and **Holli Lizee**,
managed by **Mr. Percy Zalasky**.

2003/2004 **Archbishop Jordan High School**, Sherwood Park,
with **Sean Normandeau, Rylan Martin** and **Megan Chartrand**,
managed by **Ms. Marge Hallonquist**.

2004/2005 **Bev Facey Community High School**, Sherwood Park,
with **Graham Hill, Eric Konschuh** and **Lindsay Ketler**,
managed by **Ms. Karen Joy**.

2005/2006 **Salisbury Composite High School**, Sherwood Park,
with **Eliot Buchanan, Matthew Leggott** and **Lindsay Bowthorpe**,
managed by **Mrs. Sharon Sereda**.

Geoff J. Butler Memorial Prizes:

1985/1986 **Bishop Grandin High School**, Calgary, with
Todd Michaud, Carl Chala and **Jason Copping**,
managed by **Mr. J. Audia**.

1986/1987 **Ross Sheppard Composite High School**, Edmonton, with
Chris Nichols, Shelly Simpson and **Maki Ikemura**,
managed by **Mr. E. Wasylyk**.

1987/1988 **Archbishop Jordan High School**, Sherwood Park, with
Keriley Romanufa, Cathy Brisbane and **Neil MacDonald**,
managed by **Ms. M. Hallonquist**.
William Aberhart High School, Calgary, with
Bartek Muszynski, Sean Brandenburg and **Timothy Brehaut**,
managed by **Mr. D. Margach**.

1988/1989 **Dr. E. P. Scarlett High School**, Calgary, with
Craig Story, Ronda Grey and **Parmjit Basra**,
managed by **Mr. M. Falk**.

1989/1990 **Wetaskiwin Composite High School**, Wetaskiwin, with
James Lee, Ken Milne and **Timothy Reeves**,
managed by **Mr. D. Nelson**.

1990/1991 **St. Mary's Community School**, Calgary, with
Cindy Mok, Shawn Pauliszyn and **Noelle Bacalso**,
managed by **Mr. G. Dorscher**.

1991/1992 **Victoria Composite High School**, Edmonton, with
Steven Taschuk, Sunny Ma and **Daniel MacQueen**,
managed by **Mr. R. Marian**.

1992/1993 **St. Luke's High School**, Edmonton, with
Tin-Yau Kwan, Hau-Hing Chan and **Koon-Yip Kwan**,
managed by **Ms. V. Wishen**.

1993/1994 **Queen Elizabeth High School**, Edmonton, with
Alan Riphagen, Leo Chan and **Corey Ingram**,
managed by **Mr. Bruce Kabaroff**.

1994/1995 **Vauxhall High School**, Vauxhall, with
Ken vanden Dungen, Ashley Friesen and **Kerby Redicop**,
managed by **Ms. Diane Nelson**.

1995/1996 **Bellerose High School**, St. Albert, with
Rajah Kumar, Erin Scheelar and **Jenny Peterson**,
managed by **Ms. Gwenneth Chapelsky**.

1996/1997 **James Fowler High School**, Calgary, with
Margaret Tong, Shaun Holland and **Rahim Damji**,
managed by **Mr. Patrick Ancelin**.

1997/1998 **W. P. Wagner High School**, Edmonton, with
David Wakulchyk, Justin Ng and **Jeremie Rossignol**,
managed by **Ms. Amber Steinhauer**.

1998/1999 **Father Patrick Mercredi School**, Fort McMurray,
with **Leah Ramchandar, Danica Belter** and **Ameeta Sudan**,
managed by **Mr. Ted Venne**.

1999/2000 **John G. Diefenbaker School**, Calgary,
with **Gordon Kwok, Chris Zablocki** and **Andrea Kettle**,
managed by **Mr. Terry Loschuk**.

2000/2001 **M. E. Lazerte High School**, Edmonotn,
with **Tze Luck Chia, Jamie Kalat-Malho** and **Michael Quong**,
managed by **Mr. Jim Ashton**.

2001/2002 **Jasper Place High School**, Edmonotn,
with **Summer Cowley, Julian Haagsma** and **David Lovi**,
managed by **Mr. John McNab**.

2002/2003 **St. Francis High School**, Calgary,
with **Jan Owoc, Vincent Terstappen** and **Michelle Hunter**,
managed by **Mr. Peter Schill**.

2003/2004 **Ernest Manning High School**, Calgary,
with **Michael Schleppe, Ted Bethune** and **Austin Jackson**,
managed by **Mr. Jason Manning**.

2004/2005 **Prairie Christian Academy**, Three Hills,
with **Dion Knelsen, Hao Wang** and **King Ho Chan**,
managed by **Mr. Robert Hill**.

2005/2006 **J. A. Williams High School**, Lac La Biche,
with **Michael Schaub, Shaughnessy Fula** and **Natalie Cloutier**,
managed by **Mr. Matt Dyck**.

Second Round Winners

First Prizes—Nickle Family Foundation Fellowship:

1983/1984	**Tom Harke**	McNally Composite High School, Edmonton.
1984/1985	**Rene Schipperus**	Western Canada High School, Calgary.
1985/1986	**Geoffrey Haynes**	Tempo Secondary School, Edmonton.
1986/1987	**Sam Maltby**	Bishop Carroll High School, Calgary.
1987/1988	**Chris Nichols**	Ross Sheppard Composite High School, Edmonton.
1988/1989	**Peter Yang**	Western Canada High School, Calgary.
1989/1990	**Calvin Li**	Archbishop MacDonald High School, Edmonton.
1990/1991	**Charles Cruden**	Harry Ainlay Composite High School, Edmonton.
1991/1992	**Jason Colwell**	Old Scona Academic High School, Edmonton.
1992/1993	**Peter Hwang**	Sir Winston Churchill High School, Calgary.
1993/1994	**Byung-Kyu Chun**	Edwin Parr High School, Athabasca.
1994/1995	**Derek Kisman**	Queen Elizabeth High School, Calgary.
1995/1996	**Byung-Kyu Chun**	Harry Ainlay High School, Edmonton.
1996/1997	**Byung-Kyu Chun**	Harry Ainlay High School, Edmonton.
1997/1998	**Filip Crnogorac**	Western Canada High High School, Calgary.
1998/1999	**Jack Chen**	Sir Winston Churchill High School, Calgary.
1999/2000	**Jonathan Lau**	Sir Winston Churchill High School, Calgary.
2000/2001	**Jeffrey Mo**	St. Paul's Academy, Okotoks.
2001/2002	**R. Barrington Leigh**	Old Scona Academic High School, Edmonton.
2002/2003	**R. Barrington Leigh**	Old Scona Academic High School, Edmonton.
2003/2004	**Jerry Lo**	Vernon Barford Junior High School, Edmonton.
2004/2005	**David Rhee**	McNally High School, Edmonton.
2005/2006	**David Rhee**	McNally High School, Edmonton.

Second Prizes—Peter H. Denham Memorial Fellowship:

1983/1984	**Stella Lee**	Bishop Grandin High School, Calgary.
1984/1985	**David Pollock**	McNally Composite High School, Edmonton.
1985/1986	**Thomas Yoon**	Henry Wise Wood High School, Calgary.
1986/1987	**Graham Denham**	Old Scona Academic High School, Edmonton.
1987/1988	**Sam Maltby**	Bishop Carroll High School, Calgary.
1988/1989	**Teviet Creighton**	Western Canada High School, Calgary.
	Jeffrey Kinakin	Western Canada High School, Calgary.
1989/1990	**Michael Roy**	Salisbury Composite High School, Sherwood Park.
1990/1991	**William Lee**	Harry Ainlay Composite High School, Edmonton.
1991/1992	**Jeffrey Davis**	Western Canada High School, Calgary.
1992/1993	**Robert Kry**	Western Canada High School, Calgary.
1993/1994	**William Hu**	Sir Winston Churchill High School, Calgary.
1994/1995	**Byung-Kyu Chun**	Harry Ainlay High School, Edmonton.
1995/1996	**Derek Kisman**	Queen Elizabeth High School, Calgary.

1996/1997	**Eddie Ng**	Bishop Carroll High School, Calgary.
1997/1998	**Eddie Ng**	Bishop Carroll High School, Calgary.
1998/1999	**Frank Chen**	Harry Ainlay High School, Edmonton.
1999/2000	**R. Barrington Leigh**	Vernon Barford Junior High School, Edmonton.
2000/2001	**R. Barrington Leigh**	Old Scona Academic High School, Edmonton.
	Alex Fink	Queen Elizabeth Jr/Sr High School, Calgary.
	Alvin Tan	McNally High School, Edmonton.
2001/2002	**Sumudu Fernando**	Harry Ainlay High School, Edmonton.
2002/2003	**Radoslav Marinov**	Harry Ainlay High School, Edmonton.
2003/2004	**Peter Zhang**	Sir Winston Churchill High School, Calgary.
2004/2005	**Zheng Guo**	Western Canada High School, Calgary.
2005/2006	**Boris Braverman**	Sir Winston Churchill High School, Calgary.
	Jeffrey Mo	William Aberhart High School, Calgary.

Third Prizes—Canadian Mathematical Society Fellowship:

1983/1984	**Rene Schipperus**	Western Canada High School, Calgary.
1984/1985	**Sean Park**	Western Canada High School, Calgary.
1985/1986	**Najib Ayas**	Old Scona Academic High School, Edmonton.
1986/1987	**Kevin Foltinek**	Sir Winston Churchill High School, Calgary.
1987/1988	**Lynn Stothers**	Western Canada High School, Calgary.
1989/1990	**Charles Cruden**	Harry Ainlay Composite High School, Edmonton.
1990/1991	**Michael Roy**	Salisbury Composite High School, Sherwood Park.
1991/1992	**Robert Kry**	Western Canada High School, Calgary.
1992/1993	**Bryant Swanson**	Western Canada High School, Calgary.
1993/1994	**Alistair Savage**	Western canada High School, Calgary.
1994/1995	**Lei Jia**	Harry Ainlay High School, Edmonton.
1995/1996	**Hubert Chan**	Archbishop MacDonald High School, Edmonton.
1996/1997	**Huifen Zhang**	Ross Sheppard High School, Edmonton.
1997/1998	**Gilbert Lee**	McNally High School, Edmonton.
1998/1999	**Siuki Kwan**	Western Canada High School, Calgary.
1999/2000	**Jeffrey Mo**	University Elementary School, Calgary, (Grade 5).
2001/2002	**Keith Chung**	Western Canada High School, Calgary.
2002/2003	**David Rhee**	Vernon Barford Junior High School, Edmonton.
2003/2004	**Dennis Chuang**	Strathcona-Tweedsmuir School, Okotoks.
2004/2005	**Boris Braverman**	Sir Winston Churchill High School, Calgary.
	Sarah Sun	Holy Trinity Academy, Okotoks.

Alberta Teachers' Association Grade 11 Fellowship:

1983/1984	**Kenneth Benterud**	Harry Ainlay Composite High School, Edmonton.
	Thomas Quinn	Bishop Carroll High School, Calgary.
1984/1985	**Anant Murthy**	Old Scona Academic High School, Edmonton.

1985/1986	**Jim Jenkins**	Western Canada High School, Calgary.
1986/1987	**Michael Buckley**	Western Canada High School, Calgary.
	Stephen Chen	Sir Winston Churchill High School, Calgary.
	Chris Nichols	Ross Sheppard Composite High School, Edmonton.
1987/1988	**Greg Huber**	Sir Winston Churchill High School, Calgary.
1988/1989	**Jason Colwell**	Old Scona Academic High School, Edmonton.
1989/1990	**Brent Ellingson**	Old Scona Academic High School, Edmonton.
1990/1991	**Ozzie Gelbord**	Western Canada High School, Calgary.
1991/1992	**Natalka Roshak**	Western Canada High School, Calgary.
1992/1993	**William Hu**	Sir Winston Churchill High School, Calgary.
1993/1994	**Henry Ling**	Henry Wise Wood High School, Calgary.
1994/1995	**Daniel Robbins**	Ecole Secondaire Beaumont, Beaumont.
1995/1996	**Jaisel Vadgama**	Western Canada High School, Calgary.
1996/1997	**Gilbert Lee**	McNally High School, Edmonton.
1997/1998	**Sonny Chan**	Western Canada High School, Calgary.
1998/1999	**Lindsay Johnson**	Western Canada High School, Calgary.
1999/2000	**Frederic Dupuis**	Western Canada High School, Calgary.
2000/2001	**Peter Du**	Sir Winston Churchill High School, Calgary.
2001/2002	**Alex Fink**	Queen Elizabeth Jr/Sr High School, Calgary.
2002/2003	**Dennis Chuang**	Strathcona-Tweedsmuir High School, Okotoks.
2003/2004	**Qu Chen**	Harry Ainlay High School, Edmonton.
2004/2005	**Yiyi Yang**	Western Canada High School, Calgary.
2005/2006	**Alex Sampaleanu**	St. Francis High School, Calgary.

Alberta Teachers' Association Grade 10 Fellowship:

1983/1984	**John Thompson**	Sturgeon Composite High School, Namao.
1984/1985	**Bonnie Kam**	McNally Composite High School, Edmonton.
1985/1986	**Sam Maltby**	Bishop Carroll High School, Calgary.
1986/1987	**David Koch**	Strathcona Composite High School, Edmonton.
1987/1988	**Mark Fokema**	Harry Ainlay Composite High School, Edmonton.
1988/1989	**Michael Roy**	Salisbury Composite High School, Sherwood Park.
1989/1990	**Ozzie Gelbord**	Western Canada High School, Calgary.
1990/1991	**Robert Kry**	Western Canada High School, Calgary.
1991/1992	**Eric Finley**	Harry Ainlay Composite High School, Edmonton.
1992/1993	**Matej Zamec**	Western Canada High School, Calgary.
1993/1994	**Brian Yeh**	Western Canada High School, Calgary.
1994/1995	**Talib Rajwani**	Old Scona Academic High School, Edmonton.
1995/1996	**Joshua Grosse**	Western Canada High School, Calgary.
1996/1997	**Fred Ng**	Bishop Carroll High School, Calgary.
1997/1998	**Kyle Doerksen**	Western Canada High School, Calgary.
1998/1999	**Collin Tsui**	Henry Wise Wood High School, Calgary.
1999/2000	**Dave Sohn**	Henry Wise Wood High School, Calgary.
2000/2001	**Charles Li**	Western Canada High School, Calgary.

2001/2002	**Xiao Lin**	Western Canada High School, Calgary.
2002/2003	**Eric Tran**	Western Canada High School, Calgary.
2003/2004	**Zheng Guo**	Western Canada High School, Calgary.
2004/2005	**Graham Hill**	Bev Facey High School, Sherwood Park.
2005/2006	**Michael Zhou**	Western Canada High School, Calgary.

Pacific Institute for the Mathematical Sciences Special Prizes:

1999/2000	**Alexander Fink**	Queen Elizabeth Jr/Sr High School, Calgary.
2000/2001	**Sarah Sun**	St. Mary's High School, Okotoks.
2001/2002	**Sarah Sun**	St. Mary's High School, Okotoks.
2002/2003	**Boris Braverman**	Branton Junior High School, Calgary.

About the Editor

Andy Liu received a BS degree with First Class Honors in Mathematics from McGill University in 1970. He earned his MS in Number Theory (1972) and a Doctor of Philosophy in Combinatorics (1976) from the University of Alberta, Edmonton. Dr. Liu won a Minnesota Mining & Manufacturing Teaching Fellowship in 1998. He has won numerous teaching awards including: Canadian University Professor of the Year in 1998 (awarded by the Canadian Council for the Advancement of Education and the Council for the Advancement and Support of Education); the Canadian Mathematical Society's Adrien Pouliot Education Award in 2003; and the Mathematical Association of America's Deborah and Franklin Tepper Haimo Teaching Award in 2004. Dr. Liu is very well known in problem-solving circles. He won the David Hilbert International Award in 1996 from the World Federation of National Mathematics Competitions. He was the Deputy Leader (under Murray Klamkin) of the USA Mathematical Olympiad Team from 1981–1984; he later was Leader of the Canadian IMO team. Dr. Liu has been the Editor of the problem section of the MAA's *Math Horizons* for seven years.